Comparative Effects of Radiation

Report of a Conference held in San Juan at the University of Puerto Rico, February 15-19, 1960, sponsored by
NATIONAL ACADEMY OF SCIENCES —
NATIONAL RESEARCH COUNCIL

Comparative Effects of Radiation

edited by

MILTON BURTON Chemistry Department and Radiation Laboratory, University of Notre Dame

J. S. KIRBY-SMITH Biology Division, Oak Ridge National Laboratory

JOHN L. MAGEE Chemistry Department and Radiation Laboratory, University of Notre Dame

JOHN WILEY & SONS, INC. NEW YORK · LONDON

Library of Congress Catalog Card Number: 60-15664

Printed in the United States of America

Invited Participants

ALLEN, Dr. A. O., Department of Chemistry, Brookhaven National Laboratory, Upton, Long Island, New York.

AUGENSTINE, Dr. Leroy G., Division of Biology and Medicine, United States Atomic Energy Commission, Washington 25, D. C.

BEHRENS, Dr. Harold, Universidad de Chile, Instituto de Fisica y Mathematicas, Blanco Encalada 2008, Casilla 2777, Santiago, Chile.

BENINSON, Dr. Dan, Atomic Energy Commission, Buenos Aires, Argentina.

BURTON, Professor Milton, Department of Chemistry and Radiation Project, University of Notre Dame, Notre Dame, Indiana.

CHARLESBY, Professor Arthur, Physics Branch, Royal Military College of Science, Schrivenham, Swindon, Wilts., England.

CONGER, Professor Alan D., Department of Botany, University of Florida, Gainesville, Florida.

DOLE, Professor Malcolm, Department of Chemistry, Northwestern University, Evanston, Illinois.

DOUGLAS, Professor Ross A., Faculdade de Filosofia, Ciencias e Letras, Universidade de Sao Paulo, Caixa Postal, 8105, Sao Paulo, Brazil.

EBERT, Dr. Michael, Experimental Radiopathology Research Unit, Medical Research Council, Hammersmith Hospital, Ducane Road, London W. 12, England.

FANO, Dr. Ugo, United States Department of Commerce, National Bureau of Standards, Washington 25, D. C.

FÖRSTER, Professor Theodor, Laboratorium für Physikalische Chemie und Elektrochemie der Technischen Hochschule, Stuttgart N, Wiederholdstrabe 15, Germany.

FROTA MOREIRA, Professor Manoel da, Instituto de Biofisica, Universidade de Brazil, Rio de Janeiro, Brazil.

GORDON, Dr. Solon A., Division of Biological and Medical Research, Argonne National Laboratory, 9800 Cass Avenue, Argonne, Illinois.

GRIFFING, Professor Virginia, Department of Chemistry, The Catholic University of America, Washington 25, D. C.

HAMILL, Professor W. H., Department of Chemistry and Radiation Project, University of Notre Dame, Notre Dame, Indiana.

HARDY, Professor James D., Department of Physiology, University of Pennsylvania, School of Medicine, Philadelphia 4, Pennsylvania.

HENDRICKS, Dr. Sterling B., Mineral Nutrition Laboratory, United States Department of Agriculture, Agricultural Research Service, Soil and Water Conservation Research Division, Beltsville, Maryland.

HOCHANADEL, Dr. C. J., Chemistry Department, Oak Ridge National Laboratory, Oak Ridge, Tennessee.

HOLLAENDER, Dr. Alexander, Biology Division, Oak Ridge National Laboratory, P. O. Box Y, Oak Ridge, Tennessee.

HOWARD-FLANDERS, Dr. Paul, Department of Radiology, Yale University School of Medicine, 789 Howard Avenue, New Haven 4, Connecticut.

HUTCHINSON, Professor Franklin, Biophysics Department, Box 2166, Yale Station, New Haven, Connecticut.

KALLMANN, Professor Hartmut P., Department of Physics, New York University, 110 Washington Square, E., New York 3, New York.

KASHA, Professor Michael, Department of Chemistry, The Florida State University, Tallahassee, Florida.

KIRBY-SMITH, Dr. John S., Biology Division, Oak Ridge National Laboratory, P. O. Box Y, Oak Ridge, Tennessee.

KOK, Dr. Bessel, R I A S, 7212 Bellona Avenue, Baltimore 12, Maryland.

KUPPERMANN, Professor Aron, Instituto de Energia Atomica, Cidade Universitaria, Cx. Postal 11049 - Pinheiros, Sao Paulo, Brazil. Department of Chemistry, University of Illinois, Urbana, Illinois.

LARRIMER, Dr. W. H., Division of Biology and Agriculture, National Academy of Sciences, National Research Council, 2101 Constitution Avenue, Washington 25, D. C.

LINSCHITZ, Professor Henry, Department of Chemistry, Brandeis University, Waltham 54, Massachusetts.

LIPSKY, Professor Sanford, Division of Physical Chemistry, University of Minnesota, Institute of Technology, Minneapolis 14, Minnesota.

LIVINGSTON, Professor Robert, Division of Physical Chemistry, University of Minnesota, Institute of Technology, Minneapolis 14, Minnesota.

MAGEE, Professor John L., Department of Chemistry and Radiation Project, University of Notre Dame, Notre Dame, Indiana.

MARSHAK, Dr. Alfred, Department of Radiology, Jefferson Medical Center, 11th and Walnut Streets, Philadelphia 7, Pennsylvania.

MASCARENHAS, Dr. Sergio, Escola de Engeharia de Sao Carlo, University of Sao Paulo, Sao Paulo, Brazil.

PLATT, Professor John R., Department of Physics, University of Chicago, Chicago 37, Illinois.

POLLARD, Professor E. C., Biophysics Department, The J. W. Gibbs Laboratories, Yale University, New Haven 11, Connecticut.

PULLMAN, Professor Bernard, Universite de Paris, Institut de Biologie Physico-Chimique, 13 rue Pierre Curie, Paris-V, France.

REID, Professor Cyril, Chemistry Department, University of British Columbia, Vancouver 8, B. C., Canada.

REITHEL, Dr. Francis, National Science Foundation, Washington, D. C.

RUPERT, Professor Claud S., Department of Biochemistry, The Johns Hopkins University, School of Hygiene and Public Health, 615 North Wolfe Street, Baltimore 5, Maryland.

STEINBACH, Professor H. Burr, Department of Zoology, The University of Chicago, Chicago 37, Illinois.

SVAETICHIN, Dr. Gunnar, Department of Physiology, Instituto Venezolano de Investigaciones Cientificas (I. V. I. C.), Caracas, Venezuela.

SWANSON, Professor C. P., Mergenthaler Laboratory for Biology, The Johns Hopkins University, Baltimore 18, Maryland.

SZENT-GYORGYI, Dr. Albert, Institute of Muscle Research, Marine Biological Laboratory, Woods Hole, Massachusetts.

TOTTER, Professor John R., Rogue Graseras 778, Montevideo, Uraguay. Department of Chemistry, University of Georgia, Athens, Georgia.

VAN DYKEN, Dr. A. R., Chemistry Branch, Division of Research, United States Atomic Energy Commission, Washington 25, D. C.

ZELLE, Dr. Max R., Division of Biology and Medicine, United States Atomic Energy Commission, Washington 25, D. C.

Foreword

This book is a report on the Comparative Effects of Ionizing, Ultraviolet, Visible and Near Infrared Radiations as they were examined and interpreted by a group of biologists, chemists and physicists at a Conference held at the University of Puerto Rico, February 15-19, 1960. This Conference was called under the auspices of the Committee on Photobiology, Division of Biology and Agriculture, of the National Academy of Sciences — National Research Council. Funds for support of the Conference were provided by the National Science Foundation and the Atomic Energy Commission.

The Organizing Committee which made the long-range plans for the calling of the Conference was composed of Alexander Hollaender, Chairman, and Milton Burton, A. C. Giese, Sterling B. Hendricks, Carl P. Swanson and R. E. Zirkle. The actual organization and operation of the Conference was managed by a sub-committee consisting of Milton Burton, Chairman, A. C. Giese, S. B. Hendricks, C. P. Swanson, R. E. Zirkle and J. S. Kirby-Smith, Secretary, with Alexander Hollaender as ex-officio member representing the Committee on Photobiology. The latter "Conference Committee" wishes especially to express its indebtedness to the National Academy of Sciences — National Research Council, as represented particularly by Dr. H. Burr Steinbach, Chairman of the Division of Biology and Agriculture, Dr. Frank L. Campbell, Executive Secretary, and Dr. W. H. Larrimer, Staff Officer of that Division, and to the Oak Ridge Operations Office of the Atomic Energy

Commission and its branch in San Juan. Also, the Committee wishes to express its appreciation to the Oak Ridge National Laboratory and its operator, the Union Carbide Corporation, for repeated aid and assistance and particularly for the personal services of Mr. Donald Cowen, its Public Information Director. The invaluable attention given by Mr. Cowen to many details of organization greatly expedited the activities of the Conference. It is the regrettable duty of the Conference Committee to report the death of Mrs. Edna Caplan, reporter of this and many previous conferences of like nature, who prepared the original record of the proceedings of the Conference.

Finally, the Conference Committee wishes to express its great appreciation to the officials of the University of Puerto Rico, particularly Dr. Jose Vivas, Dean of the School of Medicine, and Professor Gustavo Candeles, Chairman of the Department of Biology, who acted as hosts of the Conference. It is particularly appropriate that this Foreword is followed by the welcoming address of the Chancellor of the University of Puerto Rico, Dr. Jaime Benítez.

Milton Burton, Chairman
Conference Committee

Dr. Jaime Benítez

Chancellor of the University of Puerto Rico
February 19, 1960

Welcoming Address

Ladies and Gentlemen:

I bring you symbolic words of affection of the University of Puerto Rico. They are meant exclusively to express our pleasure at having you with us and our hopes--indeed more than hopes, our earnest request--that you will again visit us, and perhaps stay with us.

You have seen some parts of our University, and if you have travelled West you may have seen more of it with a fragment of Puerto Rico. The name "scientists," means to me men of imagination and of intuitive capacity to appreciate the greater meaning in even the smaller parts of a whole. I can, therefore, trust that these little fragments of our Island and of this University may serve to bring to your consciousness the fact that things of enormous significance for our lives are happening amazingly fast in this island and in this institution. We are gathering together forces of very diverse nature, many of them explosive, others constructive, all of them part and parcel of an extraordinary striving effort toward an accomplishment with great repercussion upon man and society in our local environment. All this is interesting in the abstract, but it is particularly exciting to us because of our personal involvement in the matter.

Sometimes I wonder whether we in Puerto Rico, which in a sense means all of Puerto Rico and in another means its intellectual, professional, and educational leadership, will be really able to cope with these tremendous forces that have focused here and which, in a certain sense, are self-proliferating and thrusting us inevitably in great leaps into the unknown. It seems to me that the possibility of mastering and understanding these forces is very intimately tied to the presence or absence here, in sufficient number, of men of intelligence, magnanimity, and alertness. Men who can operate at the frontiers of knowledge and of understanding, and who can help in the infusion of knowledge, wisdom and perspective into our populace, surrounded as it is by new developments and new realities. I assure you that nothing is more responsive to greatness of the exceptional persons I envision than the young men and women and the not so young men and women who make up, in Puerto Rico, our University. And I am convinced of the exceptional consequences in terms of spiritual meaning, growth and health resultant from contact with people really devoted to the pursuit of knowledge and to the examination and creation of new ideas.

I have repeatedly noticed the responsiveness of large audiences to grace and wisdom in the lecturers and professors who have come to visit us. I know that, insufficient and inadequate as your contacts have been, to our loss, with our Faculty and with our students, they have enriched us already. I trust that you will take this earnest request not as final words of greeting; they express rather a profound concern that we may strengthen the enlightment, the wisdom and the vigor of our own people. We hope that your present and your more prolonged future visits, will provide you with a sense of personal investment in our future and that some of you at least may join us in our common, cooperative, joyous pursuit of knowledge.

Preface

The arrangement of this book closely follows the order of presentation of the subject matter at the conference. The papers, with the exception of two which were requested shortly before the meeting in order to round out the program, were invited well in advance of the conference. The general idea behind the actual plan of operation was to relate elementary processes to subsequent physico-chemical events and final biological effects for each type of radiation. Papers were intended to be provocative as well as informative, and the stimulation of discussion was a major aim in all instances.

Even a casual perusal of the Discussions will indicate to readers familiar with such matters that this material was not left in the informal and disordered state characteristic of conversations, whether they be scientific or not. The discussants cooperated notably in the preparation of amended remarks. Further, the Editors made free use of editorial privilege in their attempt to achieve order and at the same time retain as much informality as possible. A primary objective was the retention of all remarks which show actual development of new ideas. Pertinent criticism has not been tempered.

A requirement in conflict with attainment of editorial perfection was achievement of an early publication date. The Editors believe that the value of this publication is enhanced by its relatively early availability, and that the imperfections resulting from hurried handling are tolerable.

They hope further that the result is a readable, useful report which will be of value for some years to come. For any imperfections which are the evident result of their intrusion into the free flow of ideas they apologize to participants and readers alike.

Milton Burton
John S. Kirby-Smith
John L. Magee

July, 1960

Contents

Comparative Effects of Radiation

1

C. P. Swanson
Johns Hopkins University

Contrasts of Initial Biological Effects Resultant from Different Types of Radiation

By viewing the entire radiobiological scene from a distance it is possible to gain a general, even if very blurred, picture of the initial events induced by the variety of radiations having a biological effect, and to make comparisons which may prove useful in pointing out gaps in our thinking or provide clues to future experiments. Twenty years ago the general notion regarding radiation cytology was that the field was generally devoid of interesting problems. Today, the problem is how to approach a field of research which requires the tools and thinking of almost every science.

From our distant vantage point, a number of initial biological effects seem obvious: certain physical, radiochemical and metabolic (enzymatic) events take place before radiation damage or change is fully expressed in the cell or the organism. These events, with which we are generally familiar, have been deduced from the study of relatively simple systems or from experiments in which the variables, when recognized, are minimized.

1.1. IONIZING RADIATIONS

Powers, Webb and Ehret ('60) have reported a significant piece of work on the storage, transfer and utilization of x-ray energy in dry bacterial spores. They have concluded that three distinguishable, primary classes of initial damage induced by x-rays can lead to inactivation. One of these results in irreversible change that is independent of any environmental variable; this class can be considered as being due to the immediate and initial action of x-rays on some sensitive and necessary molecule, supporting thereby the target theory. Parenthetically, it should be emphasized that the terms "direct" and "indirect" when used to describe radiation effects are meaningless in any physical sense, and should be abandoned. The other two classes are related to the oxygen levels in the cell, and constitute that action of x-rays which is reversible or modifiable. These investigators also demonstrated that any variable which depresses the biological action of x-rays has a corresponding depressant effect on the magnitude of the electron spin resonance signal.

The important point about the work of Powers, Webb and Ehret is that their experiments utilized thoroughly dehydrated spores, with radiation exposures made at extremely low temperatures. Both circumstances -- dryness and cold -- prolong the lifetime of active radicals and interfere with their diffusion and with the transfer of absorbed energy, and the question immediately posed is what relation these data have to similarly induced events taking place in a wet, metabolizing cell existing under normal conditions. For example, in the dry, frozen spores, oxygen has a pronounced post-irradiation effect by reacting with long-lived radicals, whereas in the normal cell oxygen is effective only when present during radiation because radicals of this type tend to disappear very rapidly, side reactions and diffusions occur with greater ease, and the transference of absorbed energy is possibly facilitated. These are the precise phenomena which are of significance to the well-being of the active cell. I can only believe, therefore, that the chain of events initiated by the absorption of radiant energy is far more complicated than is indicated experimentally, particularly since the initial physical events have a duration of 10^{-6} seconds or less, while the biological events, stemming from induced metastable states, require minutes or even hours before an irreversible condition has been achieved.

Such a point of view, however, is but one side of the picture, for the metabolizing cell is a well-buffered system capable of considerable repair and recovery. A variety of nutritional and environmental circumstances can either enhance or markedly depress the extent of radiation injury. For example, Wolff ('60) has demonstrated that protein synthesis, which can be varied experimentally, is part of the metabolic system necessary for the repair of broken chromosomes; Kimball, Gaither and Wilson ('59) have clearly shown that potential mutational change in Paramecium requires time in order to be fully expressed, and that cellular circumstances which inhibit division favor a return to an undamaged state; and Stapleton, among others, has been able to modify the radiation sensitivity of cells by a factor of 20 by varying the conditions under which growth is achieved and exposure is made. It would appear, then, that while the initial interaction of radiation with cellular substances probably differs very little from cell to cell under a wide variety of environmental conditions, the ultimate expression of this interaction in the form of broken chromosomes, mutations, and death can be varied profoundly. The induction of metastable states and the effect of cellular metabolism are consequently keys to our understanding of the biological effects of radiation.

It would appear, therefore, that the number of events occurring between the absorption of radiant energy and the final expression of cellular change is a function of the state of the cell during and after exposure to radiation. In the frozen, dehydrated cell the events are probably far fewer, and ramify to a lesser degree, than in the dividing or metabolizing cell. As a result, the consequences of the irreversible action of radiation, as opposed to the modifiable events, are of relatively greater import to the former than to the latter.

With ionizing radiations the crucial cellular events are nuclear rather than cytoplasmic in origin. DNA, of course is the key genetic material but, since the degree of damage is closely related to the stage of cell division, it seems very probable that this close relationship is due to the manner in which DNA, RNA, histone and complex proteins are woven together to form the fabric of the chromosome. Our knowledge in this realm of cytochemistry is slight. It is of interest to note, in this regard, that the

ESR studies of Kirby-Smith and Randolph ('60) show that protein gives a far higher resonance signal than does DNA when x-rays are used, while the opposite situation occurs with ultraviolet radiation. What this result means in terms of initial absorption and ultimate damage is problematical, but it is not too difficult to conceive of elaborate transfers of energy and of energy traps or sinks within the chromosomal fabric.

1.2. ULTRAVIOLET EFFECTS

When we turn to the initial effects of ultraviolet radiation, a consistent picture emerges, namely that the effect produced depends on the wave length employed. How energy absorption is translated into biological change remains unknown. For mutational and chromosomal effects, the action spectrum parallels the absorption curve for DNA or its precursors. Wave lengths shorter than 2500 Å and those around 2800 Å and longer are heavily absorbed in both the cytoplasmic and nuclear proteins and, consequently, affect processes (e.g., spindle formation and cell division) which depend upon protein synthesis and aggregation. Consummation of the mutational and inactivation events is not dependent upon absorption alone, however; time is required and at least a portion of the events is reversible by visible light through the mechanism of photo-reactivation. The involvement of enzymes in the photo-reactivation process will, I presume, be discussed by Dr. Rupert, but the physical mechanism of photo-reversibility of ultraviolet-induced damage remains to be explained.

We can well imagine that most, if not all, of the initial ultraviolet events are direct in the sense that the degree of hydration and the level of oxygen in the cell play little or no role in determining the amount of damage inflicted, but we have as yet very little information as to the events taking place between absorption and the final realization of, for example, a completed mutation or a broken chromosome.

Two additional pieces of information relating to the effects of ultraviolet radiation are worthy of mention. The first is that the lambda phage in E. coli, strain K-12, can be converted, with an efficiency approaching 100%, from a lysogenic to a lytic state. In Salmonella typhimurium, the phage P-22, which, like lambda, occupies a site on the bacterial chromosome, is not ultraviolet-inducible. The difference between these two systems remains to be accounted for in terms of an understandable mechanism.

The second study relates to the combined effects of ultraviolet and x-rays on the induction of chromosomal aberrations. My earlier studies (Swanson '44) had indicated that ultraviolet has a profound effect in preventing the formation of x-ray induced aberrations either when employed as a pre-treatment or when given within the space of an hour after x-radiation. Those studies were supported by subsequent investigations in Drosophilia and maize. Recently, however, Kirby-Smith and Nicoletti (unpublished) have demonstrated that light of wave lengths 2650 and 2804 Å has a synergistic rather than an inhibiting effect when combined with x-rays. The discrepancies between these two studies remain to be resolved. However, once again, as in the x-ray experiments discussed above, the degree of hydration may provide a clue because the synergistic action was found when dehydrated ungerminated pollen grains were used, while inhibition by UV occurred when the generative cells of actively growing pollen tubes were exposed.

1.3. FAR RED AND INFRARED

When we consider the biological significance of the far-red and near-infrared regions of the electromagnetic spectrum, we find that an increased biological interest is being evinced because of the role of light of this quality in various photoresponsive aspects of plant development. These are discussed in this Symposium by Dr. Hendricks, and I wish to point out only that the infrared (or far red) has a synergistic action when combined with x-rays as a pre- or post-treatment (Swanson '49). Recent findings in our laboratory confirm the finding of Moh and Withrow ('58, '59) (and, I might add, the earlier suspicions of Dr. Hendricks) that the potentiating wave length is in the neighborhood of 7500 Å. Much remains to be done to clarify the infrared-x-ray combination effects, and to pinpoint the mechanism involved, but certain aspects are suggestively similar to the photo-reactivation phenomenon and its enzymatic relations.

In conclusion, it needs to be reiterated that there is an enormous gap separating what is known of the initial physical events of radiation -- energy absorption, excited states, energy transfer, active radicals and long-lived metastable states -- and what is known of the biological manifestations stemming from such events -- mutations, altered chromosomes, cell death, etc. New and great ideas are required to bridge the gap.

DISCUSSION

EBERT: I should like to hear what the speaker has to say about the difference between frozen bacteria and unfrozen ones. I am not familiar with the situation. If you have a frozen bacterium and warm it up, does it start to metabolize at once, does it start to work where it stopped when you froze it, or is a long set of rearrangements going on? Does freezing actually just fix everything in position -- prevent processes on freezing and then permit them to proceed after defrosting? I know that slow freezing of large cells introduces difficulties. However, what happens in a quickly frozen small cell under irradiation?

SWANSON: The only thing here is imposition of a rather important time factor; i. e., in the frozen state, as compared to a wet, metabolizing cell, everything is slowed down.

EBERT: You stated that wet cells are different in every respect from frozen ones. What is the difference? Is it just that water is solid and the various consequences of that fact?

SWANSON: I am not sure that I can answer your question completely. In fact I raise the question more than I answer it, but it seems to me that when one changes the time factor in here, so that diffusion and transfer of energy can proceed, one introduces a factor which can profoundly modify the events because ramifying phenomena seem much more possible than in the frozen state. All that I said had a big question mark after it.

BURTON: We should note that any change in the time factor (resultant from freezing) has different effects on the different competing processes dependent on the different energies of activation which determine what ultimate chemical processes occur. Freezing will, therefore, inevitably affect the various process controlled by diffusion, by collisional deactivation, by free-radical combination, by abstraction of atoms by free radicals, and by excitation transfer all very differently. It is the competition of processes whose rates are affected significantly differently by cooling (and freezing) which is determinant.

KIRBY-SMITH: I should like to make a comment about Dr. Swanson's remarks on the differences between radiation effects in wet and dry systems. We have, as Dr. Swanson has indicated, observed a marked synergism in the action of combined X-ray and ultraviolet radiation on dry Tradescantia pollen which is completely opposite to Dr. Swanson's earlier results on growing Tradescantia pollen tube cultures. I believe these differences are a good example of the widely different radiobiological responses of wet and dry systems mentioned by Swanson and that no real conflict exists. Indeed, our most recent results do indicate a definite drop in the synergistic effect as the moisture content of the pollen grains is increased. The magnitude of the effect also appears to depend inversely on the interval between the X-ray and ultraviolet exposures.

POLLARD: I want to say, in regard to Dr. Ebert's point, that the unfortunate thing is that there is no quick thaw. Quick freezing is fine, but during thawing all the salt distributions in the cell are different, all the pH ranges are different, all the effects of the temperatures that come up are different, and I think, therefore, that the effects cannot be honestly compared.

CHARLESBY: One does not even have to go that far to show the difference. In certain organic-aqueous systems even a small change in concentration will drastically change the radiochemical behavior. Dilute solutions of some water-soluble polymers crosslink under irradiation but add a little more water and you get a complete reversal of this effect of irradiation. So one does not have to push this notion very far.

My second point is with regard to your statement concerning the effect when there is no viable effect. I should like to hear that amplified, because almost any change, even a small change, in environment, is capable of affecting the direct effect of radiation.

CONGER: I think there is a profound difference in the termination of events after irradiation of dry or frozen biological systems, and of wet metabolizing ones. With dry or frozen systems in general, any post-irradiation modification of amount of damage is terminated either by wetting or by warming up. However, in the wet metabolizing systems, DNA duplication, or something associated with it, seems to be the event which terminates or fixes the initial damage. I don't really see any very simple way to call this last case -- the one that is related to DNA duplication -- a simple radiochemical one.

DOLE: Probably everybody knows this but I thought I might mention it anyway. In the case of polymers, if one irradiates a polymer like polyethylene at liquid nitrogen temperature and then warms it to room temperature there is a considerable post-irradiation oxidation effect. However, if one irradiates it in the molten state and then cools it to room temperature there is no post-irradiation oxidation effect. This situation seems to me to parallel what you said about the post-irradiation oxidation effect in the dry material as compared with the wet process.

GORDON: I would like to comment on the topic to which Dr. Swanson briefly referred, namely, biological inferences from ESR spectra. What I had in mind, specifically, was work in Powers' laboratory on the X-irradiation of dry bacterial spores (B. megaterium). There are three major radical species produced in these spores following anaerobic irradiation. These are indicated by a single central line, a "triplet" at 30 gauss spacings and a "doublet" at 20 gauss. If oxygen is introduced, a fourth radical species appears. From various thermal and gas experiments, it may be inferred that the doublet species arises at the expense of the doublet and triplet. Pertinent here is the fact that the concentrations of these specific radicals particularly in thermal restoration and nitric oxide experiments, follow precisely the radiation effect on spore viability. It seems to me that this is indeed a provocative parallel between events at the molecular level of energy transfer and the subsequent biological damage.

KUPPERMANN: Dr. Swanson, I am very much intrigued by the cumulative effect of ultraviolet and x-rays or infrared and x-rays. Do you have any explanation of a molecular mechanism by which this happens or is this a phenomenon determined on a more macroscopic scale?

SWANSON: We have worked only with the end result. I do not know what the mechanism is, but I think that Dr. Hendricks probably has some information in terms of the far red system, since he has extracted that part of the system at least that is responsive to the 6500-7500 Å region.

KUPPERMANN: So that no detailed mechanistic studies are available as of now?

SWANSON: Dr. Hendricks can answer that better than I can.

HENDRICKS: Studies have been made and some of the results will be presented this afternoon. I would like to comment about the transformation of protochlorophyll to chlorophyll as temperature is reduced. The transformation requires hydrogen transfer and is second order. The transformation takes place at -80° to about one fifth of the stoichiometric amount. It is entirely suppressed at -100°.

With regard to the change with temperature of the effect on plant growth of far-red radiation mentioned by Dr. Swanson, cooling to -80° reduces the efficiency to about 5 per cent of the room-temperature value. The reversible pigment change is lost upon cooling to -180°. This behavior is typical of relaxations of aqueous systems. Two important features of aqueous systems in plants are to be considered. The first is: Solutions persist to about -40°, which is the lowest liquidus. The second is: Relaxations in most aqueous systems usually take place in the region of -180° as indicated by viscosity and heat capacity measurements.

A transition of the second kind is often found in this temperature region.

GORDON: Might I return to the ESR experiments. In dealing with ESR patterns, I think there is a risk of unwarrantedly stepping from correlation to causality. Impressive as the parallel between biological effects and change in ESR spectra upon spore irradiation is, we should remember that relatively large X-ray doses are used to produce the radical concentrations that can be picked up in paramagnetic resonance studies. For example, the Argonne experiments to which I referred involved dosages of several million rads, though qualitatively similar spectra were seen following dosages of 250 kr.

CONGER: I don't agree with that exactly. We do the same thing with seed embryos with 5,000 r. We can detect radicals at this dose, and in the same material we are doing biological experiments in the dose range from 5 kr to about 20 or 30 kr.

CHARLESBY: I am very worried about trying to read too much into the ESR. I cannot help feeling that reactions must take place which do not show up in the ESR. More than that, if one starts with some relatively simple materials and irradiates them, one gets ESR spectra which are very different in spite of only minor changes in chemical structure. A change of only one atom or side group in a thousand may give entirely different ESR spectra. Any attempt to read structural differences into the ESR spectra must be done with considerable caution.

FROTA MOREIRA: I am afraid I am going to complicate things a bit, but I should like to comment on our experience with more complex systems.

At the Institute of Biophysics we have done some work, with Dr. Marcovich from the Institute of Radium in Paris and Dr. Paes de Carvalho from the Institute of Biophysics, in which we have studied the effect of X-ray radiation on the electric response of the E. electricus electroplate. As you know the electric eel delivers electrical discharges when stimulated. Using small pieces of the electric fish main electric organ and submitting it to 35 kv X-ray irradiation, we found that the irradiated sample showed a response that fell constantly and finally disappeared. In these circumstances we were able to obtain an immediate response to the action of X-ray irradiation, which is unusual when one performs radiation experiments. I am thinking about this material being a rather good way to study initial effects.

PULLMAN: As a structural chemist I might say that I was struck by something which seemed to me to be missing or perhaps not sufficiently explicitly indicated in Dr. Swanson's paper. Namely, when speaking about the initial effect of the radiations you consider the direct effect or the indirect effects. Now, what seems to me to be essential from the biological point of view is the initial chemical changes in the biological system, -- the nature of the first chemical transformation in it. So, I think that one must distinguish between the indirect effects which need not be the initial transformation of the biological system and the initial effects which are the first chemical transformation of the biological system.

This is a gap which has to be filled before we can really understand what is happening because biological systems, and especially, say, DNA, are complex molecular structures composed of numbers of sub-units, and what is really important is to know what is the first chemical transformation which occurs in DNA. In other words, what are the sensitive atoms and groups in DNA? If one looks very carefully into the published data one can see that we seem to have some indication of what this initial biochemical transformation is. Better understanding of this point may lead to a better understanding of the later metabolic transformations. If you understand the initial chemical change in the biochemical structure under the effect of radiation then you may correlate it with the function of the structure and understand it much better.

HOLLAENDER: In our Laboratory we tried to get a nucleic acid which is highly sensitive to very small amounts of radiation. The work was done by Anderson ('54) and

Fisher et al. ('59). The most sensitive system we could find was a nuclear protein very carefully isolated from biological materials. This carefully isolated nuclear protein will respond to as little as a few roentgens and even down to one roentgen. This is the range of sensitivity in which a definite effect on rate of mitosis is shown. With the exception of the work of Forssberg and Novack ('58), this is possibly one of the most sensitive systems we have found.

ALLEN: With respect to what Dr. Hollaender just told us I should like simply to remark that twenty years ago Professor Crowther, at Reading, showed that 0.25 r had an appreciable effect on the mobility of colloidal gold particles in a gold sol. This was presumably not really a chemical change, but a change in the electrical double layer surrounding the sol particles. Maybe some of these effects are really more physical than chemical in nature.

2

Ugo Fano

National Bureau of Standards

Collective Effects in Absorption of Energy from Ionizing Radiation

2.1. SUMMARY

In the last ten years it has been realized that the energy spectrum of primary excitations* produced by fast charged particles differs substantially in the condensed and the gaseous state of matter. In the gaseous state this spectrum parallels the optical absorption spectrum and is understood to slope down, with increasing energy, rather gently at first in the far ultraviolet where most of the energy absorptions take place. Bethe ('33) has given a complete theory of collision excitations for the case of a gas of hydrogen atoms; Figure 1 shows the calculated distribution of excitation energies resulting from small-angle inelastic collisions of a fast particle. Spectra for real gases are expected to show the same qualitative features.

In solid films, on the contrary, electrons are found to produce excitation spectra characterized by a major, rather sharp, peak in the far ultraviolet, i. e., in the 10-20 ev region (except in alkali films where the peak lies below 10 ev). The experimental evidence has been reviewed adequately, whereas the theoretical evidence has been presented in fragmentary and at times confusing or controversial fashion. (Marton, et al. '55, '56; Pines '55; Uehling,

* Ionizing processes are included here in the word "excitation".

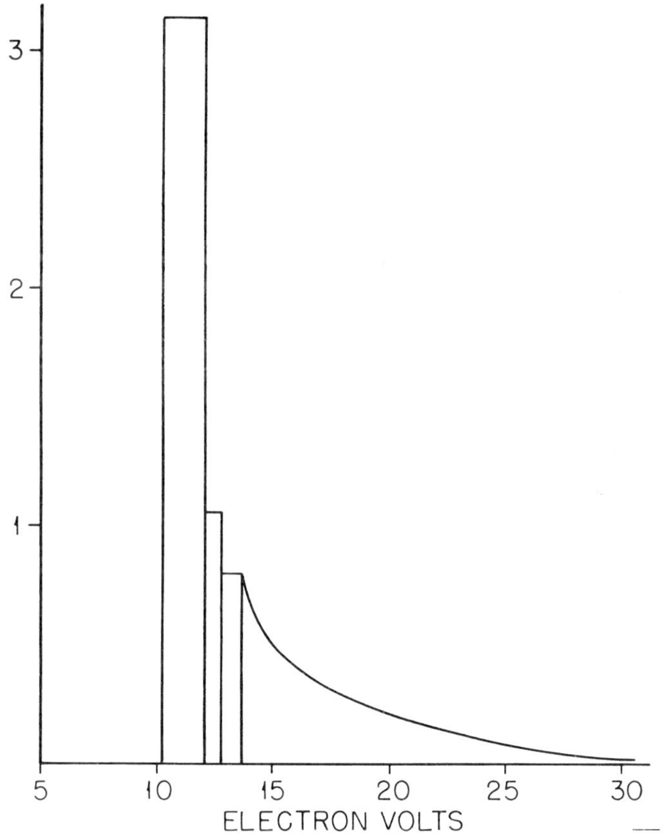

Fig. II. 1. Energy-loss spectrum for small-angle deflections of fast charged particles in gas of H atom (Bethe '33). The ordinate is number of events per unit energy interval in arbitrary units.

Editor, '60; Fano '60). Figure 2 shows typical loss spectra found for films. In the gaseous state these materials would have a complex loss spectrum corresponding to their optical absorptions at energy losses lower than, or comparable to, the observed peak and should have similar absorption at higher energies.

The peak in such a loss spectrum is understood to correspond to a collective excitation involving the electrons of

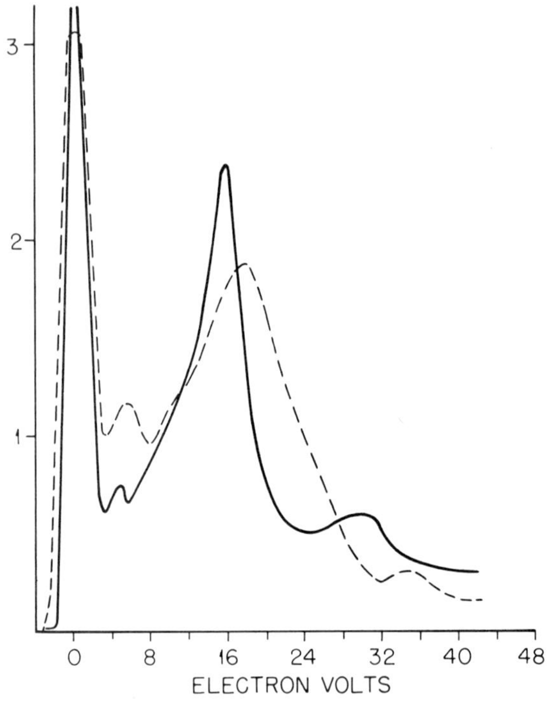

Fig. II. 2. Energy-loss spectrum for electrons in films of antimony, with solid curve, and antimony sulfide, with dashed curve (Marton, et al. '56). The ordinate is energy loss per unity energy interval in arbitrary units. The peak at zero energy is due to elastic scattering.

a rather large volume of matter, perhaps 100 Å or more in diameter. This excitation lasts, in its initial form, only for a very short time of perhaps $\sim 10^{-15}$ sec. It is not known how it evolves, e. g., whether it spreads out, or whether it becomes quickly localized within a single atom or group of atoms thus becoming similar to the excitation produced in gaseous matter. Thus, it cannot be said at this time whether the initial collective aspect of the excitation, which had been overlooked in the past, has actually a significant influence on its eventual physico-chemical effects.

2.2. DISCUSSION*

2.2.1. Experimental and theoretical foundation of discrete loss spectra of electrons in condensed materials. - Although most experimental data have been obtained with the use of metal films, it is clear from theoretical considerations that the effect is perfectly general. Some experimental data have been obtained with molecular and ionic crystals and some with amorphous materials such as plastic films. The theory is not completely worked out but the effect is understood in a general way. The loss spectra represent a collective effect and not the effects of individual atoms or molecules of the material. To a large extent any loss spectrum depends upon the density of electrons in the material in question, i. e., the number of valence electrons per unit volume. For fast particles it does not depend upon the charge and velocity of the particle.

2.2.2. Total energy loss of fast particles and the Bragg Law for condensed materials. - It is well known that the range-energy relations of high-energy particles in condensed materials can be obtained with good approximation from the range-energy relations of the same materials in the gas phase by treating the condensed phase as a compressed gas. In view of the drastic alteration of the low-energy loss spectrum of the condensed phase it seems strange that such a relationship holds. The explanation is, in the first place, that the average energy loss experienced by a charged particle per unit path length (i. e., its stopping power) depends only logarithmically on the spectrum of energy losses. Furthermore the condensation of the material influences only the low energy portion of this spectrum. This portion contains in general a majority of all collisions but a minor

* At the request of the Symposium Chairman Dr. Fano presented an extemporaneous discussion of collective effects. The first part of Chapter 2 is a brief summary of this material as prepared by him. The second part of Chapter 2 is a summary of the general discussion which followed Dr. Fano's presentation as prepared by the Editors. Although Dr. Fano was the principal contributor to this discussion, there were many questions and comments from the other participants.

fraction of the total energy loss. The residual influence of the condensation upon the stopping power is of the order of 1-2 percent and is often called the "zero-energy density effect". Departures of this order of magnitude from the Bragg Law have been detected.

2.2.3. Collective nature of excitation. - The energy losses under discussion are single events and initially there is only a single excitation although many electrons are involved. A type of state which is expected to be excited strongly is the type which has a strong dipole transition moment with respect to the ground state. Such an excited state is strongly coupled by dipole-dipole interactions between neighbors. It is clear that there are two types of polarization for such excited states: (a) the dipole moment can be parallel to the direction of propagation of the standing waves in which the states are analyzed, or (b) the dipole moment can be perpendicular to this direction. Case (a) is called longitudinal excitation and case (b) is called transverse excitation. Charged particles moving with velocities considerably less than light excite only the first type of state and electromagnetic radiation excites the second type.

In a material which is characterized by a complex frequency-dependent dielectric constant it is well known that absorption of electromagnetic radiation is proportional to the imaginary part of the dielectric constant of the appropriate frequency. It can be shown that in such a medium the probability for excitation of longitudinal waves is proportional to the imaginary part of the reciprocal of the dielectric constant. For systems in which the dielectric constant is near unity (as in a gas), these two quantities approach each other.

2.2.4. Use of correlation functions. - A technique of theoretical physics which is growing in importance at the present time is the use of correlation functions. An example of a correlation function is the probability for finding an electron at various distances from another electron at a given time. The x-ray diffraction pattern of a material is determined by such a (static) correlation function. A more complicated kind of correlation function is the probability distribution for finding an electron at various distances from another electron at later times (i.e., if an electron is at $\vec{r}_1$ at the time t_1, what is the distribution function for finding an electron at $\vec{r}_2$ at the time t_2?). The energy loss of fast charged particles can be described with the use of such a correlation function. The same results are obtained with a more conventional model which uses coupled oscillators explicity (Fano '56). In some respects the correlation function treatment is more powerful and it will most likely be used more often in the future for problems of this sort.

2.2.5. Dependence on particle energy. - The low-energy loss spectrum is insensitive to incident particle velocity (and thus to its energy) in the region above a few kilovolts for incident electrons, and probably changes only a little if this energy is reduced to a few hundred electron volts. Not much can be said for still lower incident energies.

2.2.6. Geometrical considerations. - It is clear from very general considerations that a low energy loss in a condensed medium cannot be very highly localized. It is localized to a region comparable to the de Broglie wave length associated with the momentum change Δp which an incident particle suffers when it looses an amount of energy $\Delta\varepsilon$,

$$\lambda = \frac{h}{\Delta p}.$$

For loss processes which correspond to decelerating the particle without changing its direction, it can be shown that the general relation between Δp and $\Delta\varepsilon$ is

$$\Delta p = \frac{\Delta\varepsilon}{v}$$

where v is the particle velocity. All electrons with energies above 50 kev have velocity approaching the velocity of light. For such electrons and $\Delta\varepsilon \approx 15$ ev

$$\lambda \approx 900 \text{ Å}$$

A loss process of this type is not very well localized, particularly since it occurs to a band of states which involve strong coupling between neighboring molecules.

This consideration applies only to the electronic state formed initially. How long this state lasts and how chemical effects originate are not known. The initial electronic state may last 10^{-15} to 10^{-14} seconds whereas it is clear that a time of the order of 10^{-13} seconds is required to dissociate molecules. Thus many processes may occur between formation of the initial electronic state and appearance of the first chemically altered species. It is clear that a dissociated molecule is localized (in the sense of the present discussion). If two or three dissociations are effected by the same primary process the correlation of their position is the most important matter for subsequent chemical effects.

2.2.7. Question of coherence. - In this discussion we are not considering the possible coherence between successive loss processes of fast particles. In a single loss process, a single excited electronic state is formed and in this sense the excitation of different atoms is coherent. Later if this single state converts into two or more electronic states it cannot be said that coherence remains.

2.2.8. Excitation migration. - It has been noted that an electronic state, which is delocalized and extends over a region of $(100 \text{ Å})^3$ or more, can become localized by dissociation of a molecule situated somewhere in the region of the excitation. As long as the initial state exists it has the property that it becomes more and more delocalized. This is a natural consequence of the fact that it is a wave packet and spreading is an inherent property of wave packets. If it breaks up into other electronic states (i. e., two or more) they may also be delocalized to some extent. No general principles can be utilized to say where chemical effects will originate.

2.2.9. Radiation chemistry. - The similarity of radiolysis of materials in gas and liquid states has been noted a number of times. Collective effects per se are not necessarily expected to introduce great differences. The molecular processes themselves (such as dissociations and rearrangements) may well be very similar in the two phases. It is clear that track effects are more important in the liquid. It may be that another general principle applies as a result of the altered energy-loss spectrum of the condensed state, particularly the shifting to high energies of the low-energy region. Such a shift suggests that initial chemical effects may be reduced in the condensed state, since there is a strong implication that more energy should be wasted as low-grade heating. This is a particularly important consideration both for radiation chemistry and for radiation biology.

3

Sterling B. Hendricks
Mineral Nutrition Laboratory
U.S. Department of Agriculture

Control of Plant Growth by Phytochrome

A photoresponsive pigment, which we call phytochrome, controls the growth of all seed plants and is evident in some responses of at least one algal species (Haupt, '58), several liverworts, and many ferns (Mohr, '56). A striking display of phytochrome action is the photoperiodic control of flowering which was discovered by W. W. Garner and H. A. Allard ('20). In the ensuing four decades studies in the U. S. Department of Agriculture by Garner and Allard and by H. A. Borthwick and his associates led to an essential understanding of this phenomenon* and other responses subject to the same control. We have summarized the advances on several occasions (Borthwick et al., '56, '48; Hendricks, in press; Hendricks and Borthwick, '55), but the continued and now rapid progress resulting from separation and partial purification of phytochrome gives new insight into the several phenomena meriting interim examination before concepts crystallize.

---

* The concepts presented here rest upon the work in the U. S. Department of Agriculture by H. A. Allard, H. A. Borthwick, W. L. Butler, R. J. Downs, W. W. Garner, P. H. Heinze, S. B. Hendricks, H. Mohr, S. Nakayama, K. H. Norris, M. W. Parker, A. A. Piringer, N. J. Scully, H. W. Siegelman, E. H. Toole, and V. K. Toole. Indebtedness is expressed to R. H. Flint and E. D. MacAlister ('37) who measured action spectra for lettuce seed germination.

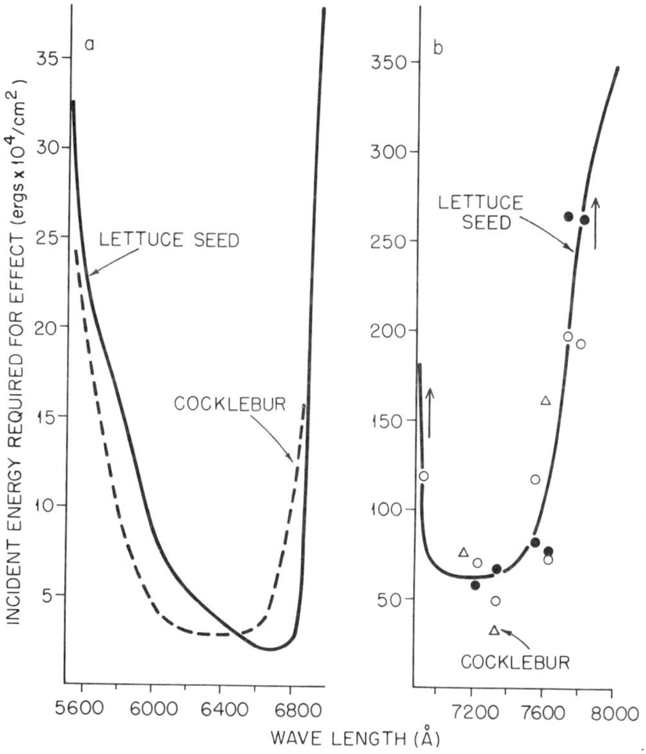

Fig. III. 1. (a) Comparison of action spectra for inhibition of floral initiation of cocklebur and promotion of 50% germination of lettuce seed in the region of 5600 to 6900 Å. The slit width for the former was about 150 Å at 6400 and about 100 Å for the latter. (b) Action spectra following irradiation with red radiation (5800 to 6800 Å). The effect is promotion of floral initiation of cocklebur and inhibition of 50% germination of lettuce seeds which is opposite to that in the region of 5600 to 6900 Å.

3.1. PHOTORESPONSIVENESS AND ACTION SPECTRA

Action spectra for flowering of cocklebur, Xanthium pensylvanicum (Parker et al., '46) and germination of lettuce (Borthwick et al., '54) and wild peppergrass seeds, Lepidium virginicum (Toole et al., '55), Figures 1A and B,

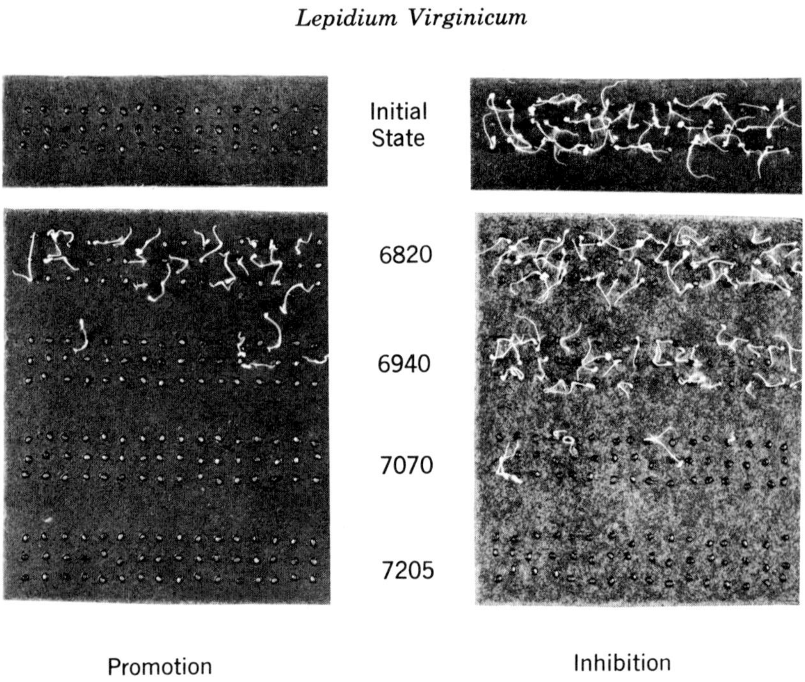

Fig. III. 2. Responses of wild peppergrass seed to irradiation in the region of 6700 to 7400 Å for 4 minutes followed by a 3-day period in darkness to permit display of germination. Seed on the left were held imbibed in darkness until placed in the spectrum, while those on the right after holding in darkness were exposed to red radiation (5800 to 6900 Å) for 5 minutes before being placed in the spectrum. The lots marked "initial state" were representative of the seed irradiated in the spectrum as they developed in 3 days of darkness.

which are typical for many responses, illustrate the diversity and some essential features of photoresponsiveness. The seeds after 24 hours of imbibition of water in darkness were exposed in a spectrum (Figure 2) for short intervals (seconds to minutes) and then returned to darkness. Germinating seeds were counted after several days and incident energies in various spectral regions required for a given degree of germination were obtained from a graded series either for inhibition or promotion (Figures 3A and B). Action spectra for flowering were measured by the control exerted by irradiation on recently expanded leaves near the middle of one or more long nights (14 hours for cocklebur), after which the plants were returned to noninductive conditions for two weeks for development adequate to display the degree of floral initiation (Figure 4).

The photoreversibility evidenced by the action spectra (Figure 1) (Borthwick et al., '52), is the key to the subject. It is strikingly shown by simple experiments as illustrated by reversibility in germination of pine and lettuce seed and stem elongation of young bean plants (Figures 5a, b, c).

Action spectra, reversibility, and related observations lead directly to the following simple generalizations of great significance.

1. A number of apparently unrelated photoresponses of plants have the same causation. This equivalence includes flowering responses of short- and long-night plants, which respond oppositely to the same stimulus.
2. The controlling photoreaction is photoreversible. The implication is that there are two forms of the effective pigment with absorption maxima near 6600 Å and 7350 Å, respectively.
3. Biological action depends upon the extent of pigment conversion.
4. Photoreversibility allows the photoreaction to be examined by a null method based on biological assay but not biologically dependent.

These generalizations are of two kinds -- the one bearing on biology, the other on the nature of the pigment.

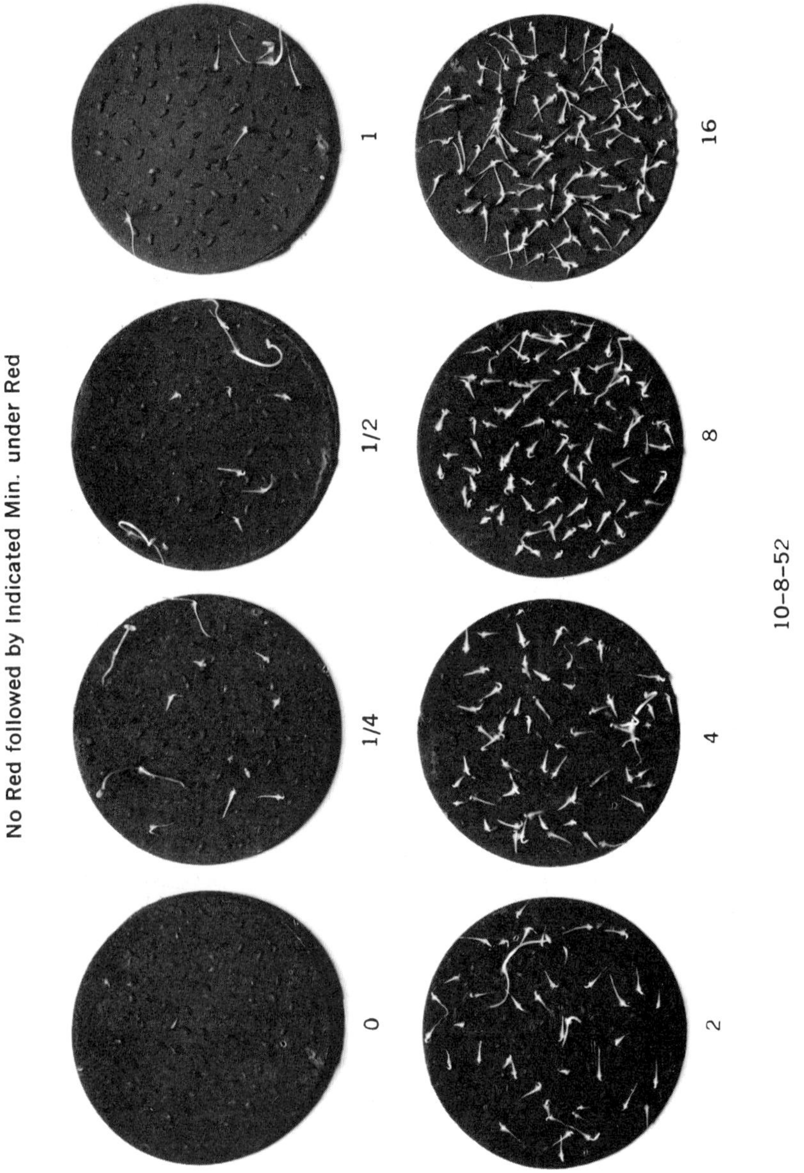
No Red followed by Indicated Min. under Red
0
1/4
1/2
1
2
4
8
16
10-8-52

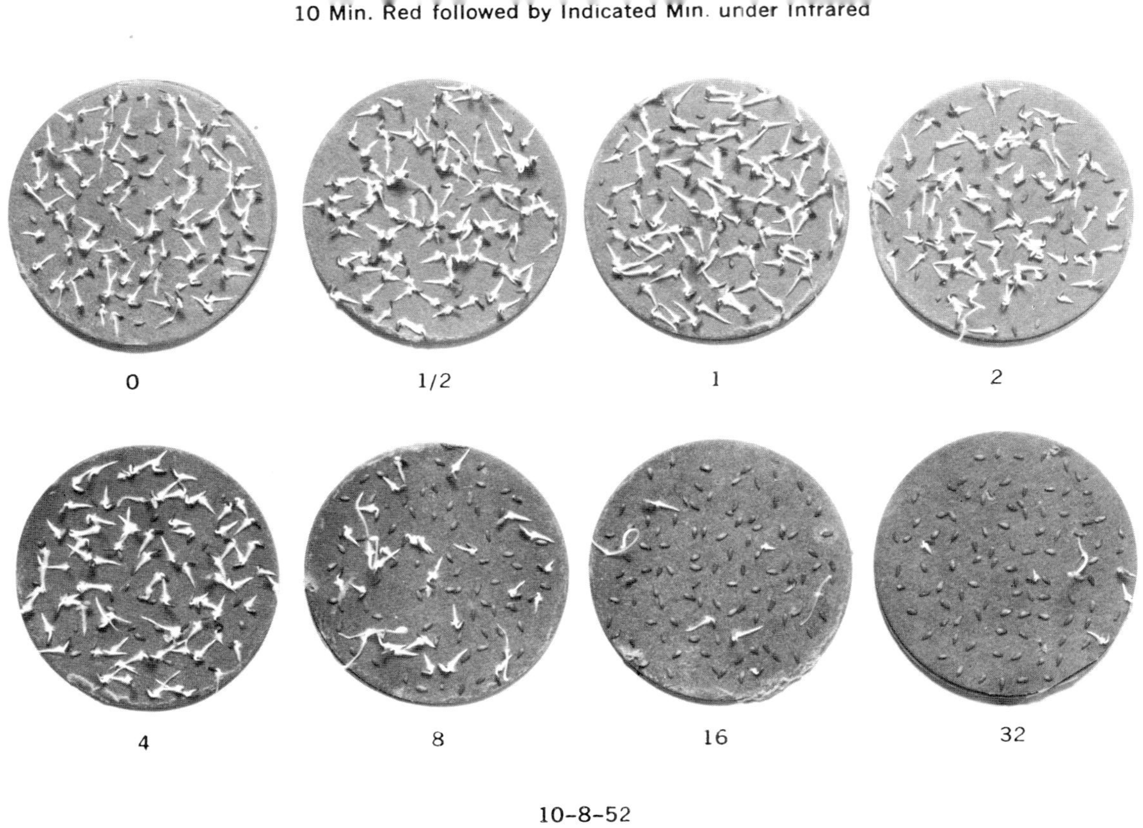

Fig. III. 3. (a) Promotion of germination of lettuce seed by irradiation in the region of 5800 to 6900 Å (red) with a low intensity source for a series of times, and (b) inhibition in the region of 7000 to 8000 Å (far-red).

3.2. THE PIGMENT, PHYTOCHROME, AND THE PHOTOREACTION

The photoreaction can be written as

$$\text{Phytochrome r (6600 Å max)} \underset{\text{Far-red radiation}}{\overset{\text{Red radiation}}{\rightleftarrows}} \text{Phytochrome fr (7350 Å max)}$$

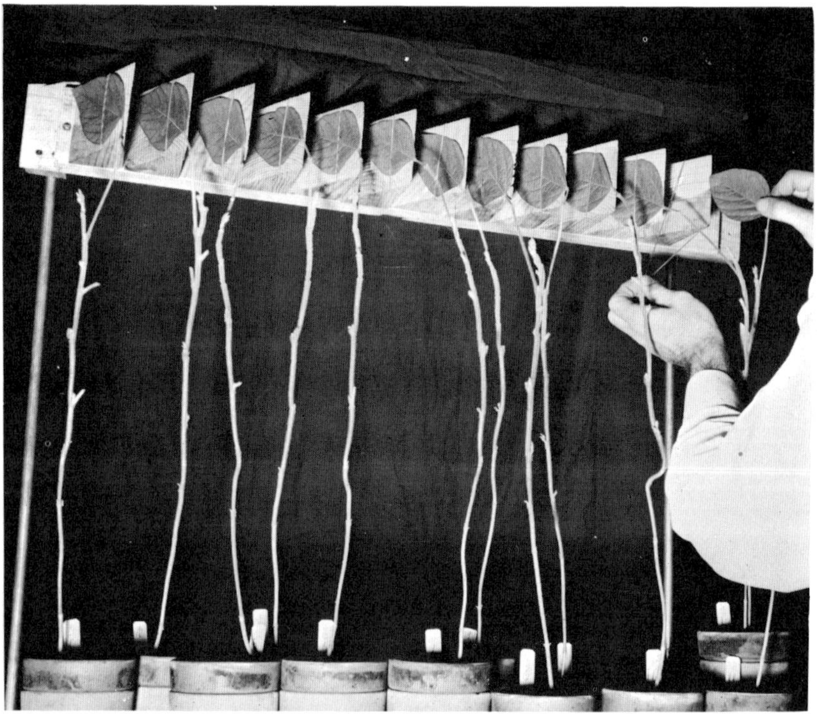

Fig. III. 4. Leaves of soybean plants being placed at the focus of the spectrograph. The actual operation was done in darkness near the middle of 14-hour nights, after which the leaves were exposed in the several spectral regions for various times (seconds to minutes) and then returned to darkness.

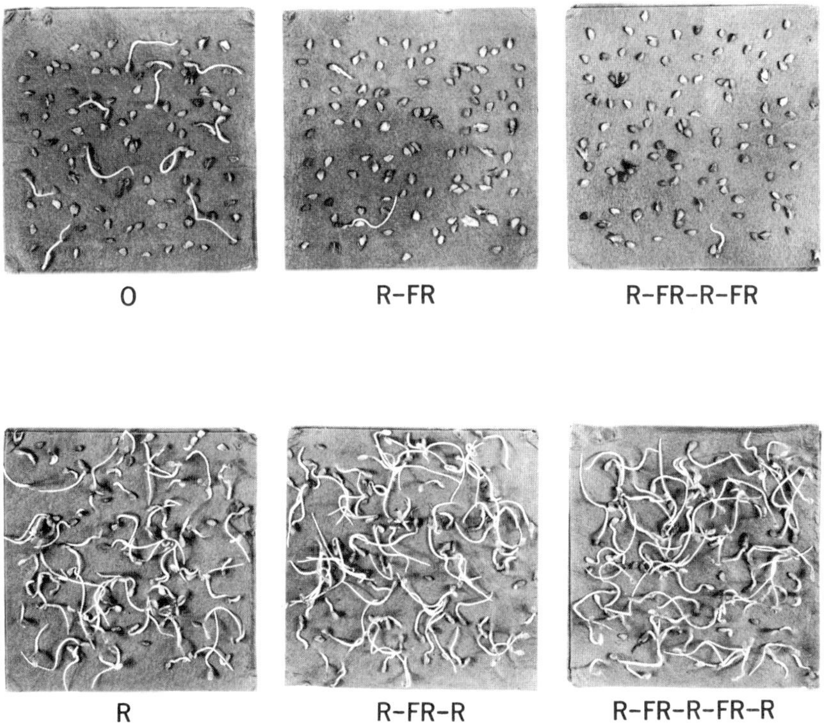

Fig. III. 5a. Reversibility of responses to red and far-red radiation. Pinus virginiana seed in darkness (O), with 5 minutes red radiation (R), followed by 5 minutes far-red radiation (R-FR), and then by 5 minutes red radiation (R-FR-R), etc. Lots of seed of each type were returned to darkness for development of the germination response.

Values of $\varepsilon\phi X$, where ε is the molar absorption coefficient ϕ is the quantum efficiency for conversion, and X is fraction of the incident energy reaching the point of absorption can be obtained (Hendricks et al. '56) by the method of Warburg ('48) for measuring $\varepsilon\phi X$ of carbon monoxidecyto-

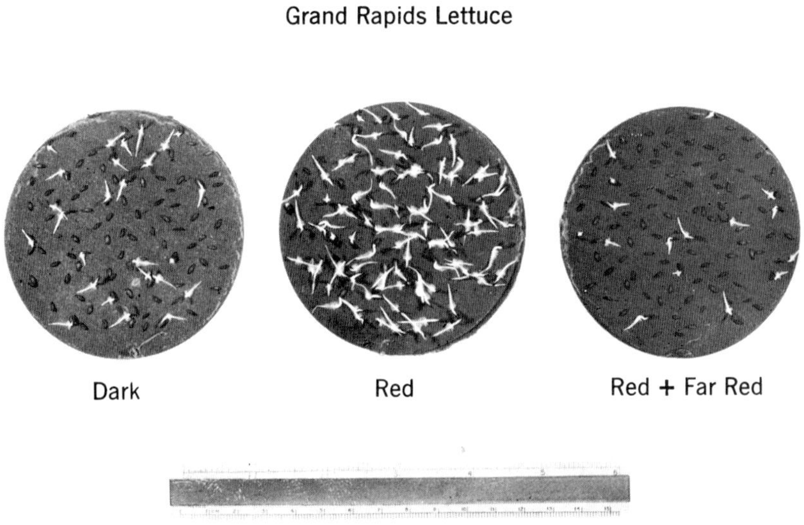

Fig. III. 5b. Photoreversibility of responses to red and far-red radiation. Lettuce seed (variety Grand Rapids).

chrome oxidase. The method depends on determination of the ratios of energies required for two levels of response. The two levels of response are obtained by displacing the reaction to the right successively to two stationary states which we may designate as a and b and then to the extreme right; thereafter the conditions are adjusted successively to bring the system back to the states b and a. This procedure amounts to measuring both the degree of pigment change and the response corresponding to the degree of change.

Measurement of response to radiation at two temperatures maintained only for the short period of irradiation shows that the temperature coefficients for the rates of change are 1.0, which permits use of first-order kinetics. The maximum values of $\varepsilon \phi X$ measured for seed germina-

Fig. III. 5c. Photoreversibility of responses to red and far-red radiation. Stem elongation of pinto beans. In this case the plant on the left was taken directly to darkness after each light period (12 hours) in a growth room giving predominantly red radiation; the center plant was irradiated for 5 minutes with far-red radiation and then placed in darkness; and the one on the right received the far-red radiation followed by 5 minutes red radiation. Treatment was for 4 days followed by 3 12-hour day and night periods for development.

tion and stem elongation are near 3×10^7 cm^2 $mole^{-1}$ and are the same within a factor of 5 for the two pigment forms. The quantum efficiency is not likely to exceed 1.0 and X cannot have a value greater than 1.0. The molar extinction coefficients of phytochrome r and fr, then, are of the order of 10^5.

Even though the two forms of phytochrome have maximum possible absorbancies, they are not evident, in albino tissue of barley and maize (which display their presence by stem elongation responses), either to the eye or to usual spectrophotometers. Biological activity must be controlled by a very small amount of material, an estimate being $< 10^{-7}$ molar in the absorbing group.

Half-maximum biological responses are obtained in the range 0.001 to 0.02 joule/cm^2 for red radiation irrespective of the size of the object; variance of response is about the same per unit area of object irrespective of how many objects occupy a unit area. This implies display of control on a cellular basis with effectiveness of 10% to 50% conversion to phytochrome fr for half-maximum response. The energy ranges are of the same magnitude for all responsive plant parts but, within their variations, if response to red radiation is depressed, the far-red radiation requirement is enhanced.

With the generality of a reversible photoreaction in mind, let us examine some cases in which the reversibility fails either immediately or after the lapse of about 30 minutes between red and far-red irradiations. All cases of photoperiodic control of flowering are of this type. An extreme example is afforded by control of flowering of the Japanese morning glory, Pharbitis nil, which has been studied in detail by S. Nakayama ('58) in Japan and with H. A. Borthwick's group (unpublished).

Figures 6a, b, c show action spectra for floral induction of Pharbitis nil at the start and near the center of dark periods adequate for flowering. At the start of the dark period the usual reversibility is clearly displayed with the isosbestic point near 6950 Å. Near the middle of the dark period the action with a maximum near 6600 Å (red) cannot be reversed; radiation in the region of 7350 Å (far-red) has the same type of action at an adequate energy level as that at 6600 Å. The same energy is required, however, to accomplish a given action of red radiation as when the system is reversible and is within a three-fold factor of that required for promotion of lettuce seed to 50% germination. The form of the pigment is changed and its action is ac-

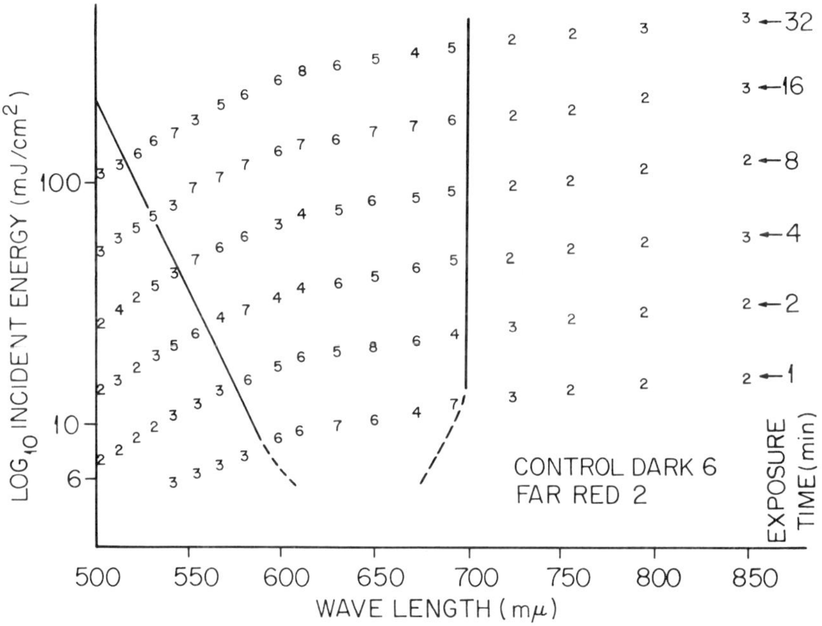

Fig. III. 6a. Action spectrum for floral induction of Pharbitus nil. Plants were grown in continuous fluorescent light for 4 days and then exposed in the spectrum for various times as indicated on the right side of the figures. During treatment plants were subject to 8-hour days and 16-hour nights. After treatment they were returned to continuous light for 10 days and then examined. The number of flowers evident at the terminal meristem is recorded in the figures. This figure is for material placed in the spectra at the start of the dark period.

complished but the reversibility is lacking.

3.3. HIGH-ENERGY SYSTEMS

Production of anthocyanins in apples (Siegelman and Hendricks, '58), red cabbage (Siegelman and Hendricks, '57), and many other plant parts is either completely determined or is enhanced by visible radiation. Formation

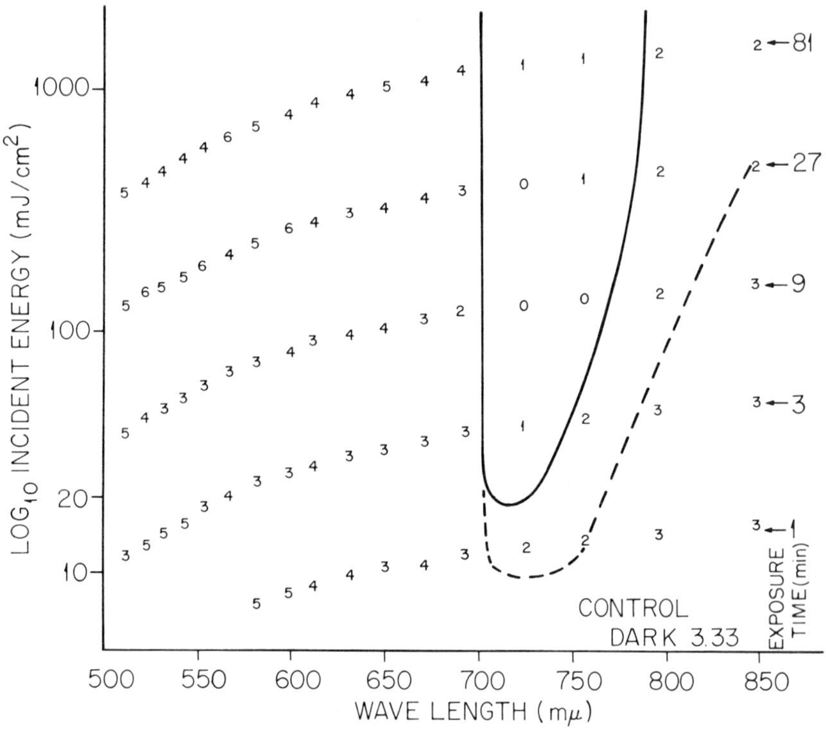

Fig. III. 6b. Irradiated with far-red radiation at the start of the dark period and then placed in the spectrum (cf. Fig. 6a for description).

of significant amounts of anthocyanin, however, requires energies of the order of 1,000-fold the half-maximum energies for growth regulation. Such a high-energy requirement makes the measurement of action spectra for anthocyanin formation far more difficult and tedious than for growth regulation.

Some action spectra for anthocyanin formation, structure elongation, and control of seed germination for high energies are shown in Figures 7a and 7b (Hendricks and Borthwick, '59; Hendricks et al., '60). An evident feature is marked effectiveness of radiation in the region of great-

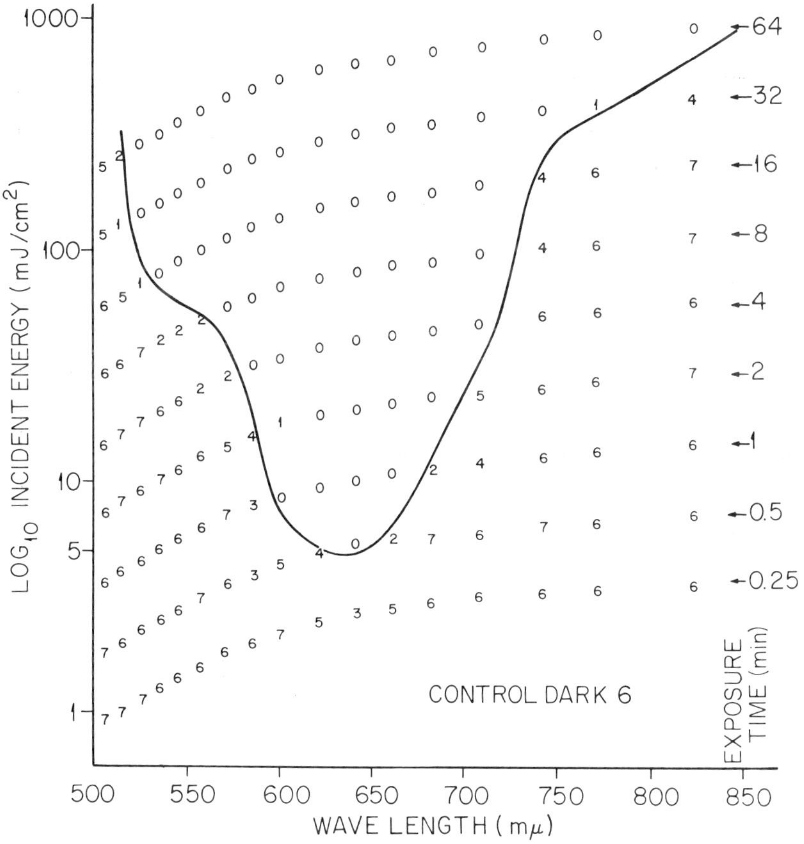

Fig. III. 6c. Placed in the spectrum near the middle of the dark period (cf. Fig. 6a for description).

est action of phytochrome.

Two photoreactions acting in sequence are found in anthocyanin synthesis. The second of these is more clearly the conversion of phytochrome. It is illustrated for anthocyanin synthesis in red cabbage seedlings by responses of dark-grown seedlings (Figure 8) in a spectrum at an energy level of 0.1 joule cm^{-2}. In the case of sorghum seedlings anthocyanin synthesis, potentiated by a two-hour ex-

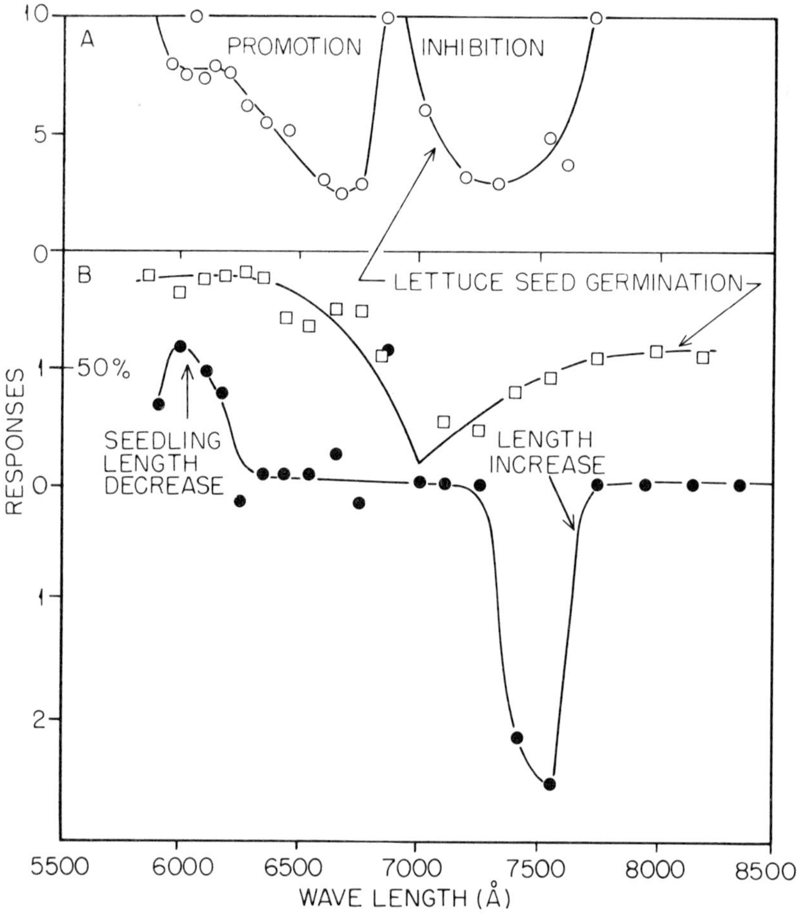

Fig. III. 7a. A. Portions of the action spectra for Grand Rapids variety lettuce seed germination for short times of exposure with low energy. The curve marked promotion is expressed in terms of incident energy (units of 10^4 ergs cm^{-2}) required for 50% germination. Curve for inhibition is expressed in incident energy (units of 2×10^5 ergs cm^{-2}) required for 50% inhibition. B. Action spectra of same material for 12-hour irradiation with high (5×10^7 ergs cm^{-2}) incident radiant flux: (a) percent germination; (b) change in radicle length in mm.

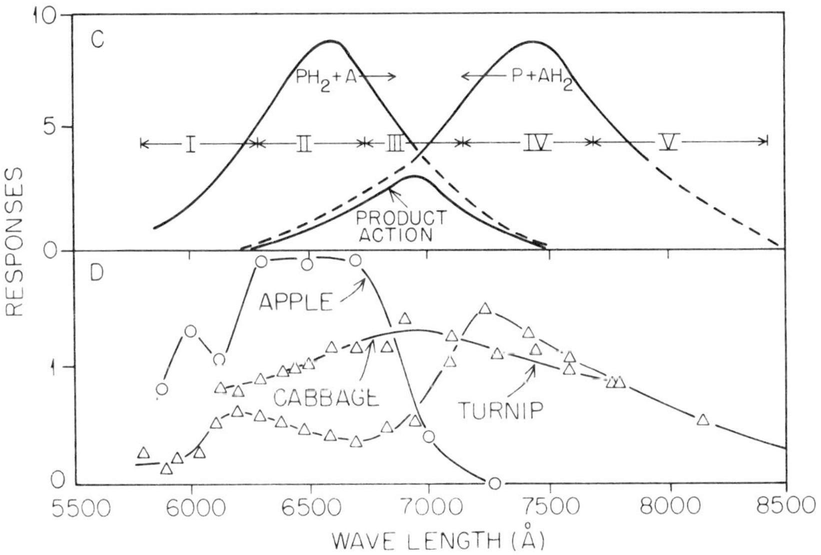

Fig. III. 7b. C. Schematic indication of molar absorptivities, expressed in terms of molar absorption coefficient (units of 10^{-7} cm^2 $mole^{-1}$), for the two forms of phytochrome and the probabilities for their continuous excitation. D. Portions of the action spectra for anthocyanin formation in turnip and red cabbage seedlings, and in apple skin. Curves are plotted in units of 10^{-9} moles of anthocyanin per 0.5 cm^2 for apple skin and per seedling for turnip and cabbage. Incident energies (ergs cm^{-2}) were 5×10^7 for cabbage, 10^8 for turnip and 2.6×10^8 for apple.

posure to high intensity radiation, can, after four additional hours, be blocked by conversion of phytochrome to the red absorbing form (r) or be permitted by phytochrome fr.

That the high energy photoreaction also involves phytochrome is shown not only by the region of action but also by the long wave-length limit for action which extends in turnip seedlings to at least 7600 Å. Continued excitation of phytochrome r or fr is apparently required. Note, for instance, that the action maximum for the high-energy action for anthocyanin formation in red cabbage is at the isosbestic point for conversion of phytochrome. This

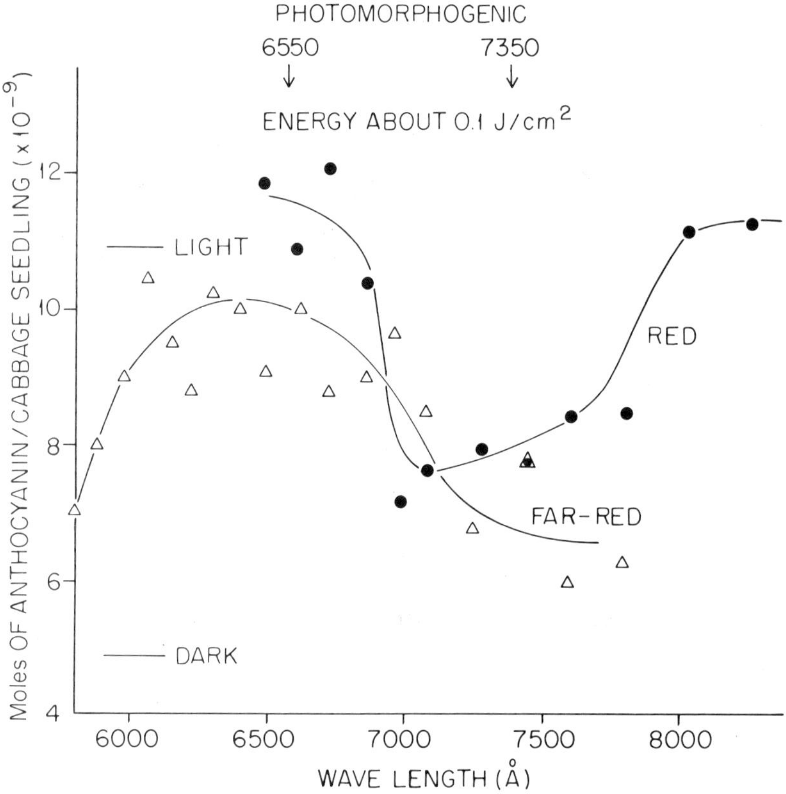

Fig. III. 8. Action spectrum for the change of phytochrome partially controlling anthocyanin synthesis in red cabbage seedlings. Dark-grown seedlings were irradiated with either far-red or red radiation and then placed in the spectrum.

action spectrum can be calculated from the phytochrome action spectrum assuming its mediation in energy transfer to anthocyanin synthesis. Similarly, action spectra for anthocyanin synthesis in apple and red cabbage can be calculated assuming principal excitation of phytochrome r for the former and cytochrome fr for the latter. An assumed quantum efficiency of a few tenths for anthocyanin formation leads to a calculated phytochrome concentration of the order of 10^{-7} molar.

3.4. PHYTOCHROME ASSAY AND SEPARATION

Physical detection and separation of phytochrome were accomplished in June 1959 by a team consisting of W. L. Butler, K. H. Norris, H. W. Siegelman and S. B. Hendricks ('59). The argument was that, if the phytochrome concentration is of the order of 10^{-7} molar in some tissue, then it can be detected with a differential spectrophotometer for the latter permits measurement of differences in transmission in the region of the absorption maxima of phytochrome r and fr as the form of the pigment is changed. The system noise would preferably be of the order of .001 O. D. at an O. D. of 3.

Mr. Gerry Birth (in press) has designed a suitable difference spectrophotometer which, used with a 2 cm-thick sample of dark-grown turnip cotyledons, indicates the presence of phytochrome by a $\Delta(\Delta O.D.)$ value of several hundredths as the phytochrome is interconverted, corresponding to $(O.D._{6600} - O.D._{7350})$, far-red irradiated, minus $(O.D._{6600} - O.D._{7350})$, red irradiated. Limited surveys indicate that shoots of dark-grown maise seedlings, four to seven days old, are suitable as a phytochrome source. The pigment is readily detectable in cauliflower florets, avocado fruit, zucchini squash fruit, etiolated graminous seedlings, and many other plants and plant parts.

The big question, of course, concerned maintenance of reversibility of phytochrome in vitro. The fact that such reversibility is maintained (cf. Figure 9) has opened up an entirely new area for investigation, which has been barely touched at this time. It has been clearly shown by biochemical observations that phytochrome is a soluble protein of the cell cytoplasm. Its concentration in dark-grown maize seedlings is of the order of 1/3000 of the protein. The concentration relative to protein has been increased to the order of 1/300 by salting protein out of solutions at appropriate $(NH_4)_2SO_4$ concentrations. Phytochrome is entirely stable in solutions at 0° and has been maintained for six months at -15°. Further purification is in progress.

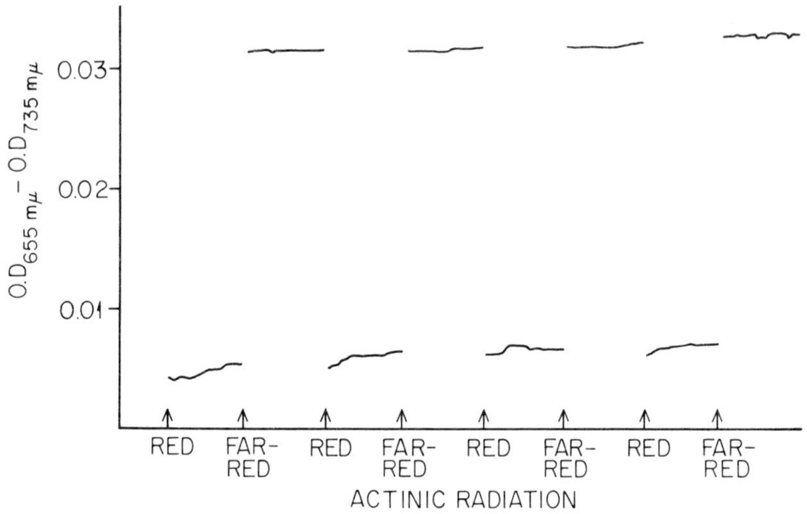

Fig. III. 9. Reversibility indicated on the differential spectrophotometer of the phytochrome response in a solution extracted from maize seedlings.

3.5. OBSERVATIONS ON PHYTOCHROME WITH THE DIFFERENCE SPECTROPHOTOMETER

Relative absorbancies of phytochrome r and fr and quantum efficiencies for interconversion can readily be measured with the difference spectrophotometer in vivo and in vitro. The relative absorbancies are equal, within 5%, at the respective absorption maxima as shown by comparison of $\Delta(O.D._{6600} - O.D._{8000})$ and $\Delta(O.D._{7350} - O.D._{8000})$ as phytochrome is changed from the red to the far-red absorbing form.

Relative quantum efficiencies for conversion are obtained by comparison of $\Delta(\Delta O.D.)$ values from 6600 to 7350 Å sources with the value from a mixed source of known intensity as afforded by an incandescent filament. The quantum efficiency for the conversion Pr→Pfr is three times that in the reverse direction. This relative efficiency is the same in more than ten materials examined in vivo and for solutions from maize seedlings, cauliflower florets, and spinach leaves.

Phytochrome conversion is first order in both directions. Reversibility is maintained upon dialysis and after freezing. It decreases as the temperature is lowered to -70°, where it is about one-tenth the value at 0°, and is absent at -180°. Irradiation at -180° followed by warming in darkness does not cause pigment conversion. Photoreversibility, however, is recovered without loss as the system warms.

These several observations are completely in harmony with the pigment conversion being a molecular rearrangement.

3.6. FURTHER CONSIDERATIONS OF IRREVERSIBLE SYSTEMS: THE BIOLOGICALLY ACTIVE FORM OF PHYTOCHROME

If the pigment conversion is a molecular rearrangement, how can it become irreversible? How can varying effectiveness for action of red and far-red radiation arise? The results of action of red radiation near the middle of the dark period on floral induction of Pharbitis nil clearly shows conversion of phytochrome r and attainment of biological effectiveness but far-red radiation does not reverse the action. An interpretation is that phytochrome r can pass directly to an active but non-reversible form or can do so indirectly by first changing to phytochrome fr.

The sequence of change is further suggested by following $\Delta(\Delta O.D.)$ values for dark-grown maize seedlings upon first exposure to light such as normal laboratory illumination for a few minutes. The value of $\Delta(\Delta O.D.)$ decreases in a period of four hours to less than half the initial value even though the seedlings are returned to darkness. If the tissue is maintained or extracted at 2° upon removing from darkness $\Delta(\Delta O.D.)$ values are constant in light or darkness. These observations are consonant with marked stability of phytochrome fr, as the biologically effective pigment form, and its change into another effective form which is not photoreversible.

3.7. THE REACTION SCHEME

A pattern of the change of phytochrome consistent with the several observations is

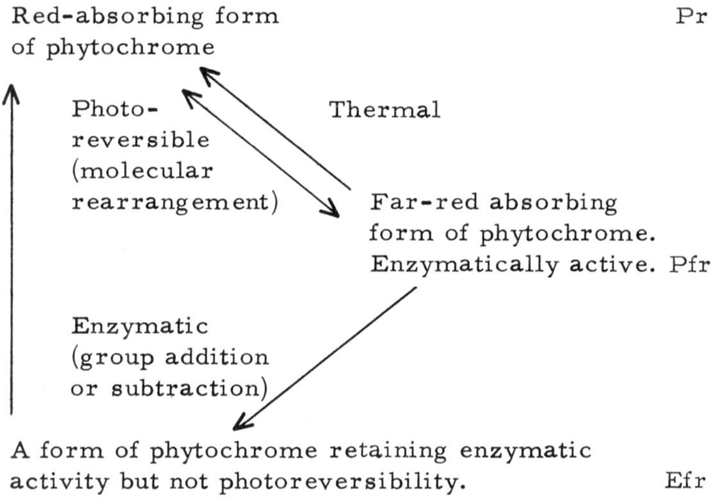

The manner and time in which Efr returns to Pr are unknown.

3.8. THE IMMEDIATE FUTURE

This presentation has emphasized very simple deductions from clear-cut observations. Our knowledge, in reality, is far more extensive and we now have reasonable insight into the four following major questions.

1. What is the probable nature of prosthetic groups of phytochrome and of the molecular rearrangement and subsequent group change?

2. What is the nature of the enzymatic action determining biological change arising from phytochrome fr?

3. How can so many diverse responses arise from one action?

4. How do rhythmic changes arise?

A remark on the third question, which concerns a principal basis of life is appropriate at this time. Living materials have reaction sequences such as the glycolytic pathway, the respiratory chain, the tricarboxylic acid cycle, and others, that are interlinked or connected at some points. Reactions at these points can become rate-limiting for the living system as a whole; i. e., they constitute bottlenecks (Krebs and Kornberg, '57). The whole system is controlled at the interlinking points. To be less general, more than 80% of all carbon in a plant passes through a coenzyme A linkage before evolution as CO_2 or incorporation into relatively invariant structures. A control of any-one of the chief reactions of coenzyme A reactions could be the answer to the third question.

3.9. SUMMARY

Many aspects of plant growth, including floral initiation and development, etiolation, seed germination, anthocyanin synthesis, bud dormancies, etc., are controlled by the action of a protein with an absorption maximum near 7350Å. This pigment, which is called phytochrome fr, changes in darkness to an inactive form, phytochrome r, with an absorption maximum near 6600 Å. Phytochrome r and fr are photochemically interconvertible and can be detected by this property in plant tissue.

Phytochrome has been separated from maize shoots and several other tissues by methods of protein chemistry using photo-interconvertibility as the basis of assay.

Phytochrome fr can also undergo change to a photo-irreversible form which is active in control of plant development. Continued photo-excitation at the wave length either of the isosbestic point or of the photo-irreversible form is effective for synthesis of anthocyanin, control of seed germination and other aspects of plant development.

DISCUSSION

HUTCHINSON: I was wondering if you could get any changes in spectrum by heating.

HENDRICKS: The reversibility of the systems is destroyed by heating to 50° in vitro. As the system is cooled to 0° the degree of reversitibility is unchanged from the room temperature value. About 0.8 of the reversible change is lost upon freezing and this is further reduced upon cooling to minus 70° C.

PLATT: This is freezing during irradiation?

HENDRICKS: No, it is first frozen and then irradiated.

PLATT: Then you warm it up again?

HENDRICKS: If you wish the system can be warmed up and the relaxation followed by means of the spectrometer.

PLATT: I don't know what Dr. Hutchinson wanted to hear from you, but I would like to know of there is a thermal dependence on conversion of one pigment to form another pigment.

HENDRICKS: No, the temperature coefficient is quite precisely 1.0.* The energy required for pigment conversion is constant as the temperature is lowered from 35° C to 0° C. Upon freezing, however, only about two-tenths of the reversibility is retained and this is reduced to about .05 at -70° C. The reversible change is absent at -100° C. If the pigment is irradiated with red light at -100° C and

* Editors' note: Kineticists frequently refer to the ratio of rate constants at two temperatures (i. e., k_2/k_1) as the temperature coefficient. This is the sense in which Dr. Hendricks uses the term. Acually, of course, from the physicists' viewpoint, he means that the temperature coefficient, $d \ln k/dT$, is zero.

then warmed up in darkness it remains in the red absorbing form. After the warming it is changed by red radiation, indicating that it was not permanently altered by cooling. In other words, it is not possible that irradiation at a low temperature potentiates a change that can be realized by warming. The radiation is absorbed at -100° C but must be degraded without an accompanying photochemical change.

REID: All of this sounds so similar to some of the things we know about the isomerization of the retinene that I would like to know what the absorption spectrum of carotenoids associated with proteins in plants is like. We know that visual carotenoids shift their absorptions far into the red in many cases, in association with protein.

HENDRICKS: Carotenoids associated with chlorophyll in plastids appear to absorb at the normal absorption maxima of beta carotene.

REID: But I would suggest that this is free carotene and that perhaps one thousandth of the carotene may be associated with the proteins in something like retinene.

HENDRICKS: That is avoiding the issue.

REID: In rhodopsin there is a shift of more than 1,000 Å.

HENDRICKS: It is true as you mention that the absorption maximum of retinene is greatly shifted by conjugation with protein.

REID: But you know that both are present together.

HENDRICKS: Of course. What is the situation for bacteria chlorophyll? Is not the shift of the absorption maximum to the region of 9000 Å a result of conjugation with carotenoid?

LINSCHITZ: No. In rhodopsin the protein plays no

particular role as protein, in shifting the absorption spectrum of the carotenoid. The protein has only the function of making available an NH_2 group which then forms a Schiff's base with the aldehyde group of retinene and finally adds a proton. One can get exactly the same kind of spectral shift in retinene just by forming such Schiff's bases with small amine molecules in acid media. There is no property which is unique for protein that would give us this large shift.

PLATT: Dr. Reid said there was a shift larger than 1,000 Å in his protein complex. I don't remember anything like that in Schiff's base.

LINSCHITZ: The absorption spectra of simple protonated retinene Schiff's bases are identical with the rhodopsin spectrum. The shift is from 3800 Å to 5000 Å.

KASHA: There have been some photochemical researches done in Israel by Y. Hirshberg and E. Fisher ('54) which involve photo-transformation and photo-reversal of spiro compounds. Those are geometric isomers of different kinds, and it may be interesting to consider if similar molecules have anything to do with the pigment change reported by Dr. Hendricks.

LINSCHITZ: Could I ask you about the possibility that there is a reversible photochemical redox reaction? It has been shown in experiments in rigid glasses that the irradiation of either quinone or hydroquinone yields the same semiquinone. When the glass is warmed up, the system goes back to the respective original reduced or oxidized form. This is well established, and one can get such a relation between oxidized and reduced form in many redox couples.

HENDRICKS: Yes. That is the way to describe a model system for the change of hydroquinone to the semiquinone.

LINSCHITZ: The formation of semiquinone from hydroquinone is not quite the same as its formation from quinone, because proton shifts are also involved. Illumination of QH_2 gives QH_2^+ by direct electron-ejection, and involves no large motions of any heavy particles. Formation of QH_2^+ from excited Q requires electron transfer from the solvent, together with proton transfers. These must be bimolecular dark reactions.

PLATT: I am very sympathetic to all these exotic changes of form such as cis-trans and Schiff's base and semiquinone forms. Nevertheless anthocyanin has a tremendous change of spectrum just in going from the acid form to the base form, that is, from the blue of cornflowers to the red of roses. It is a shift of the order of 1,000 Å which depends only on protonation.

HENDRICKS: The protonation results in loss of the double bond and breaks the conjugation which shifts the absorption spectrum by the order of 1000 Å.

PLATT: The acid-base possibility is just less exotic than those other proposals.

KASHA: But this color change is not due to a photochemical reaction. Of course, there are photochemical reactions in which protons are ejected, so that it is not impossible that Dr. Platt's remark is pertinent.

HENDRICKS: We have often speculated that the pigment is phycocyanin. In the phycocyanin formula one ring is doubly conjugated to the remainder of the system. This could be changed upon addition of hydrogen without breaking the overall conjugation.

GORDON: A brief comment about Dr. Linschitz' suggestion of a quinone or analogous reduction in the pigment conversion. Isn't that type of reaction improbable here, Dr. Hendricks? Isn't the photoreversibility of the pigment complex unimpaired in the presence of dithionite?

HENDRICKS: Yes, it goes freely in the presence of dithionite and also in the presence of an oxidant such as dihydroascorbic acid.

LINSCHITZ: Wouldn't you think that these reactions would be very highly specific?

HENDRICKS: Because of the possibility of marked specificity we have tried a number of oxidants and reductants, but none has had an effect.

LINSCHITZ: I mean that if you are going to maintain high quantum efficiencies in this reaction, involving excited states and the constant probability of energy losses by thermal conversion, that it is quite likely that the reaction in vivo must maintain very high specificity. I don't think that the fact that dithionate or oxygen have no effect really disproves the reversible redox mechanism.

GORDON: A speculation might be worthwile on the basis of the red-far red response of mitochondrial ATP generation. It might be proposed that this phosphorylative mechanism for the channeling and transfer of respiratory energy was the determinative biochemical transformation affected by the pigment complex. In other words, that the various photosensitive phenomena you describe are determined, essentially, through metabolic shifts introduced by a change in the availability of high-energy phosphate.

4

Claud S. Rupert
Johns Hopkins University

The Mechanism of Photoreactivation

Photoreactivation is a marked reduction in the biological effects of 200-300 mμ ultraviolet radiation produced by a post-irradiation treatment with longer wave lengths. This effect was first recognized less than a dozen years ago (Kelner '49) although indications of it were probably observed earlier and analogous effects in photochemistry have been known since early in the century (Sheppard '27). The subject has been ably reviewed within the past two years (Jagger '58), so that we start any discussion of its mechanism with a good idea of the present state of knowledge.

The questions we shall be asking originated from experimental study of intact organisms, but they are questions we wish to answer in terms of the physics and chemistry of molecular processes. The risk of over-simplification in this endeavor is as obvious as it is unavoidable. It is acknowledged here at the outset to avoid repeating it continually in what follows.

4.1. PHOTOREACTIVABLE ULTRAVIOLET DAMAGE

An inquiry into the mechanism of photoreactivation necessarily starts with the nature of the ultraviolet damage which it overcomes. It was first noticed some thirty years ago that action spectra for ultraviolet effects in organisms have a strong resemblance to the absorption

* This work was supported in part by Senior Research Fellowship, SF-312, from the U. S. Public Health Service.

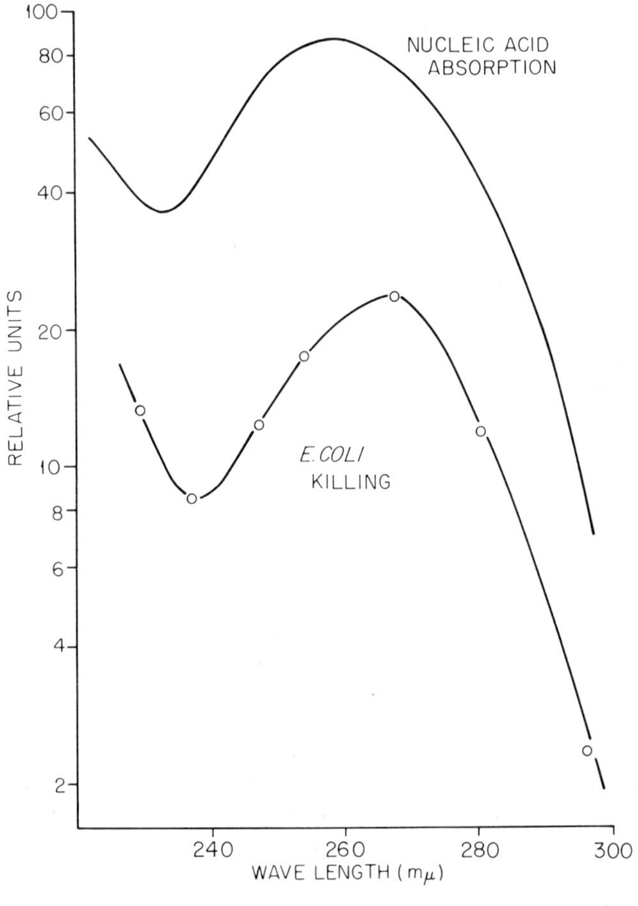

Fig. IV. 1. Similarity of action spectrum for UV killing of E. coli (Gates '30) to absorption spectrum of nucleic acids.

spectra of nucleic acids (Figure 1) (Gates '28, '30, Zelle and Hollaender '55). There are exceptions, and some action spectra have extra humps or broadenings around the aromatic absorption regions of proteins at 280 mμ (e. g., Setlow '57), but in general this evidence seems to say that the damaging radiant energy is absorbed in nucleic acids -- with occasional help from proteins -- or else in materials having similar absorption.

Nucleic acids not only absorb ultraviolet radiation, but they are damaged by it. Bacterial transforming DNA, which conveys genetic traits of the cells from which it was extracted to other strains of the same species, is inactivated by this radiation, and its radiosensitivity is the same when irradiated inside the donor cells before extraction as when irradiated outside the cells after purification (S. H. Goodgal*, unpublished data). Infectious RNA from tobacco mosaic virus is also inactivated by ultraviolet (Siegel et al. '57). Both kinds of nucleic acid can be found in photoreactivating systems, including bacteriophage with only DNA (Dulbecco '50), plant viruses containing only RNA (Bawden and Kleczkowski '53), and infectious tobacco mosaic virus RNA itself (Bawden and Kleczkowski '59). It is possible, therefore, that either type of nucleic acid can serve as the site of photoreactivable damage, although other cellular components cannot be summarily excluded.

This evidence is not particularly amplified by studies on cells. In certain cells nuclear damage has appeared photoreactivable while cytoplasmic damage was not (von Borstel and Wolff '55, Brandt and Giese '56), suggesting DNA but not RNA involvement. On the other hand, more recent work has shown photoreversal of ultraviolet effects in cell fragments lacking nuclei (Skreb and Errera '57, Pierce and Giese '57).

4.2. PHOTORECOVERY PROCESSES

Some feature of a cellular environment is evidently requited for photoreactivation. Irradiated virus or infectious RNA must infect a host cell before any recovery from ultraviolet effects can be produced by light, and in the only case accurately measured, the action spectrum for photoreactivation of a bacteriophage in its host cell was the same as that for photoreactivation of the host cell itself (Jagger and

* The author is indebted to Dr. David Pittman of Southern Illinois University, Dr. Jerome Schiff of Brandeis University, and Dr. S. H. Goodgal, currently at the Pasteur Institute, for permission to quote their experimental results in advance of publication.

Latarjet '56). These examples strongly suggest a cellular mechanism acting on nucleic acid, since DNA is the major material transferred to a cell upon bacteriophage infection, and RNA is apparently the only material involved in infection by dismantled tobacco mosaic virus.

The principal hypotheses which have been advanced to explain the recovery during illumination are, first, that a cell poison is produced by the ultraviolet radiation, and that this poison is destroyed during photoreactivation. By this hypothesis photoreactivation prevents the essential cell damage by ultraviolet. It accounts very well for many quantitative and qualitative features of photoreactivation (e. g., Novick and Szilard '49), but efforts to demonstrate such poisons by independent means have failed (Jagger '58). A second hypothesis proposes that new metabolic paths are opened during illumination which bypass processes blocked by ultraviolet damage. However, cells which have been irradiated and then photoreactivated through one or more successive cycles have essentially unaltered sensitivity to further ultraviolet irradiation (Lennox et al. '54, Jagger '58), a fact which rather discourages this view. Both these hypotheses become strained when applied to free nucleic acid. A third hypothesis presumes that the primary ultraviolet damage is itself repaired during illumination. This is the simplest assumption of all, and since no really good evidence is known against it, it is the currently favored mechanism. Its position is very much strengthened by the fact that such a process can be carried out with cell-free components in vitro.

4.3. PHOTOREACTIVATION OF DNA

Ultraviolet inactivated transforming DNA does not recover its activity if exposed to light alone. However, when it is mixed with extracts of photoreactivable cells -- baker's yeast or Escherichia coli B -- partial restoration occurs upon illumination (Goodgal et al. '57, Rupert et al. '58, Rupert '60). Recovery ceases when the illumination is interrupted. An extract of non-photoreactivable cells -- Hemophilus influenzae, the species from which the transforming DNA was taken -- is ineffective (Goodgal et al. '57). The increase in transforming activity is obtained only when the DNA has first been ultraviolet inactivated, and occurs then only if the extracts are mixed with the DNA at the time of illumination. The time rate of recovery increases with temperature up to a point, but sufficient heating will inactivate the extracts.

The yeast and E. coli extracts evidently act on the same kind of damage, since DNA which has been maximally restored by one is not further restored by the other, and both can give recovery to about the same degree. They have differences, however. The E. coli system (Figure 2) consists of two components, one dialyzable and heat stable, the other non-dialyzable and heat labile. The dialyzable component can be replaced by reduced DPN, reduced glutathione, or better still, by both (Rupert, unpublished experiments). The maximum level of recovery is limited when the amount of dialyzable component is insufficient, but after maximum recovery under such conditions, further recovery will occur if more dialyzable component is added. Evidently something of small molecular weight is being used up. However, this something is not stoichiometrically related to the amount of DNA repaired. Increasing the amount of irradiated DNA in a reaction mixture with a certain limited concentration of dialyzable material gives, not a lower level of recovery as would be expected if something had been exhausted, but a slightly higher one. Such a result makes it unlikely that the dialyzable component is consumed directly as a reactant in the repair process. It may rather be overcoming some kind of inhibition in the extract. Unfortunately, it has not been possible to purify this system significantly because of nucleases which become activated in the process and destroy transforming DNA, so that this point remains unsettled.

The yeast system (Figure 3) is much simpler to manipulate. It apparently consists of a single, non-dialyzable component which is not used up in the restoration (Rupert '60). A single aliquot of yeast extract, at a rate-limiting concentration, will photoreactivate two successive samples of irradiated DNA at about the same rate and to about the

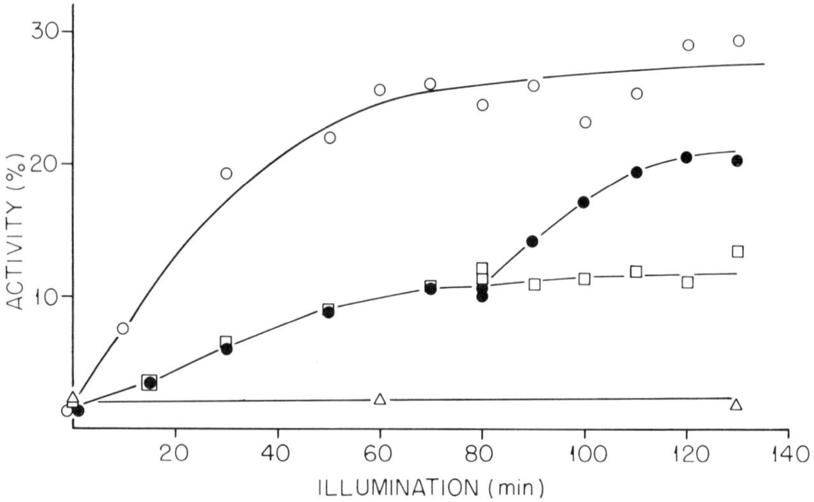

Fig. IV. 2. Photoreactivation of ultraviolet inactivated transforming DNA by components of E. coli extract (Rupert '60).

- ○ Dialyzed extract plus concentrated dialysate.
- ● Dialyzed extract with dialysate added after 80 minutes illumination.
- □ Dialyzed extract with buffer added after 80 minutes.
- △ Dialysate only.

same extent. The active agent appears in a single, compact fraction after 100-fold purification. It is inactivated progressively by trypsin and chymotrypsin at rates proportional to these enzymes' concentrations, it can be salted out with ammonium sulfate and recovered quantitatively, it is adsorbed to alumina Cγ gel and calcium phosphate gel, and it can be eluted from them in active form. In short, it is apparently a single protein with catalytic activity -- an enzyme.

4.4. THE YEAST PR ENZYME

This enzyme behaves in a conventional manner, except

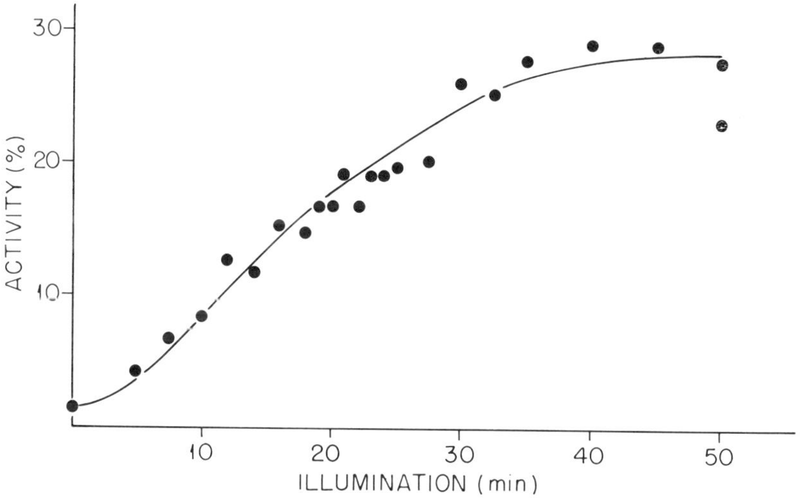

Fig. IV. 3. Photoreactivation of ultraviolet inactivated transforming DNA by yeast photoreactivating enzyme (PRE) 40-fold purified from crude extract.

that it is involved in a photochemical reaction. It forms a complex with its substrate, as can be shown by ultracentrifugation (C. S. Rupert, to be published). The enzyme itself sediments relatively slowly and consequently is recovered only slightly depleted from the upper layers of a centrifuge tube after several hours at 100,000 g. The presence of unirradiated DNA does not alter this behavior, although the DNA sediments relatively rapidly to the bottom of the tube. Centrifugation with irradiated DNA in the dark, however, causes the enzyme to follow the DNA down the tube and disappear from the upper layers. Illuminating the mixture of enzyme and irradiated DNA for a sufficient period before centrifugation removes the ultraviolet lesions with which the enzyme is complexed so that it is once more left behind in the top of the tube.

At constant temperature and light intensity, the reaction velocity is described (at least approximately) by the Michaelis-Menten equation, since a Lineweaver-Burk plot gives a straight line with a slope and intercept both proportional to the enzyme concentration (C. S. Rupert, to be published). Adding unirradiated non-transforming DNA (e. g., from calf thymus) to a reaction mixture has no effect on the photoreactivation of transforming DNA, but addition of irradiated DNA gives competitive inhibition. This inhibition increases with the amount of competing DNA and with the ultraviolet dose to which it has been subjected. It decreases (and ultimately disappears) if the competing DNA is photoreactivated sufficiently before testing to repair its ultraviolet lesions.

The above facts mean that we can consider the enzyme E and its substrate S (the photoreactivable lesions of DNA) to interact by the conventional scheme:

$$E + S \underset{k_2}{\overset{k_1}{\rightleftharpoons}} ES \xrightarrow{k_3(I)} E + P,$$

where the product, P, is a repaired DNA structure. The only unusual feature is the dependence of k_3 on the light intensity, I. In the dark, $k_3 = 0$.

4.5. OTHER PHOTOENZYMES

This is not a very surprising result. Any light-driven enzyme reaction proceeds most efficiently if the enzyme combines with its substrate before the light absorption, as this one does. Otherwise deactivation of an excited molecule can occur before the opportunity to form a complex presents itself -- although this tendency may be less important in some highly organized structure than in solution. One might reasonably expect to find other enzymes of this same general type taking part in the normal photochemistry of living things.

One such enzyme has already been known for a number of years. The protein opsin from the rods of the vertebrate retina (Wald '53) can be considered as an isomerase

(Hubbard '56). This system is highly specialized as a radiation detector but, as shown in Figure 4, it closely parallels the photoreactivating system in its formal behavior.

Retinene (vitamin A aldehyde) has several cis-trans isomeric forms. When the all-trans isomer is exposed to radiation from a band centered around 385 mμ, it is converted to a mixture of isomers. Two of these, "neoretinene b" and "isoretinene a", combine with opsin in the dark to form an enzyme-substrate complex (rhodopsin or isorhodopsin), in which the absorption band is shifted over 100 mμ toward the red. Illumination in this new absorption region converts these two isomers back to the all-trans form, with dissociation of the product from the protein.

Correspondingly, a normal DNA structure irradiated in a band centered around 260 mμ develops several kinds of ultraviolet lesions. Part of these are not photoreactivable, since transforming activity cannot be restored to the 100% level, but those which are reactivable combine with the photoreactivating enzyme to give a complex. Illuminating this complex with wave lengths some 100 mμ longer than those causing the damage regenerates the original structure and frees the enzyme.

This parallelism obviously breaks down at some point -- one does not know just where -- because retinene is chemically quite different from DNA. The chromophore absorbing the reactivating light could be either part of the substrate (as in the visual system) or part of the enzyme. The repair reaction could be either an isomerization or something else. It might be noted, however, that if there is really no component used up during photoreactivation by the yeast system, then -- since there is no evidence that ultraviolet inactivation of DNA depends on the presence of any special compound, the process occurring just as readily intracellularly as in dilute saline -- the damage and repair processes could only be isomerizations or reactions involving ubiquitous material like water. It so happens that both these particular reactions occur with pyrimidine nucleotides irradiated in dilute solution, and both are reversible (Wang '58, '59a, '59b).

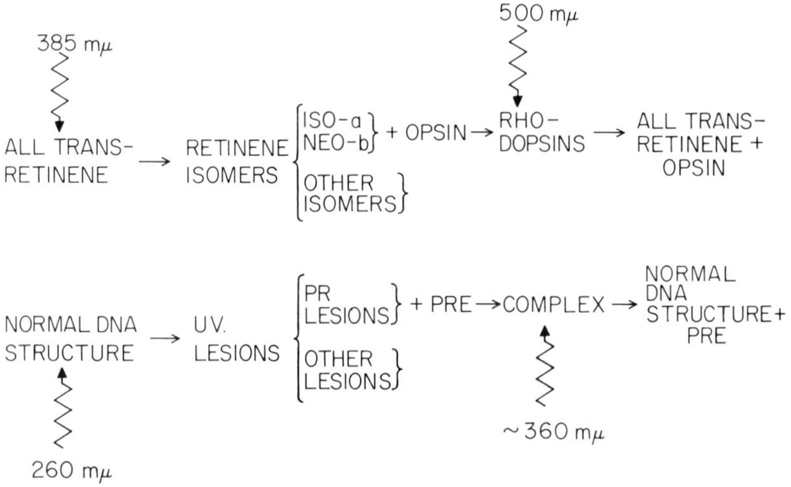

Fig. IV. 4. Formal parallelism between visual and photoreactivation systems.

4.6. SOLAR RADIATION DAMAGE

An interesting feature of the photoreactivating enzyme is its repair of radiation damage to the genetic substance of cells, using a longer wavelength radiation to drive the reaction. The most obvious utility of such a mechanism is to cause the longer wavelengths of sunlight to repair shorter wavelength damage to DNA, and such a process is actually carried out in vitro as shown in Figure 5. An hour's exposure to a near-zenith sun at sea level inactivates transforming DNA by an amount roughly equivalent to 3500 ergs mm^{-2} of 254 mμ radiation. This damage is not only photoreactivable, but it is photoreactivated in sunlight in the same quartz container simply by adding a small amount of the enzyme. This enzyme concentration is too low to give appreciable optical shielding, so that the rise in activity represents a net repair in the face of continuing damage.

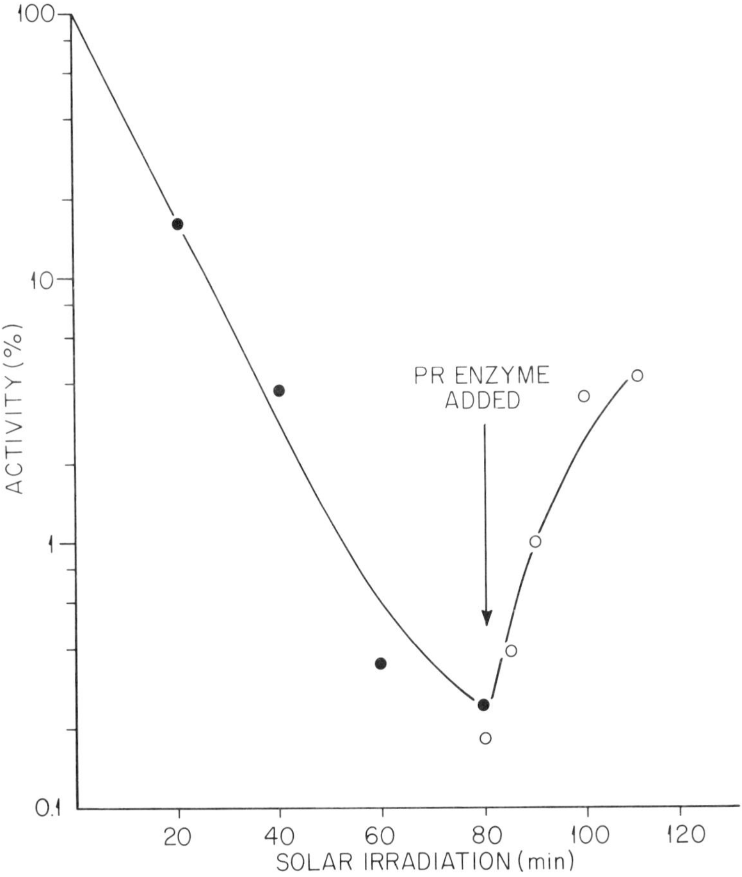

Fig. IV. 5. Solar inactivation of transforming DNA followed by solar reactivation in the presence of yeast PRE.

The widespread occurrence of photoreactivation -- in all orders of living things tested (Jagger '58) -- would be understandable if the ability of living things to repair solar radiation damage to their DNA gave them a competitive advantage, either now or at some earlier time in phylogenetic history.

4.7. ADDITIONAL PHOTOREACTIVATION MECHANISMS

Before closing we should consider some features of photoreactivation in which this enzyme is probably not involved.

By observing competitive inhibition of transforming DNA repair, one can show that the enzyme acts on irradiated DNA of all types (including Kornberg's synthetic DNA) (Bessman et al. '58). In each case an inhibition is produced by irradiated nucleic acid which is not produced by unirradiated material, and the inhibition can be removed by preillumination of the competing material with the enzyme. By this test the enzyme is apparently inactive on RNA. Hence, in any photoreactivation of ultraviolet damage to RNA (e. g., the RNA plant viruses, infectious TMV RNA, and probably certain extranuclear damage to cells) our enzyme cannot account for recovery.

It is also not concerned with photoreactivation of the most common type of ultraviolet-induced cytochrome deficiency in yeasts. The action spectrum for this ultraviolet effect has a typical nucleic acid-like shape (Raut and Simpson '55), although genetic studies indicate that the deficiency represents loss or inactivation of some non-chromosomal element (Ephrussi and Hottinguer '51, Raut '54). This effect is normally photoreactivable (Pittman et al. '59), but in strains carrying a particular gene mutation it is not. These same cells, however, are still reactivable as far as their survival is concerned (Pittman et al. '60), and their extracts will photoreactivate transforming DNA (C. S. Rupert and D. Pittman, unpublished experiments). We must conclude that more than one kind of repair is encompassed in what we call photoreactivation. Both the loss of the extrachromosomal elements in yeast and the inactivation of chloroplast precursors in Euglena gracilis (which shows a nucleoprotein action spectrum and is photoreactivable) (Lyman, Epstein and Schiff '59) and unpublished experiments) seem likely examples of RNA rather than DNA damage.

4.8. SUMMARY AND APPRAISAL

It is possible that we understand the photoreactivation process in a superficial way. The evolution of some mechanism to counter solar radiation damage in primitive organisms would be natural, and if photoreactivation is the remnant of such a mechanism, then it is clear why repair, rather than replacement, is essential for DNA -- which carries the cell's master file of genetic information. Evidently a mechanism has developed for this repair, driven by wavelengths of radiant energy which normally arrive at the same time as the damaging radiation. Some critical pieces of RNA may also require repair instead of replacement, but damage to many other cellular components would probably be satisfactorily countered by a rate of synthesis sufficient to compensate for destruction. The selective pressure favoring repair mechanisms over simple replacement would be less acute in these cases.

This interpretation plausibly accounts for the major features of photoreactivation without contradicting any known facts. No direct experiment, however, links the ability of a present-day organism to endure sunlight with its capacity for photoreactivation, and doubters are certainly entitled to make themselves heard in the discussion.

Assuming that we accept this superficial interpretation, we, nevertheless, understand very little about the photoreactivation mechanism on a deeper level. The actual means of RNA recovery is unknown, and even for DNA, where we have an enzyme, the photochemistry itself with the role played by the protein is a blank. One can guess intelligently, and one of the things this meeting can do is to explore the possibilities based on chemical knowledge, but this is not the same thing as understanding.

What I have given is in no way a framework for the general discussion. I hope that it is plausible enough to serve as the beginning for it, but I also hope that it is sufficiently controversial to excite some disagreements, for here is a subject which deserves all the vigorous debate which a diversified meeting of this kind can provide.

DISCUSSION

KASHA: The word "lesion" can be interpreted in several ways. One way DNA inactivation might occur would be local destruction of, say, one pyrimidine, which, however, may still leave the pyrimidine in the helical DNA structure. Is it possible to consider an enzyme reaction at a "lesion" without some major change in the structure of the DNA itself? In other words, I would think that the ribophosphate helices might keep the pyrimidine protected from external reaction, unless the DNA broke open at that point.

RUPERT: By "lesion" I mean a local alteration in the DNA molecule that disrupts its biological functioning. It might involve a change in one or several neighboring nucleotides.

It is certainly true that the Watson-Crick structure tends to bury the purine and pyrimidine rings in the center of the helix, but they are not completely hidden from the outside. A deep groove winds around the helix through which parts of them are accessible. It is also possible, as you suggested, that the helical structure is disrupted at a lesion. Marmur and Doty ('59) found a small drop in the heat denaturation temperature of DNA irradiated with approximately the same UV doses we use, and this must mean some attenuation of the hydrogen bonding that maintains the structure. In practice accessibility of the ultraviolet lesions to enzyme action doesn't seem to be a serious problem. If we use competitive inhibition of transforming DNA repair as a criterion, the enzyme interacts with UV lesions in Watson-Crick DNA about as readily as with lesions in denatured DNA, or in Sinsheimer's single-stranded ϕX-174 DNA.

PULLMAN: Electronically, the most susceptible point would be in the pyrimidine ring and probably at its C_5-C_6 bond. This bond is also at the external face of the helix so I think it can be attacked in spite of the fact that it is a helix.

PLATT: I thought some of the earlier reports on photoreactivation had the reactivation spectra in the visible region.

RUPERT: In most cases the reactivating light lies around the borderline between the visible and near ultra-violet, but only a few good action spectra have ever been determined. The one for E. coli cells has its maximum around 360 mμ and trails off weakly through the violet and blue-violet. The one for Streptomyces griseus spores peaks around 435 mμ, extending fairly symmetrically on both sides of this value. According to Jagger's review article ('58), most systems have their maxima around 380 mμ. The most effective wavelengths for our photo-reactivating enzyme are not accurately known, but they are in the usual spectral region.

PLATT: Is the absorption spectrum of the initial transforming principle changed by the irradiation?

RUPERT: We see no change, and I think there is good reason why we should not at this level of irradiation without taking special pains. Even at 10 or 20-fold higher UV doses, the gross properties of the DNA are very little altered, so that hardly as much as one nucleotide in ten could be affected. That means that at our doses, less than 1 percent (and perhaps very much less than this) has been changed. As one ordinarily determines optical densities with, say, a Beckman DU, a change in absorption corresponding to the usual hyperchromic effect, or even a rather big change in absorption spectrum, involving only this fraction of the material would be hardly perceptible. With the best modern techniques for detecting small absorbance changes it might be worth looking for, but we haven't done this.

PLATT: Is your PRE species specific?

RUPERT: No. We obtain the enzyme from yeast and use it to restore irradiated Hemophilus influenzae DNA. And then, working in this system, we can get a competitive inhibition from irradiated calf thymus or bacteriophage DNA which is eliminated if the enzyme is allowed to act on these materials in the light before the Hemophilus DNA is added. Yeast PRE evidently acts on ultraviolet damage in DNA from all these species.

PLATT: Are there any non-photoreactivable mutants?

RUPERT: There supposedly are. Several yeast strains from the Southern Illinois University collection did not photoreactivate when sent to Dr. Albert Kelner for test. That was quite a while ago. Dr. Pittman is trying to track these down now. There are, of course, whole species of cells which show no photoreactivation.

KASHA: I would like to comment on a paradox related to the long wavelength reactivation. One of the weakest links in the DNA structure photochemically speaking, is probably one of the pyrimidines. The aqueous photochemistry of some pyrimidines has been studied, and the final products have been isolated and identified (Wang '58).

But these products always absorb at a much shorter wavelength than the pyrimidine itself does, so we have a mystery here, which really requires some new molecular species to be found which can satisfy the longer wavelength absorption requirement.

PLATT: It seems to me that it cannot be as simple as a hydrated pyrimidine because of the variation in action spectrum from one species to another.

RUPERT: Most PR action spectra, so far as they are known, are about the same. Where differences occur, they could perhaps be due to auxiliary pigments on the protein which extend the active wavelength region beyond that where the unaided protein-substrate complex would absorb. Unfortunately we don't have even the prospect of checking this by direct observation of absorption spectra at the present time. The enzyme in our reaction mixtures must be at a lower concentration than the substrate because Michaelis-Menten kinetics apply. If what I surmised is true, that something less than 1 nucleotide per 100 is involved in photoreactivable lesions, and if one assumes that the protein has a molecular weight somewhere between 10,000 and 100,000, some limits can be set on the mass of enzyme present. It turns out that at least 10^5 or 10^6 fold purification of crude yeast extract would be necessary to get the pure protein. We have been able to go to 100 fold, which is just nothing. And this is the situation with the PR system we can manipulate best.

KASHA: I wonder whether anyone has considered trying photochemical experiments on synthetic polynucleotides where one might have a chance of actually preparing a large amount of the photo-inactivated material.

RUPERT: The synthetic polyribonucleotides are all inactive; i. e., after irradiation they do not competitively inhibit the photoreactivation of irradiated transforming DNA. That includes a two-stranded polyadenylic-polyuridylic acid supplied by Dr. Davies of NIH and a two-stranded polyadenylic-uridylic-cytidylic-guanilic acid supplied by Dr. Rich at MIT.

KASHA: I meant: to irradiate a polynucleotide and then break it down and try to separate out some more complex product.

RUPERT: The only way I know to link such studies with the phenomenon we have been studying is to see, first, that our enzyme will combine with such a radiation product so that we get a competitive inhibition, and, second, that the enzyme can act on the product in the light (but not in the dark) to eliminate this inhibitory power. Otherwise what is your criterion for picking out from all the heterogeneous radiation damage that which is concerned with photoreactivation? You must connect somehow with a photoreactivating system.

Now, if we irradiate synthetic polynucleotides and put them into our photoreactivation mixture we make no change whatever in the reaction rate. (The only exception is Kornberg's synthetic DNA containing all four bases. It acts like natural DNA, although his polydeoxyadenylic-thymidylic acid does not.) It seems to me that whatever ultraviolet lesions are present in these "non-competing" polynucleotides are either of a character different from lesions in DNA, or else they are in structural surroundings where the enzyme cannot act on them.

KIRBY-SMITH: Did you find any DNase per se?

RUPERT: The yeast PRE preparations show no DNase activity. However, in the messy E. coli system we have never been able to get away from DNase. We hold it only rather uncertainly in check with inhibitors naturally present in the crude cell extract.

SZENT-GYORGYI: I want to get your idea about reactivating light. What can it do?

RUPERT: If I had to guess I would surmise it supplies the energy needed to drive the repair reaction, in a way somewhat analogous to the retinene-opsin system. The damaged system is raised to an excited state from which it can pass over into the repaired form (maybe an isomerization or maybe something else) while complexed with the enzyme. Following this, there is a dissociation of the product from the protein because no strong affinity exists between the repaired structure and the enzyme. One cannot say at the present time whether the reactivating light is absorbed in a chromophore carried by the protein or in the ultraviolet lesions of DNA themselves, with their absorption spectra possibly modified by attachment of the protein.

SZENT-GYORGYI: How long can you wait before you reactivate?

RUPERT: We have waited for as long as six weeks with the material frozen. Usually we leave it liquid at 5°C over a number of days. We just irradiate a batch of DNA and work with it until we run out, and then irradiate more.

These photoreactivable lesions in DNA are very stable structures. A sample of irradiated non-transforming DNA can be heated to 100°C without losing its competitive inhibitory power. Tested in this way, photoreactivable lesions are stable to 100° for at least 40 minutes, and to pH 2.7 or pH 12 overnight at room temperature. This again makes me feel that a certain amount of energy must be injected in order to flip them back to the normal repaired DNA structure, this energy normally being supplied by light.

REID: There is some evidence, is there not, of being able to do the photoreactivation before you do the damage? If this is genuine how will it fit into the picture?

RUPERT: That apparently does not fit into this picture at all. John Jagger at Oak Ridge, who has recently been working on this so-called photoprotection, feels it involves some entirely different mechanism. It occurs under narrowly circumscribed conditions with some strains of cells and not with others, although both may be capable of photoreactivation. And it occurs in E. coli B/r if they are plated with salt, but disappears if they are not, while their photoreactivation is independent of this detail. It also requires light doses very much higher than the doses needed for photoreactivation, even up to the point that the light itself is beginning to decrease cell survival.

HUTCHINSON: I just wondered if some chemist could not suggest a method for adding water between those five and six positions which could be applied to intact DNA. Put some of the DNA in your system and see if you get competitive inhibition. Let the biochemist create the photoreactivable damage in DNA instead of having the ultraviolet treatment do so.

RUPERT: Wang has added water chemically across the 5-6 double bond in 1-3 dinethyl uracil.

HUTCHINSON: Can it be done in the intact DNA?

ALLEN: Do I understand that you require a dialysate preparation or some other reducing agent?

RUPERT: In the case of the E. coli system we do. In the case of the yeast system there is no evidence of any dialyzable component. You can precipitate the enzyme with ammonium sulfate, redissolve and dialyze, or adsorb to calcium phosphate gel, elute and dialyze and the resulting active product behaves just the same as the starting material. In the E. coli system on the other hand, if you simply dialyze the enzyme preparation, the maximum level of recovery it will produce gets lower and lower.

ALLEN: Does that suggest that you are dealing with oxidation and reduction?

RUPERT: It might if there were any stoichiometric relation between the amount of dialysate used up and the amount of DNA repaired. But I think a serious discrepancy shows up here, as I tried to point out in my paper. Possibly I covered this part too rapidly. The point is this: In the E. coli system, the maximum level to which recovery proceeds (the recovery "plateau") is lowered by decreasing the amount of dialyzable material present. This result suggests that some small-molecule component is being used up. But if, with a deficiency of dialyzable material, you then double the amount of irradiated transforming DNA in the mixture, the recovery level is about the same or even a little higher, instead of being somewhat lower, as you might expect.

Also, (and I hadn't mentioned this) we find you cannot exhaust the system by letting it act on irradiated non-transforming DNA before the transforming DNA is added. If you add some other irradiated DNA to the E. coli enzyme with a limiting amount of the dialyzable component present, illuminate the mixture until you would expect this DNA to be maximally repaired, and then add irradiated transforming DNA, the latter is repaired just as completely as if the first DNA had not been added. The recovery plateau will not be as high when there is a deficiency of dialyzable material as it is when there is an excess, but it is not affected by whether the system has already repaired a previous batch of DNA.

There is just no clear relationship between the quantity of DNA repaired and the amount of dialysate apparently consumed. I am afraid that something about working with a crude cell extract has been misleading us in the E. coli story. The dialyzable component could be entirely extraneous to the actual photoreactivation.

LINSCHITZ: How about saying that the light is actually making the reducing agent in the enzyme which is then replaced right at the place you want the repair?

RUPERT: All right. But so far no evidence suggests that anything is used up during photoreactivation with the yeast system. One can use the same aliquot of enzyme over again, putting more DNA in it, without seeming to exhaust anything. Any postulates about mechanism should be consistent with this fact.

REID: I have already been doing some speculating on this question of whether there was an oxidation reaction involved, and it seems to me that you can quite generally say that it looks as if damage is probably an oxidation, and the reaction is probably a reduction which might well occur via the triplet state of the reactivating molecule. We do know that triplet states are fairly good scavengers of oxidants or oxidizing materials and in this way you might be able to reuse the same molecule but nevertheless have the competing reaction which would partially use it up, but not stoichiometrically.

RUPERT: I have done one experiment to test the effect of oxygen on our photoreactivation system. I am not very happy about the design of it, but maybe somebody here can help me.

To show how recovery proceeds in the absence of oxygen, about 0.4 ml of reaction mixture was put in a chamber of about 13 ml volume which was flushed over a period of an hour in the dark with 4.5 liters of nitrogen previously bubbled through alkaline pyrogallic acid. Two of these anaerobic chambers were used along with one left open to the air. This latter reaction mixture was illuminated and sampled at intervals to get a complete recovery curve (like Figure 2 of the paper). One of the anaerobic mixtures was illuminated long enough to get partial recovery, and the other long enough to get maximum recovery. When the contents were tested for transforming activity their points fell right on the recovery curve obtained for the aerobic mixture. This experiment was done for both the yeast and E. coli photoreactivation systems with the same result.

The point that I don't like is this. We are dealing here with a substrate that can't be over about 10^{-8} M and I am not sure how we can satisfy ourselves that something equivalent to this concentration of oxygen is not being supplied by the components of the "semi-crude" cell extract which is present in the mixture.

HOLLAENDER: I believe you can use sodium hydrosulfite in reasonable concentrations. It is not toxic to E. coli and as far as I know it is not toxic to yeast in concentration in which it protects very much better than oxygen.

RUPERT: We'll have to try that. There is no clear-cut oxygen effect in photoreactivation of intact cells. In the best studies, luminescent bacteria were employed to test for the absence of oxygen, the oxygen tension being decreased until luminescence disappeared. It made no difference in the photoreactivation of the yeast and E. coli cells being studied in these experiments. These findings taken together may not be totally conclusive, but they surely argue against molecular oxygen being involved in the recovery process.

HENDRICKS: Is there a chance that these changes are similar to those studied by Vernon? He found that photooxidations involving DPNH give hydrogen peroxide depending upon the aerobic or anaerobic character of the system. Oxygen does not have to be present. The reactivation with an absorption maximum at 3600 Å could involve a flavin. Many of the other reactivations with action spectra extending to 5000 Å could also involve flavins. Is this type of action excluded in your system?

RUPERT: No, I don't think you can say that at all. When a substrate is present at less than 10^{-8} molar, I am not sure how you can say that any particular thing is or is not present in significant amounts, unless you actually have a crystalline enzyme preparation. If we purify cell extracts 100-fold or so, we still have an awful lot of junk present in great excess over the reactivating enzyme.

HENDRICKS: Does not dialysis also have these characteristics?

RUPERT: With the yeast system, dialysis does not have any effect. We have dialyzed as long as 65 hours against a couple of thousand volumes of buffer on a shaker without any change in activity, compared with the undialyzed material held at the same temperature for the same length of time.

HENDRICKS: Can't FMV be removed by dialysis against cyanide?

RUPERT: There may be tricks we don't know that would work on something like this. All we can say is that whatever is necessary for PR activity does not dialyze away in dilute phosphate or tris buffers or dilute salt solutions.

5

Michael Kasha
Florida State University

Ultraviolet Radiation Effects: Molecular Photochemistry

Radiation chemical and biological effects may involve some electronic excitation, and ionization, processes in common even if contrasting radiation sources are used. At the outset it is evident that four different disciplines have overlapping contributions to make to the understanding of this field. They are: molecular electronic spectroscopy, photochemical kinetics, organic photochemistry, and radiation chemistry. Each of these has its own language, characteristic methods of investigation, and seeks a characteristic fragment of knowledge -- and thus a large communication barrier exists. The great interest in, and the great urgency of, radiobiological problems requires some major attempts at a deliberate union of these disciplines -- perhaps, one could say, a van de Graaff wedding.

In this presentation, instead of making a comprehensive survey of the voluminous empirical literature in the field of ultraviolet photochemical effects, a few central problems have been singled out in order to raise the question of how much is known today about detailed electronic mechanisms of photochemical processes. Such a study could be called molecular photochemistry, when upon interpreting the photochemical reactions of a molecular system we take specific cognizance of the molecular skeletal and electronic structure in relation to function -- analogous to the focus implied in the term molecular biology.

Photochemical kinetic studies constitute a major part of the researches in photochemistry undertaken by physical chemists. Much of this research is on a stoichiometric level and may give little information on the intramolecular electronic processes occurring, although the advent of flash spectroscopic techniques is rapidly changing the picture (e.g., Bridge and Porter '58; Porter and Strachan '58). On the other hand, the powerful and systematic methods of organic chemistry certainly have much to offer in the unraveling of the intricate molecular mechanism of photochemical reactions (e.g. the studies on a uracil by Wang et al, '56; Wang '58). Unfortunately, however, spectroscopic theory has not been applied often in the elucidation of such mechanisms, partly because its own nomenclature details have served to make it somewhat inscrutable, but largely because of actual limitation in the development of the theory. However, the last decade has brought great extensions in the understanding of molecular electronic spectra, so it is worthwhile to inquire if any connections between theory and experiment can now be made.

5.1. THE PHOTOCHEMICAL LAWS

5.1.1. Zeroth Law:* The Activation Principle. -

Grotthus and Draper, over a century ago, established the principle which may be called the photochemical activation principle: Only the light that is absorbed by a molecule is effective in producing a photochemical change. In earlier times, when the distinction between scattering phenomena and quantum transitions was unclear, this principle had a central role in the interpretation of photochemical reactions. In the perspective of the quantum age, the utility of this principle has diminished. It seems appropriate to recognize the activation principle as a zeroth law -- a now self-evident starting point for any photochemical interpretation.

---

* Editors' note: This law has been sometimes called the first law of photochemistry (cf. G. K. Rollefson and M. Burton, Photochemistry and the Mechanism of Chemical Reactions, Prentice-Hall, Inc., New York 1939).

5.1.2. The Quantum Activation Principle (Photochemical Equivalence). - Stark ('08)and Einstein ('12) applied the quantum hypothesis to the consideration of photochemical reactions, and were led to the enunciation of a principle which may be called the quantum activation principle, and which may be stated: Each quantum of radiation absorbed by a molecule activates one molecule in the primary step of a photochemical process. Stark laid particular emphasis on distinguishing between the primary act and secondary steps in the reaction. The principle as stated implies that a limiting quantum yield of unity can be expected for photochemical reactions, if the complications of stoichiometry of secondary reactions, chain mechanisms etc., are excluded (cf. Livingston '55); hence the idea of "photochemical equivalence" arose, in analogy to electrochemical equivalence.

The principle as stated included no intramolecular limitations on intrinsic validity. However, as revealed especially in the great Faraday Society Discussion ('26) on photochemistry, numerous kinds of exceptions were found. In reviewing the limitations of the principle, Allmand ('26) listed as one of the main conditions required by Einstein's thermodynamic, and his Bohr-model, proof of the principle that "all molecules are at the same energy level before reacting and are hence equally reactive."

The Einstein derivations apply to a two-state atomic energy-level diagram consisting of two electronic states of atomic character, and cannot be expected to be valid for molecules generally, if radiationless transitions to a third, or additional, state distribute the excitation energy before photochemical reaction. This leads to the necessity of considering the Stark-Einstein principle as a limiting case, and of considering the nature of the excitation of polyatomic molecules explicitly, in formulating photochemical laws. Possibly the neglect of such a refinement is at the root of the quantum yield debate among photobiological researchers.

5.1.3. Principle of Bipartition of Molecular Excitation. - Studies of electronic excitation in polyatomic molecules reveal that in general a series of extremely rapid physical events take place before luminescence emission or photochemical reaction can occur. The most important of these are first, **internal conversion** in which states of like multiplicity combine radiationlessly, leading to the rapid population of the lowest excited singlet state of a normal molecule; and second, **intersystem crossing**, leading to the competitive excitation of the lowest triplet excited state of a normal molecule. The consequence of these uniquely polyatomic phenomena may be stated as a principle of bipartition of molecular excitation: **Each quantum of radiation absorbed by a normal polyatomic molecule has a probability of bipartition between the lowest excited singlet and triplet states of the molecule.** The probability of the bipartition of excitation in a molecule is quantitatively determined by the intersystem crossing ratio χ (Kasha '50). The limiting magnitude of this ratio can be readily determined by spectroscopic and quantum yield measurements (Kasha '60a).

For many molecules, experimental studies show that triplet states of molecules can be excited with a quantum yield (see next section) which is comparable in magnitude with the quantum yield of singlet excitation. The intersystem crossing ratio is determined mainly by the intramolecular spin-orbital interaction in the molecule, and is a molecular constant. However, intermolecular spin-orbital interaction, which is generally weaker, can modify the intersystem crossing ratio in some exceptional situations.

If a photochemical reaction occurs for a molecule only if in a singlet excited state, or only if in a triplet excited state, then the existence of a finite intersystem crossing ratio suggests that limiting non-integral quantum yields are predicated by intramolecular photophysical processes in the molecule. Only in the case that the quantum yield of triplet excitation (as defined below) reaches unity would the limiting case of the Stark-Einstein law as outlined by Allmand ('26) be attained.

5.1.4. Two-State and Three-State Quantum Yields. - Recognizing that the ground singlet, and the lowest triplet and singlet excited states are the three most important

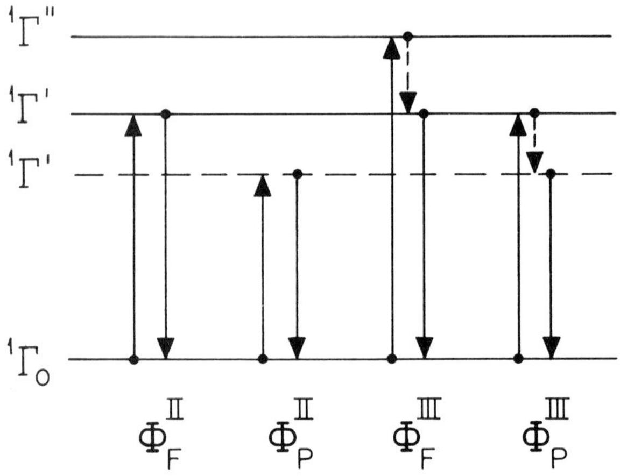

Fig. V.I. Molecular energy level diagrams defining two-state and three-state quantum yields.

states in photochemistry of most molecules, a special problem in defining quantum yields arises. As will be apparent from Figure 1, two states alone may be involved in determining the luminescence (or photochemical) quantum yield for a given excitation. However, the case of three states being involved, with a highly efficient radiationless transition between the two upper states, is a common one.

A two-state quantum yield is defined as the number of molecules emitting or reacting at a particular excited state divided by the number of quanta absorbed in going from the ground to that same excited state.

A three-state quantum yield is defined as the number of molecules emitting or reacting at a particular excited state divided by the number of quanta absorbed in going from the ground to some other excited state, above the excited state in question.

These definitions have considerable practical importance because the range of values of the total quantum yield, the relations holding between quantum yield and observed lifetime of an excited state, etc. are different for the two definitions. Since triplet state quantum yields are almost invariably measured as three-state quantum yields (because of the vanishingly weak direct absorption from the ground singlet), a word of caution concerning the use of normal (two-state) quantum yield relationships seems not out of place. The formulation of the quantum yield-lifetime relations is straightforward and will be given in another place.

5.2. MOLECULAR EXCITATION

5.2.1. Typical Molecular Orbitals. -

To begin a discussion of the relationship of electronic spectroscopy to organic photochemistry a brief examination of the nature of molecular orbitals will be made. Diagrams of such orbitals seldom see the light of day, and the repetition of the statement that molecular orbitals can be approximated by linear combinations of atomic orbitals does not automatically help to promote a fuller understanding.

To show the three types of orbitals normally dealt with in the middle ultraviolet excitation of polyatomic molecules, the formaldehyde molecule may be considered. Figure 2 shows the five possible types of orbitals superposed on the plane of the molecule. (Solid line represents a positive contour for the orbital wave function; dashed line, negative.) The sigma-orbitals are cylindrically symmetrical. The features of particular interest are: (a) the antibonding sigma or σ^*-orbital has an excitation node between the C and O atoms; (b) since the O atom is more electronegative than C, the σ-orbital has a larger magnitude at the O atom than at the C atom. (However, the sigma orbitals are strongly localized between the atoms, so this displacement of electronic amplitude is small; the localized C-H sigma orbitals are not shown).

The pi-orbitals have a plane of antisymmetry coinciding with the skeletal plane of the molecule. The features of particular interest are: (a) the antibonding pi or π^*-orbital has an excitation node between the C and O atoms; (b) because of the larger electronegativity of the O vs. the C atom, and because of the high mobility of the pi-orbital electrons, a considerable charge displacement from the C to the O atoms is present in the π-orbital; (c) as a requirement of orthogonality between orbitals, the π^*-orbital

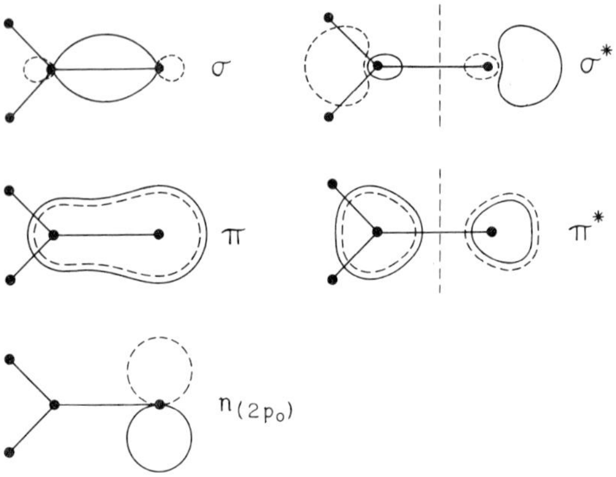

Fig. V. 2. Classification of types of molecular orbitals (formaldehyde skeleton) by symmetry properties and bonding character.

consequently must have a charge displacement from the O to the C atom.

The n-orbital or non-bonding orbital shown is the $2p_0$ orbital, whose axis lies in the molecular plane. This is essentially an atomic orbital, localized on the O atom. The more firmly bonded spherical $2s_0$ orbital is not shown as it probably does not enter into transitions in the ultraviolet range under consideration.

Theoretical calculations (Parr and Pariser '55) indicate that a contribution of 1.3 D to the dipole moment can arise from the charge polarization in the π-orbital of formaldehyde, and that an additional contribution of 1.5 D can be contributed by the charge distribution of the lone pair or n-orbital electrons, thus accounting approximately for the observed dipole moment of the molecule. The square of the orbital wavefunctions described above yields the probability distribution of electronic charge for an orbital.

For a larger polyatomic molecule, instead of obtaining just two orbitals of each symmetry type, a number increasing with the number of atoms in the molecule is obtained. Figure 3 shows the molecular π-orbitals and the n-orbital for the pyridine molecule. Three bonding π-orbitals, and

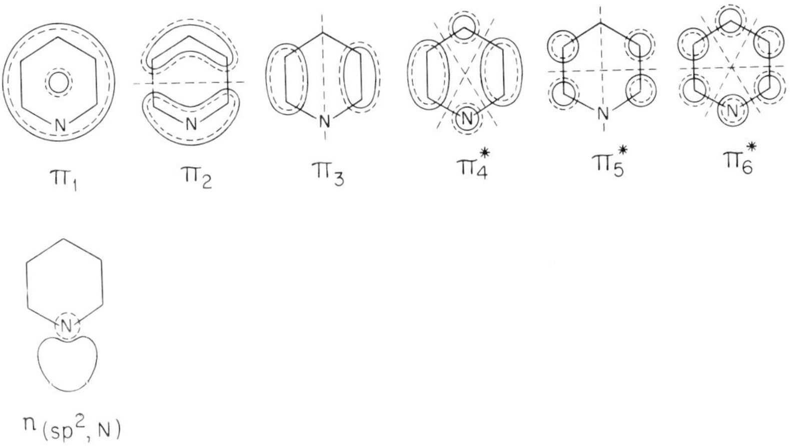

Fig. V. 3. Classification of molecular orbitals (pyridine skeleton) showing π-orbital and n-orbital nodal characteristics.

three antibonding π* -orbitals exist. All of these orbitals have as a plane of antisymmetry the molecular plane. In addition, with increasing orbital energy there is an increase in the number of nodes perpendicular to the molecular plane. The orbital numbering is in accord with self-consistent-field energy calculations (McWeeny and Peacock '57). The n-orbital in pyridine is an approximately sp^2-hybrid orbital on the N atom. It is highly directional and has an axis in the plane of the molecule.

The principal reason for presenting here the orbital diagrams for the pyridine molecule is to emphasize the delocalization of electronic distribution for all of the π-orbitals. Each orbital $\pi_1 \ldots\ldots \pi_6{}^*$ represents the one-electron orbital wavefunction, whose square represents the probability charge distribution for a single electron (the relatively small effect of the electronegativity of the N vs. the C atom is not shown). Consequently, in any photochemical mechanism, arguments based on a specific localization of a π-electron on one atom would not be justified unless a particular cause for such localization could be established. Usually, theory would suggest at most a tendency, which could be exaggerated by the perturbation induced by an approaching reagent.

5.2.2. Typical Molecular Energy Level Diagrams. - Consideration of the available orbitals and the number of electrons involved in molecular formation leads to the formulation of configurations and states of the molecule. For example, in the ground state of pyridine the electron configuration would be $\pi_1^2\pi_2^2\pi_3^2n^2$. The details of molecular state formation will not be given here (cf. Kasha '60). If formaldehyde or pyridine is considered, it is clear that, if an electron from the closed-shell ground-state configuration is to be excited, then a choice of promotion types is available. An n-orbital electron promoted to a π^*-orbital will yield singlet or triplet n, π^* states, while a π-orbital electron promoted to a π^*-orbital will give singlet or triplet π, π^* states.

Figure 4 shows the two most typical molecular energy level diagrams, found in numerous examples. The left diagram applies to many hetero-atom molecules (most aldehydes, ketones, quinones, thioketones, nitro-, nitroso-compounds, etc.). The right diagram applies to unsaturated and aromatic hydrocarbon molecules, which contain no n-orbitals.

5.2.3. Typical Molecular Emissions. - The phenanthrene molecule conforms to the right diagram of Figure 4. As shown in the total emission spectrum, Figure 5, a relatively strong $^1\Gamma'_{\pi,\pi^*} \rightarrow {}^1\Gamma_0$ fluorescence emission is observed in the near ultraviolet, partially in the visible region. But the $^3\Gamma'_{\pi,\pi^*} \rightarrow {}^1\Gamma_0$ phosphorescence emission, while having an integrated emission quantum yield which is about one fourth of that of the fluorescence emission, is nevertheless finite (cf. Kasha '60a).

If the corrected relative quantum yields of these emissions are determined, the intersystem crossing ratio $\chi = \Phi_P/\Phi_F$ can be obtained. The luminescence measurement made in the absence of intermolecular quenching by the techniques of rigid glass solution spectroscopy at 77°K thus offers a limiting value of the quantum yields of excitation of lowest triplet and singlet excited states. The total

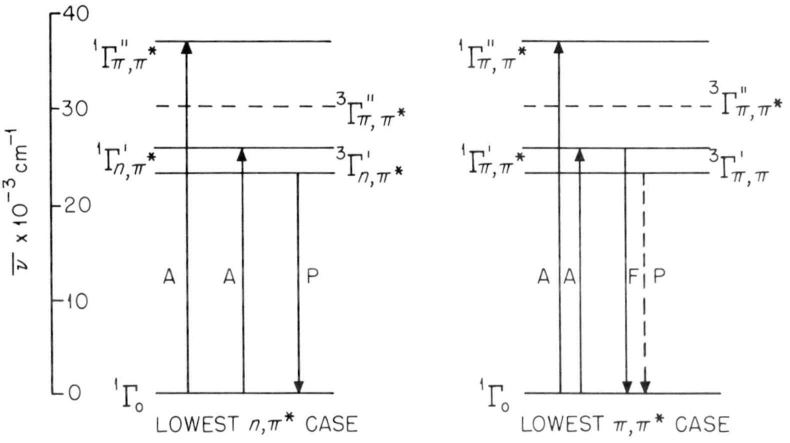

Fig. V. 4. Typical molecular electronic energy level diagrams showing transitions between lowest excited states.

luminescence yield approaches unity in many molecules in rigid glass solution at 77°K.

The benzophenone molecule conforms to the left diagram of Figure 4. As is evident from the total emission spectrum, Figure 6, a unique $^3\Gamma'_{n\pi^*} \rightarrow {}^1\Gamma_0$ emission is observed (cf. Kasha '60a). Since this luminescence at 77°K has a quantum yield approaching unity, an efficient photochemical reaction might be anticipated if the $^3\Gamma'_{n\pi^*}$ excited state were involved (the limiting case of the Stark-Einstein principle being applicable). However, the details of the excitation process (cf. Kasha '60a) do not eliminate reaction at the singlet excited state, $^1\Gamma'_{n\pi^*}$, in cases where the left diagram of Figure 4 applies. Only a detailed study of the rates of all the competing photophysical intramolecular steps and intermolecular photochemical steps can determine the probable reaction path. If a photochemical step has a rate constant comparable with or greater than the rate constant for intersystem crossing (cf. Kasha '50), then reaction at the excited singlet state will occur. The competition with intersystem crossing, however, will probably lead to non-integral quantum yields in the excited singlet reaction case.

5.3. REACTIVITY OF EXCITED STATES

Triatomic and slightly more complex polyatomic molecules frequently are found to undergo relatively large changes in shape (i. e., atomic configuration) upon electronic excitation. In the recent Symposium on the Structure and Reactivity of Electronically Excited Species ('58) much attention was paid to such changes in shape of simple polyatomic molecules.

In larger polyatomic molecules, the excitation of one electron in a configuration in which many electrons remain in bonding orbitals, is not likely to result in any considerable changes in molecular geometry. As interesting as such small changes which do occur are to spectroscopists analyzing band structure, they usually would be of little influence in determining photochemical reactivity.

The main problem in theoretical molecular photochemistry of large polyatomic systems (aside from energetic and quantum yield aspects) is to relate the electronic distribution functions deduced by the spectroscopist and quantum chemist to the structural analysis and mechanistic thinking of the organic photochemist.

In selecting examples for discussion of electronic mechanisms in molecular photochemistry, a few cases have been chosen in which the simplicity of the reaction allows a relatively clear statement of at least the main features of the mechanism.

5.3.1. Reactivity of $n\pi^*$ Excited States.

As an example of an attempt to relate spectroscopic interpretation to organic photochemistry, the photochemical reaction of ketones in solution in alcohols is particularly interesting. The optical excitation is well defined, and the structure of the reaction products is known. Figure 7 indicates a set of mechanistic primary steps which are an attempt to transpose the information gleaned from the molecular orbital diagrams of Figure 2 for the formaldehyde skeleton. The first phase of the discussion should apply equally well to

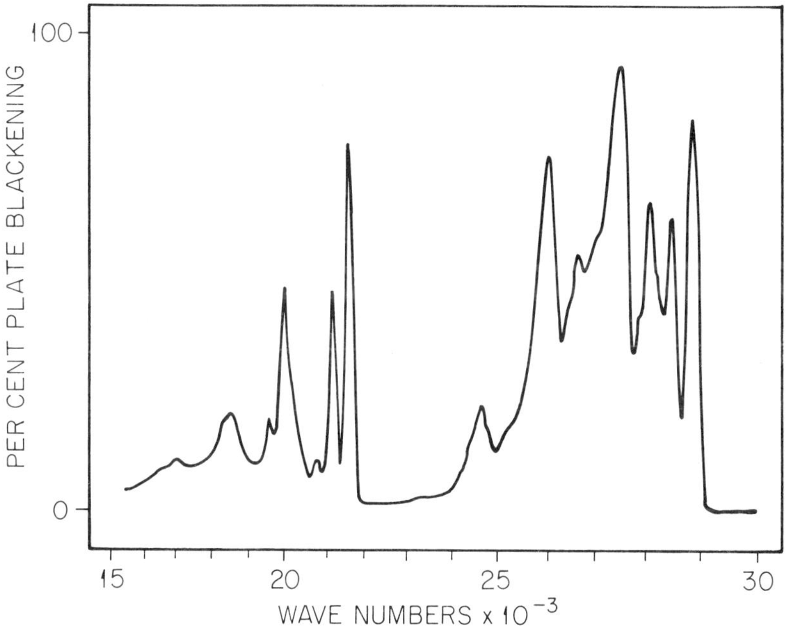

Fig. V. 5. Typical total luminescence spectrum of a molecule with lowest π, π* excited states (phenanthrene, EPA rigid glass solution, 77°K; fluorescence at right, phosphorescence at left).

any aliphatic aldehyde or ketone.

According to the orbital configuration of an isolated or aliphatic carbonyl group in both the lowest singlet and triplet excited states: $\sigma^2\pi^2n^1(\pi^*)^1$, an n → π* orbital promotion does not remove electrons forming the double bond. Thus in Figure 7 the structures b representing the excited configuration, show the double bond intact. The two structures b of Figure 7 represent extreme limits of the formal charge assignment in the excited molecule. The upper structure b of Figure 7 shows one half of an antibond (the antibond —+— shown, thus, as enclosed in parenthesis), representing the promoted electron in the π* -orbital. This would put a formal charge of (-)1/2 on the C, and (+)1/2 on the O atom. The lower structure b of Figure 7 represents the other ex-

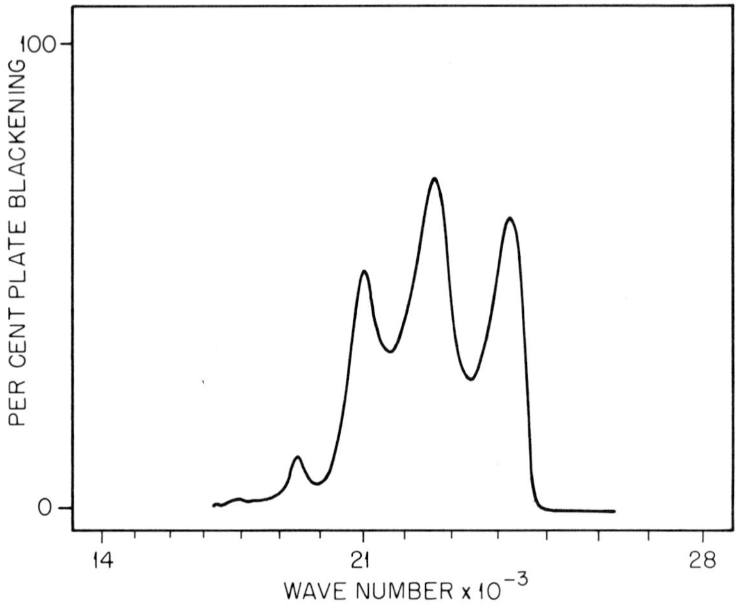

Fig. V. 6. Typical total luminescence spectrum of a molecule with lowest n, π* excited states (benzophenone, EPA rigid glass solution, 77°K; fluorescence absent, phosphorescence shown).

treme limit of a full (-)1 electronic charge on the C, and (+)1 on the O atom. The actual charge distribution of the electron in the π* -orbital must lie between the two structures b of Figure 7, since we know that the π* -orbital has its center of gravity displaced toward the C atom, as discussed previously, and illustrated for the π* -orbital in Figure 2. Obviously, the dipole moment of the n, π* singlet (and triplet) excited state of an isolated carbonyl would tend to reverse the dipole moment of the ground state. Having a formal (+) charge on an O atom is an unfamiliar situation, but one must remember that some 75-95 kcal mole^{-1} of excitation energy is involved in the n → π* promotion excitation in aldehydes and ketones.

The deduction which one might make is that the n-orbital

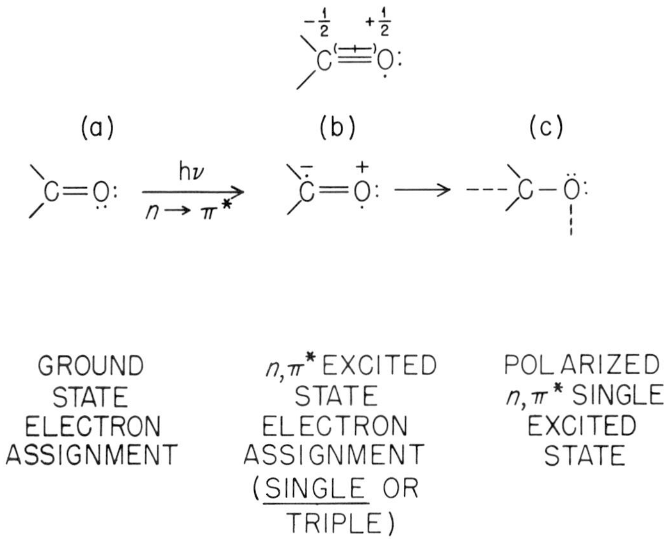

Fig. V. 7. Schematic electron assignment diagrams for n, π* excited states associated with carbonyl groups.

in the excited states will be a highly localized site of photochemical reactivity, but that once its electron deficiency were satisfied, the electronic structure of the carbonyl group would rearrange (i. e., polarize under the perturbation of the reagent) as shown in structure C of Figure 7, thereby forcing localization onto the C atom of the electron which had been promoted. One might say that the tendency of the electron in the π* -orbital to concentrate on the C atom is enhanced by the reagent perturbation, until the second point of reactivity on the C atom is saturated by bond formation. However, if a reagent fails to appear at the O atom during the lifetime of the n,π* excited state, the molecule returns to its ground state either radiatively or nonradiatively without reacting (presuming no spontaneous decomposition occurs, as seems to be the case in many condensed systems). If the reagent is an H atom originating from a hydrogen bond to the O atom, the failure of a following reaction to occur at the C atom during the lifetime of the

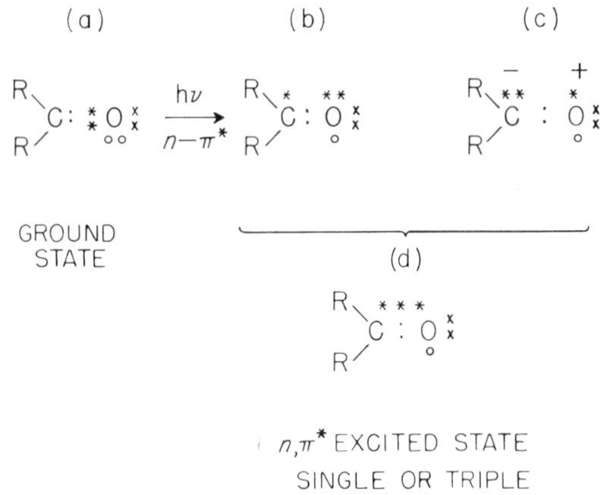

Fig. V. 8. Schematic valence bond diagrams for n, π* excited states associated with carbonyl groups (after Wheland).

n,π* excited state could lead to a rapid reversal of the hydrogen atom abstraction.

The electronic structures of the n,π* excited states of aliphatic carbonyl compounds have been discussed from the point of view of valence bond resonance structures by Wheland ('55). Figure 8 is a representation of his diagrams conforming to the n-orbital assignments of Figure 7 of this paper. The following code identifying orbitals is used: (•) electron in a σ-orbital, (*) electron in a π-orbital or a π*-orbital, (o) electron in a $2p_0$ n-orbital, (x) electron in a $2s_0$ n-orbital. The precise meaning of Wheland's structures loses contact with the orbital assignments of Figure 2 of this paper, as is clear if one asks the significance of assigning a (*) orbital (i. e., π or π*-orbital) electron to a C or O atom. However, it is of interest to point out that the three-electron-bond structure proposed by Wheland (lower right of Figure 8) is an exact formal analog to one of the limiting structures, upper b of Figure 7, described above. Thus, a three-electron bond structure is equivalent to the orbital configuration $\sigma^2\pi^2n^1(\pi^*)^1$; i. e., one double

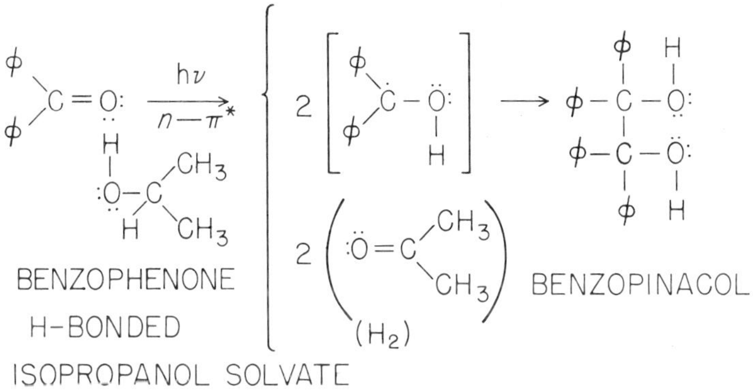

Fig. V. 9. Example of the hydrogen abstracting power of an n, π* excited state of a ketone (benzophenone).

bond, half of an antibond and an n-orbital half-filled.

It is important to point out that the n, π* excited state configuration described above corresponds fairly closely to what a chemist describes as a biradical structure: one electron is localized on an atom in an n-orbital, and another electron is in an independent (orthogonal) orbital, the π*-orbital, which is spatially rather isolated from the n-orbital (although spread out over part of the volume space of both the C and O atoms, and having a center of gravity displaced toward the C atom). Nevertheless, this "biradical" description fits both the singlet and the triplet n, π* excited states (cf. Kasha '60a). Hence one can expect that a highly reactive molecule is generated when the n → π* promotion takes place in carbonyl molecules.

Aromatic carbonyl n, π* excited states can be described somewhat similarly, except that now many more π*-orbitals may exist, and each will involve delocalization over all atoms "formally conjugated by double bonds". However, the n-orbital will preserve its essentially localized character, and the lowest empty π*-orbital, reached by the lowest energy n → π* promotion, will always have a node between the C and O atom just as in the simple isolated carbonyl group. Thus, to a degree, the electronic configuration of the excited state of an aromatic carbonyl compound will

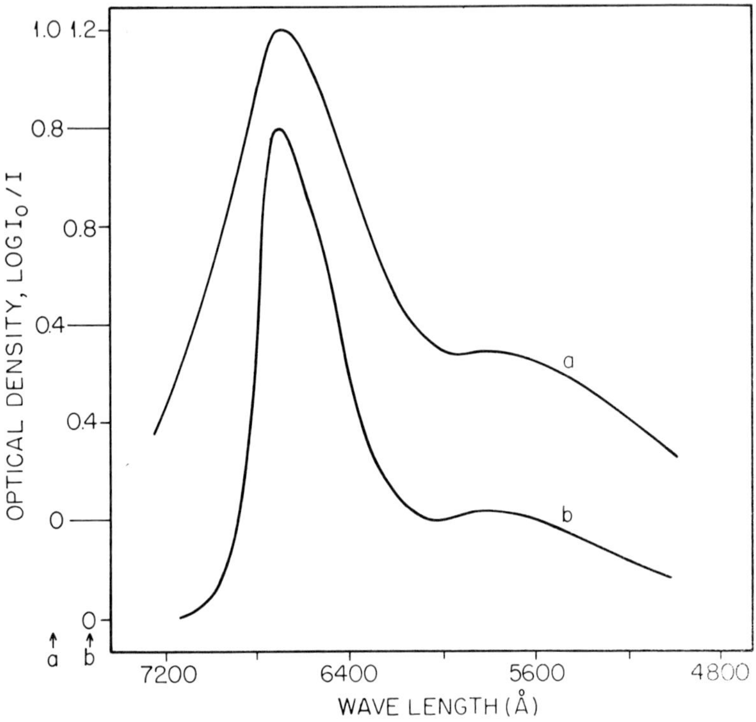

Fig. V. 10. Comparison of absorption curves for $(p\text{-}CH_3C_6H_4)_3N^+$ produced by two methods. Curve a, by chemical oxidation (aqueous room temp.) (Granick and Michaelis); curve b, by photooxidation (90°K, EPA rigid glass solution). (After Lewis and Lipkin).

have similarities to the simple isolated carbonyl case.

The photochemistry of aliphatic and aromatic carbonyl compounds in solution has attracted much attention. Here we shall consider only the case of the conversion of benzophenone to benzopinacol in an alcoholic solution. Plotnikow ('20) and Bäckström ('34) have described this reaction in terms of the hydrogen abstracting power of excited states of ketones and postulated a biradical intermediate for the reaction (cf. Bäckström and Sandros '58, Bäckström et al. '58). The sequence of steps which could now be used

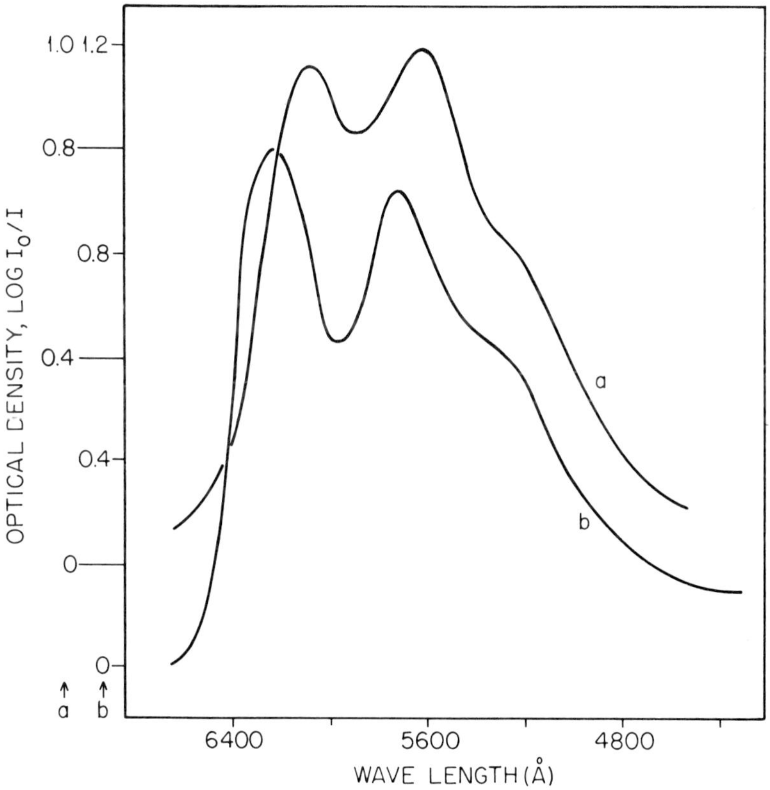

Fig. V. 11. Comparison of absorption curves for Wurster's blue produced by two methods. Curve a, by chemical oxidation (aqueous, room temp.) (Michaelis, Schubert, and Granick); curve b, by photooxidation (90°K, EPA rigid glass solution). (After Lewis and Lipkin).

to describe this reaction in isopropanol is illustrated in Figure 9. It is known from spectral behavior (Brealey and Kasha '55) that benzophenone in alcoholic solution exists as a hydrogen-bonded solvate. This puts the H-atom of the alcohol at precisely the most reactive point of the benzophenone molecule in its n, π* excited state. An intermediate such as shown in the central portion of Figure 9 would be appropriate. The alcoholic radical fragment would be stabilized by elimination of H atoms (whose fate is not known, but probably yielding H_2). Two radical intermediates of the benzophenone are then able to condense to form the benzopinacol product, which forms with a high quantum yield (cf. Bäckström et al. '58).

Bäckström and coworkers ('58) favor identification of the triplet excited state of benzophenone with the biradical intermediate. It is possible that in a bimolecular condensation of the type described, the longer-lifetime of the triplet state would favor a higher quantum yield. However, in the present view, the biradical description fits both the lowest singlet and triplet excited states, so a simple choice cannot be made without further rate-constant data. Bäckström and coworkers refer to the Lewis and Kasha ('44) structural description of the triplet state of ketones. For example, their structure for the triplet excited state of acetone (with n-orbital electrons now shown) is

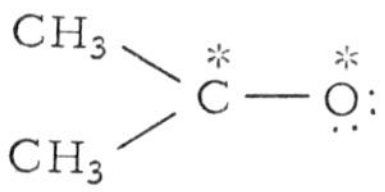

in which the electrons with parallel spins are designated by asterisks. Such a structure would describe a $\pi \rightarrow \pi^*$ promotion with an excited state configuration $\sigma^2\pi^1 n^2(\pi^*)^1$. Since the excitation is known to be of $n \rightarrow \pi^*$ type, and since the electrons in a π-orbital and π^*-orbital are delocalized on the C and O atoms, the Lewis and Kasha structure seems to be inappropriate today.

Other molecular types, such as N-heterocyclics, which have important n, π^* lowest excited states should also show specific reactions which should be interpretable in terms of electronic distributions in the excited state. However, each molecular bonding situation will have characteristic features which will distinguish the reactive intermediates. It is of great interest to investigate the reactions of the simple N-heterocyclics from this point of view.

5.3.2. **Reactivity of π, π^* triplet excited states.** - The relatively high quantum yield of triplet excitation in many molecules, even when n, π^* states are not available to facilitate (cf. Kasha '60a) the population of the triplet excited state, suggests that a specific reactivity of triplet states could be detected. Photochemists have often adduced biradical excited intermediates in the past, and Lewis and Kasha ('44) have identified these with the lowest triplet excited state of the molecule. One might expect triplet states to be photochemically important, both from the point of view of the long intrinsic mean lifetimes, and from the point of view of possible specific reactions of a biradical-like state. The question is, however, to what extent does a triplet excited state of a molecule have biradical characteristics. Lewis and Kasha described triplet states of molecules by means of resonance structures, such as the following ones for benzene and anthracene

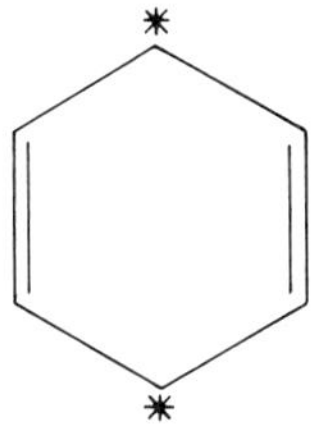

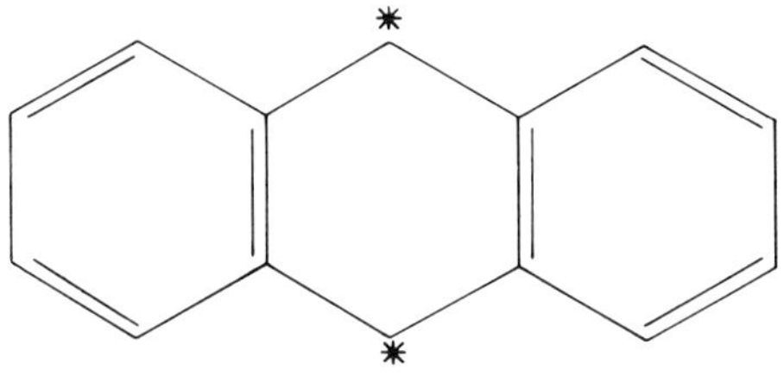

in which the electrons with parallel spin are designated by asterisks. Of course, these resonance structures do not mean that the electrons with parallel spins are actually localized as shown, since one would expect a set of such structures would be required to describe properly the actual lowest triplet excited state. However, Lewis and Kasha ('44) presented the thesis that actual localization was implied by their analysis of the possible resonance energy of the triplet state (the validity of this argument is obscure). Lewis and Kasha ('44), particularly in the case of anthracene, favored the structure shown above on the basis of the known photochemistry of anthracene.

However, the molecular orbital description does not at

first sight seem to indicate that such a localization of electrons is possible in the triplet state. The orbital wavefunction for the π-electrons of a molecule, e.g., as shown by the orbital diagrams for pyridine (Figure 3), indicates that each π-electron is spread over the whole molecular skeleton. It is of great interest, however, to inquire if theoretical calculations indicate a distinctive electronic distribution for the lowest singlet vs. the triplet excited state of a molecule. Such differences might indicate a unique tendency toward polarization of the excited triplet state under the influence of a perturbing reagent. It is to be expected that the electron pair giving rise to a triplet state should be more readily polarized.

Calculations based on the simplest molecular orbital method do not distinguish between singlet and triplet excited state energies or wavefunctions. In other words, in the primitive calculations the electronic distributions for excited singlet and triplet states would be predicted to be identical. However, more refined calculations, e.g., those of Pariser ('56) on naphthalene, yield different singlet and triplet state energies for the same electronic configuration -- as required in real molecules -- and give necessarily different electronic wavefunctions for singlet and triplet excited states. Such calculations should be extended until sufficient numbers of cases are available for comparisons with photochemical reactivity. It is possible that the best index of polarization tendency of electrons giving rise to triplet states would be an electron-pair spin-density calculation for aromatic molecules. Such calculations are now possible, but have not yet been carried out.

Emphasis has been given in this section to electronic distributions within molecules and their influence on photochemical reactivity as regards specific product structural determinativeness. There are several reports in the literature of photochemical reactions distinguishing excited singlet and triplet states, where structural effects were not noted.

5.4. PHOTOIONIZATION IN CONDENSED SYSTEMS

Very seldom is a photochemical reaction carried out in such a manner and on such a system that one knows precisely the nature of the elementary primary process involved. Such is the case, however, in the photooxidation experiments of Lewis and Lipkin ('42), which remain today a photochemical classic.

Lewis and Lipkin irradiated numerous N-heteroaromatic molecules in dilute solutions in rigid EPA glass solution at 90°K, using ultraviolet light. What is noteworthy about their work is their proof of the identity of the spectra of the colored species produced by ultraviolet radiation, with the spectra produced by Michaelis and coworkers by chemical and electrochemical methods. Moreover, Michaelis and coworkers, using potentiometric methods, proved that a one electron oxidation was involved in the production of the colored species.

Figure 10 shows these spectral results for the colored ion-radical produced by the removal of one electron from tri-p-tolylamine

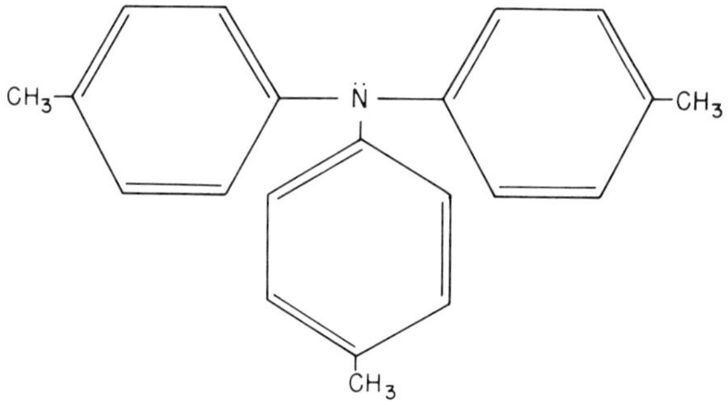

It is evident that the spectrum obtained by Granick and Michaelis ('40) for aqueous solutions at room temperature bears a close resemblance to that obtained by Lewis and

Lipkin for EPA rigid glass solution at 90°K, despite the difference in the solvent systems.

Figure 11 shows an additional example, which emphasizes the spectral characterization. This case is for the ion-radical (Wurster's blue) produced by the removal of one electron from N, N, N', N'-tetramethylparaphenylenediamine

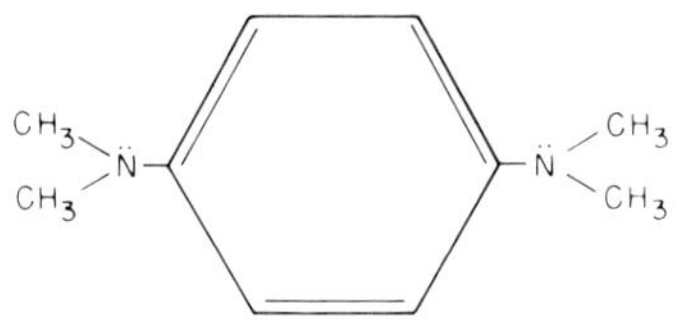

Again, the spectrum obtained by Michaelis, Schubert, and Granick ('39) for aqueous solutions at room temperature bears a striking resemblance to the Lewis and Lipkin spectrum obtained for the colored species in EPA rigid glass solution at 90°K.

The physical processes involved in the photooxidation process studied by Lewis and Lipkin have been the subject of further investigation. Debye and Edwards ('52) have demonstrated that the decay law for the recombination luminescence of the trapped electron and the ion radical corresponds to a second order process, as required. Linschitz et al. '54, '54) have extended the photochemical studies with some keynote experiments, in which the nature of the trapped electron was investigated, and demonstrated that a triplet excited state of the ion-radical precursor was produced in their cases after the recombination.

Nevertheless, some important energetic aspects of the photoionization process for the molecules in the rigid glass system remain to be investigated. The ionization potentials of the gaseous heteroaromatic molecules are in the range 8-12 ev. In the rigid glass solutions, the photoionization process occurs at a greatly lowered energy, probably at 3-5 ev. All of the workers in this field stress the sensitivity to the nature of the solvent, of the condensed phase photoionization. Since the solvent acts as a trap for the ejected electron in the primary process, the energy of the photoionization should be lowered from the ionization potential by the electron trapping energy of the acceptor molecule. Monochromatic photoionization threshold studies would be of great interest, especially in relation to nature of the solvent used. In essence, what is a simple photoionization process in the gaseous molecule becomes an intermolecular charge transfer process in the condensed phase. The presence of a solvent-dependent photoionization continuum underlying the normal absorption bands is suggested by this view-point and is subject to experimental test.

The possibility of low-energy photoionization processes of heteroaromatic molecules in solution is a topic of considerable importance in photochemistry and radiation chemistry, and research in this field should be greatly extended.

5.5. CONCLUSION

In this paper the discussion has been limited to some fundamental aspects of molecular photochemistry. The complexity of the problems dealt with in organic photochemistry (e.g., see Barton '59) may possibly serve to make the chemist impatient with such a discussion as given here, especially when armed with his typically successful methodological understanding of his reactions. However, as Barton and others have pointed out, when dealing with the high energy process of photochemical transformation, a new set of reaction rules must apply. It is certainly possible that the nature of such rules, and even their precise form, may be greatly enhanced by contributions from molecular spectroscopy.

In selecting photochemical literature for discussion, especial care was taken to list some of the key papers in the field. These should provide a good introduction to some of the main lines of research in molecular photochemistry.

5.6. SUMMARY

A discussion is presented relating details of molecular electronic excitation to molecular photochemistry. A new photochemical principle is proposed, which applies explicitly to polyatomic molecules, a principle of bipartition of molecular excitation: Each quantum of radiation absorbed by a normal polyatomic molecule has a probability of bipartition between the lowest excited singlet and triplet states of the molecule. It is shown that this bipartition can be experimentally related to the intersystem crossing ratio, a measurable molecular constant, which indicates the limiting, generally non-integral quantum yields that can be attained for a photochemical reaction at a specific molecular excited state.

Some elementary photochemical processes are discussed in detail, including (a) reactivity of n, π^* excited states, (b) reactivity of π, π^* triplet excited states, and (c) photoionization in condensed systems.

DISCUSSION

KASHA: There were some additional points which I did not include in my lecture, but on which I wish to make brief comments.

1. Space and time do not permit a discussion of the role of n-π transitions in N-heterocyclics; in another place I am presenting a discussion of this topic (Kasha '60).

An especially interesting topic in N-heterocyclics is the splitting of electronic energy levels of these molecules upon formation of hydrogen-bonded dimers. This work, in progress in my laboratory at the Florida State University (with Mr. A. El Bayoumy), shows the role of the exciton phenomenon in such situations. I shall perhaps have the opportunity to elaborate on this point after Dr. Förster's lecture.

2. In purines and pyrimidines there is suggestive evidence that the lowest energy allowed electronic transition is of n-π type. Dr. Alex Rich (MIT) and I have been doing some experimental work on crystalline purines, and on oriented DNA fibres, in polarized light, in order to establish this point. Also, in synthetic polynucleotides we have noted a hyperchromic region, on the long wavelength side of the 2600 Å absorption band, upon formation of the helical from the random coil molecule. This result would agree with the n-π transition interpretation, according to Tinoco's recent theory (Univ. of Calif., in press '60) of hypochromism in the π-π region. (Rich and I expect to publish these results shortly.)

3. The existence of n-π transitions as the lowest allowed electronic transitions indicates their probable key role in the photochemistry and radiation chemistry of DNA.

4. Direct evidence for energy migration in ultraviolet irradiation studies of DNA has been recently found (L. Lerman and L. Tolmach ('59), J. Marmur and K. Berns, unpublished). Especially in these latter studies, it was found that native DNA gave evidence for facile energy transfer and

was more sensitive to ultraviolet radiation effects, than was denatured DNA.

RUPERT: Setlow at Yale found that unoriented DNA irradiated in the dry state formed a gel in water instead of redissolving, as though some crosslinking were produced by the ultraviolet. Have you tested the solubility of your oriented material after irradiation with polarized UV perpendicular or parallel to the orientation?

KASHA: No, we haven't. The light intensity used in polarized light absorption studies is very weak, however. Your remark reminds me of another point I forgot to mention. There was originally planned for today's session a talk on the photochemistry of interfaces. I recall some early experiments of Terenin's on irradiation of molecular layers, in which he found that new higher molecular weight products formed when one N-heteroaromatic compound was condensed on top of another, and irradiated with ultraviolet. That work was terminated too early. It is a very important field to continue. Terenin did not prove precisely what species he had produced, but they were certainly of a greatly increased molecular weight.

In listening to Rupert's discussion of DNA photo-lesions yesterday, with their photoreversibility it seemed to me that to get a longer wave length absorption one has to consider the possibility that the bases between layers may be condensing, rather than fissioning.

AUGENSTINE: I want to ask a question about a point discussed sometime earlier. You indicated that, in compounds similar to the purines and pyrimidines, you know that the n-π transition is a very probable one and in your lecture you stated that in such a case there is a high probability of transition from excited singlet to the triplet, with consequent phosphorescence. This could make a very important difference, in the actual reactivity of irradiated DNA. What is known about conditions for deciding among these various possibilities? In other words, what is known about what is required for the excited singlet to react, or to have a transition from the excited singlet to the triplet?

KASHA: I will start with an oblique answer. The four common bases present in DNA differ considerably in the extent that the n-π transition sticks out at long wavelength (or low energy side). It is possible that in a particular hydrogen-bonded nucleotide, the n-π transition does not at all determine the final excitation of the molecule. This is something which has to be determined experimentally in the future. If, however, there is a lowest n-π excited singlet, then it is of great interest to know whether the base will react as a singlet, or undergo transition to a triplet first. In my paper I suggested that, if there is a possibility of hydrogen abstraction (or some as yet unknown characteristic reaction of n-π singlet excited states of N-heterocyclics), then a triplet state may never be produced. Lacking this, the triplet state is usually favored with n, π lowest singlet excited states.

Thus, the exact environment of the N-heterocyclic molecule, as well as the relative energies of the n-π and π-π transitions will determine whether or not reaction at the excited singlet will occur.

PULLMAN: May I ask a question on the point about which Kasha is speaking concerning the n-π transition? Which of the lone-pairs is involved? One has lone-pairs on nitrogen and lone-pairs on oxygens. The n-π transition is due to which lone pair?

KASHA: In the biological pyrimidines there are available quite different lone-pairs from those in the simple unsubstituted pyrimidine itself. In pyrimidine free base the fairly strong absorption band at about 3500 Å is due to an n-π transition that originates from the nitrogen lone pairs. But in the biological pyrimidines it is most likely that oxygen lone pairs are involved.

PULLMAN: In the pyrimidines, but not the purines?

KASHA: Not in the purines.

PULLMAN: This might be possible to some extent because one has a few nitrogens in purine, and the mobility and the ionization potentials of lone pairs on these different nitrogens will be different.

For instance, one can predict that the most mobile lone pair on nitrogen of the purine base depends upon the base and of course in adenine is going to be N_1.

The lone pair on the nitrogen N_1 will be more mobile than the lone pair on the N_7, and biochemical conclusions depend on such differences. It is possible also that the formation of N oxides can be interpreted on this basis.

KASHA: There is a nitrogen heterocyclic containing four nitrogens on which photochemical reactivity occurs at one of the nitrogens. This suggests that a photochemical selectivity may also depend on which lone pair is being excited.

FANO: As Professor Kasha pointed out, the exciton coupling, or exciton model, of molecular excitation is closely related to the diffused excitation of condensed matter which I was discussing yesterday. I should like to illustrate this relationship by pointing out similarities and differences between the two situations, and specifically to mention four separate points.

1. In the extended isotropic medium I was considering yesterday, one must take into account the phase-gradient (i. e., the modulation in space) of an exciting disturbance, whether it be light or the field of a fast charged particle, even though this gradient is small in accordance with the smallness of the momentum transferred in the excitation. This gradient has, however, a negligible influence when one deals, like Dr. Kasha, with a rather small molecule. Thus the excitation of condensed matter involves a geometrical element, namely, the direction of the phase gradient (i. e., of the momentum transfer), which is absent in the excitation of small molecules. The mutual orientation of the phase gradient and of the electric dipole oscillation in the excitation determines the classification of the excitations of extended media into longitudinal and transverse ones; this distinction does not arise in small molecules.

2. Consider now, in Dr. Kasha's problem, a situation where a monomer has two types of excitation, e.g., one π-π and one n-π. As one proceeds to the dimer, each of these levels splits into two, as described by Kasha, and one level of each pair keeps most, or all, of the oscillator strength of the original (π-π or n-π) level. It was understood here that the level split, within each pair, is much smaller than the separation between π-π and n-π. Suppose now, for the sake of the argument (and we'll return to this in item 3 below), that the interaction between the two monomers of the dimer increases to the point where it causes a level split comparable to, or larger than, the separation of the π-π and n-π monomer levels. We shall, then, still have a total of four levels but they will no longer be classified into two pairs, one π-π and one n-π. I would then expect that a single level among the four, namely the highest, would take up most or all of the oscillator strength pertaining to the two initial levels (π-π and n-π). This effect results from the fact that the wave function of this particular level is a coherent superposition with appropriate phases of the four wave functions of the π-π and n-π excitations of the two monomers. This formation of a new state, which takes up the dipole oscillator strength of different levels of different chemical groups, is the characteristic of the plasma-type excitation phenomenon I was discussing yesterday. The mixing of two levels (π-π and n-π) of two monomers, in the fictitious example I was just describing, constitutes a first step in that direction.

3. The splitting of the monomer levels owing to dipole interaction in the examples quoted by Dr. Kasha gives a good indication of the magnitude of the same effect in the circumstances I was considering yesterday. Let us take as a basis that dimerization causes a split of 800 cm^{-1}, i.e., of 0.1 ev. (I hope that this figure is representative.) Now, it must be considered that the phenomena discussed by Dr. Kasha concern the excitation of a chromophoric group within each molecule and that this excitation involves presumably an oscillator strength of the order of one electron per group. On the other hand, the effect I was discussing yesterday involves the combined oscillator strength of <u>all</u> valence elec-

trons within each molecular group. The number of these electrons is nearly one half of the molecular weight, i.e., it is of the order of 10-100 for groups such as Kasha's monomers. Therefore, the electrostatic interaction, which is ~ 0.1 ev in Kasha's situation should be boosted by a factor of 10-100 in the situation I was considering. This large boost results from the difference between the optical or near-ultraviolet range of the spectra, where a minute fraction of the electrons is effective, and the far ultraviolet (10-12 ev range) where all valence electrons are loosened up. All of us are so conditioned by common experience with the optical properties of matter in the visible or near-visible range that we have trouble in visualizing the corresponding properties in the vacuum ultraviolet. This is, I believe, why many of us hesitate to reconcile ourselves to the suggestion that the exciton coupling responsible for 0.1 ev splits in Kasha's phenomenon has a far more drastic influence in the higher excitations.

4. Situations intermediate between that of Kasha's dimers and that of extended matter are encountered in the excitation by near UV of long stacks of chromophoric groups, such as are contained in the nucleic acid of genetic material. Interesting results regarding the spread of excitation along such a stack have been obtained, if I remember correctly, in studies with the "transforming principle" by Dr. L. S. Lerman at Colorado Medical School. This information is based on recollection from a visit I made there about one year ago.

LIPSKY: I have two questions to ask of Dr. Kasha. The first one involves the absence of n-π singlet fluorescence. Is this independent of exciting wavelength? More specifically, do you see any fluroescence if you excite in the long wavelength edge of the n-π singlet absorption band?

I have one other question. This is concerned with the splitting that you observe in the spectrum of the dimer. Are there examples in which the long wavelength component of the split band disappears completely leaving you only with a blue shift? If the answer to this is yes, then do you know whether these monomers are more strongly concentration quenched than those for which the splitting is observed?

KASHA: The common dimerizational effect is to yield a blue-shift, because it most often happens that dimers occur with transition dipoles parallel. This occurs in most dyestuff dimers which involve planar dyes, and in some nitrogen heterocyclic dimers. Thus, one always finds in such cases that the long wavelength component is nearly absent.

Strong fluorescence quenching is caused by such dimerization; it is a phenomenon long recognized in the literature.*

CHARLESBY: I have two points to raise. The first one concerns the fact that benzophenone catalyzes the photochemical crosslinking of polyethylene. I am wondering if you have any idea of the mechanism. The second point concerns the fact that if one irradiates a number of polymers at a low temperature, I was thinking particularly of polystyrene, one can change the color. On warming up, a certain amount of reaction is taken place, with a certain amount of light given off. This emission may occur quite a considerable time afterward. This role is ascribed to electrons being knocked off and bounding back again as they are warmed up. Do you have any comments on that?

KASHA: On the first point, the benzophenone crosslinking which Oster has patented ('59), but as far as I have information has not explained: I think that the explanation is simply that hydrogen abstraction by the n-π excited state of benzophenone is the primary process occurring, with radical crosslinking in the hydrocarbon chain following. This seems to be closely parallel to the hydrogen abstracting properties of the excited ketones and quinones as described by photochemists, and elaborated in my paper.

CHARLESBY: It is the idea that one benzophenone ab-

*See McRae and Kasha ('58) for references, and a discussion of the relation of the spectral shifts to fluorescence quenching.

stracts two hydrogens, and does not take any further part in the process?

KASHA: Yes, that is the simplest possibility. I wonder if Dr. Linschitz would care to comment on the second point. He has recently worked in the field.

LINSCHITZ: I agree with the idea that the emitted light results from recombination of positive ions and electrons, and that these may be responsible for the color in the irradiated polymer. In fact, we have shown that this recombination very often gives triplet excited states rather than singlet, the emitted light having precisely the triplet-singlet spectrum.

FÖRSTER: I should like to ask Dr. Kasha's opinion about the different intrinsic reactivities of corresponding singlet and triplet states. In other words, how does one deduce such difference, because it is determined not only by electronic distribution but also by spin?

KASHA: This is the point on which I especially wanted to elaborate, but it was partly in the form of a question that I made my comment. In the molecular orbital calculations, e g., those of Pullman, no distinction whatsoever is made in either energy or wave functions of corresponding singlet and triplet. In some more elaborate calculations, (e. g., by Pariser) one actually does calculate different wave functions and energies for singlet and triplet excited states. But it seems to me that these differences are fairly small as far as the electronic distributions are concerned, and I raised the question: is there another kind of calculation, which would show differences in terms of polarization of excited state by a reagent? I think that the statement made sometimes that triplet states really are no different in their reactivity from singlet states is probably not right, and I think that besides the lifetime difference there should be really inherent theoretical positional difference for molecular reactivity, and organic photochemists and others could do well to look for specific products which are geometrically determined by such positional differences.

6

Bernard Pullman
Alberte Pullman
Université de Paris

Structural Aspects of Radiation Effects on Nucleic Acids and Related Substances

6.1. THE DATA*

The main chemical effects produced by ionizing radiations on nucleic acids in aqueous solution, appear to be†:

deamination and dehydroxylation,
rupture of hydrogen bonds,
fission of sugar base linkage,
liberation of purine bases,
breakdown of the pyrimidine bases,
oxidation of the sugar moiety, and
breakage of the nucleotide chain and liberation of inorganic phosphates.

In the presence of oxygen the irradiation leads to the formation of hydroperoxides of the pyrimidine bases (Scholes and Weiss '59; Weiss '58).

In this paper, we are essentially interested in the interpretation of the differences in behavior, with respect to radiation damage, of the purine and pyrimidine constituents of the nucleic acids. In this respect the general results

* This work was sponsored by the U.S. Public Health Service Grant CY-3073.

† Butler '56, '59; Dale '55; Barron '54; Swallow '56; Scholes, Stein and Weiss '49.

mentioned seem to indicate:

(a) the greater general sensitivity to radiation damage of the pyrimidine constituents of the nucleic acids. The purines appear to be more stable than the pyrimidines.

(b) the greater ability of the glycosidic linkage of the purine bases compared to that of the pyrimidine bases. This conclusion based on the liberation of purine bases may not be unequivocal because the pyrimidine bases, which might have been liberated also, are more sensitive to destruction. Nevertheless, as we shall see later, the conclusion is probably correct.

These two general conclusions, and in particular the first one, receive further confirmation and development from the study of the effect of radiations on the corresponding free bases, their nucleosides and nucleotides as related in the following three paragraphs.

1. An investigation (Scholes and Weiss '59, Weiss '58) of the effect of ionizing radiations, in the presence of oxygen, on these different types of compounds has confirmed the resistance of purines to form hydroperoxides. As to the pyrimidines, it indicates that thymine and uracil are much more sensitive than cytosine. The structure of the hydroperoxide has been established (Scholes and Weiss '59,

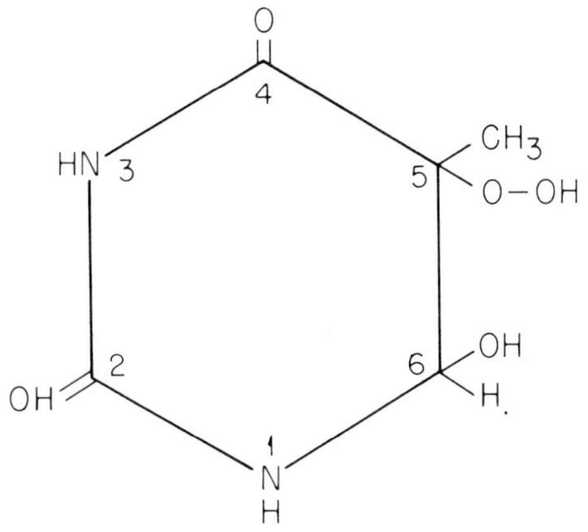

Weiss '58, Ekert and Monier '59): it corresponds to the addition of OH and O_2H radicals to the 5-6 bond of the pyrimidine, as illustrated in the formula established for the hydroperoxide of thymine, which appears to be particularly stable,

2. The effect of ultraviolet light of $\lambda <$ 3100 Å, in the presence of H_2O_2, indicates also the resistance of the purines (adenine) and the destruction of pyrimidines (thymine). Because, under these conditions, the effect of the irradiation is due essentially to the indirect action of the OH and O_2H radicals, generated by the splitting of H_2O_2, this investigation is related to the preceeding one.

3. According to Scholes and Weiss ('50) strong irradiation may also cause splitting of the purine bases. The adenine ring appears nevertheless to be much more resistant than the guanine one. Thus, irradiation with high-energy electrons (ca. 4×10^6 r) of an aqueous solution of guanylic acid or guanosine, leads to the formation of guanine (breakage of the glycosidic link) and to rupture of the imidazole ring of guanine with the subsequent formation of 2, 4 diamino-5 formamido-6 hydroxypyrimidine (Hems '58),

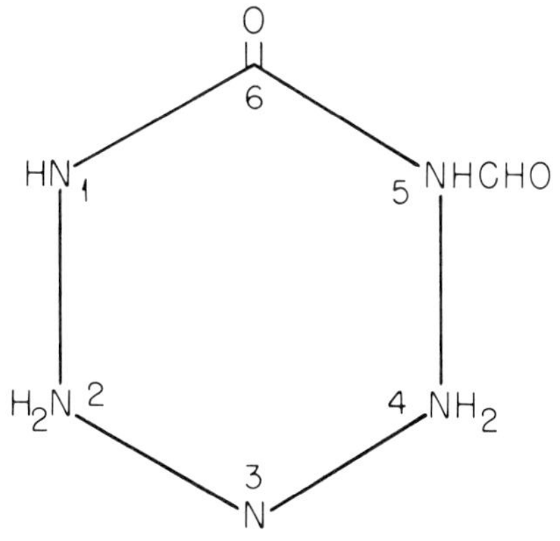

Under the same conditions, according to Hems and Eidinoff ('58), the corresponding derivatives of adenine yield only the free base: the glycosidic linkage is broken but the base resists destruction.

It seems justified to conclude from these different studies that the relative resistance of the principal purine and pyrimidine bases to destruction by the direct and indirect effects of ionizing radiations is in the order: adenine > guanine >> cytosine > uracil > thymine. It must nevertheless be remembered that the relative resistance of the purine or pyrimidine bases among themselves has not been determined in the nucleic acids but only on their separated components. It cannot be excluded that in these acids this order may be changed.

Some of the essential results, related to the irradiation of nucleic acids by ultraviolet light, lead to conclusions which are similar to those inferred from the effect of ionizing radiations. Thus, all observers agree that purines are much more resistant to ultraviolet irradiation than the pyrimidines, this statement being valid for the effect both on nucleic acids and on the free bases, their nucleosides or nucleotides (Errera '52, '55; Christensen and Giese '54; Shugar and Wierzchowski '58). A closer examination of the available data reveals some useful complementary information. For example, as far as the purines are concerned, the last authors cited not only confirm the greater stability toward UV radiations of adenine over guanine but also indicate that the examination of available qualitative results suggests that these two purines are more stable than xanthine and hypoxanthine. It appears difficult to ascertain the relative resistance to ultraviolet radiations of uric acid, with respect to that of the other purines; nevertheless it seems that this compound is rather labile, its irradiation giving rise to a variety of products. Although one cannot be certain that the relative stability of these complementary purines towards ionizing radiations would parallel their relative resistance towards ultraviolet radiations, such correspondence does seem highly probable. It would thus appear that the two purines which form part of the structure of nucleic acids represent relatively more stable entities than their metabolic degradation derivatives (xanthine, uric acid) or their immediate precursors on the pathway of the de novo synthesis of these acids (hypoxanthine).

As concerns the photochemical behavior of the pyrimidine bases they exhibit some analogies, but also some differences, with respect to their behavior towards ionizing radiations. The essential feature seems to be that, in aqueous solutions, the pyrimidines may be divided into two groups:

(a) compounds which, like uracil and cytosine (and their nucleosides or nucleotides) contain an unsubstituted 5-6 bond. The photolysis of these compounds is to a lesser or greater degree reversible and is accompanied by the uptake of water at this 5-6 bond and an acid-base catalysed removal of it.

(b) derivatives substituted in the 5 or 6 positions (e.g., thymine, 5-methylcytosine and orotic acid), the irradiation of which leads irreversibly to the destruction of the pyrimidine ring and the formation of a variety of products. The uptake of a water molecule is probably not involved in these cases (Scholes and Weiss '59). Sufficiently strong irradiation ($\lambda < 2537$ Å) may, of course, result also in the disruption of the pyrimidine ring of uracil and cytosine.

The structure of the reversible irradiation product has been established, in particular in the case of 1, 3 dimethyluracil, as a 6-hydroxyhydroderivative (Wierzchowski and Shugar '57; Wang '59)

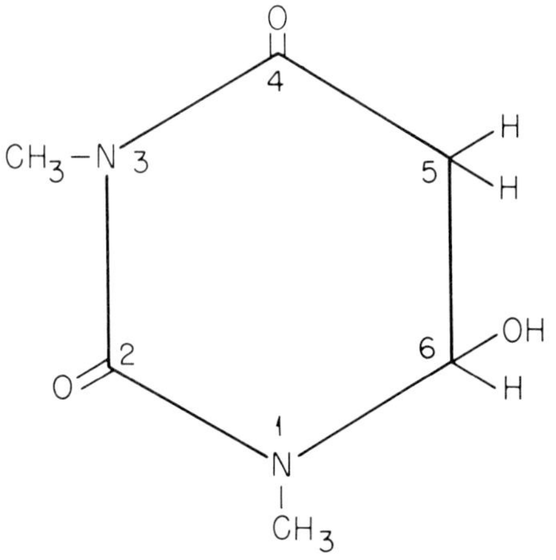

The different nature of the irradiation product, as well as the distinction between 5-6 unsubstituted or substituted derivatives, prevents a direct comparison with the results obtained from the study of the effect of the ionizing radiations. Nevertheless, under these restrictions, certain similarities can again be observed, particularly in regard to the relative stabilities of the different pyrimidines. For example, according to Shugar and Wierzchowski ('58) and Wierzchowski and Shugar ('57), the quantum yield of photolysis of uracil (or its derivatives) is much greater than that of cytosine (or its derivatives). In connection with this observation it must also be remarked that the photoproducts of cytosine are considerably less stable than those of uracil. Finally, it must be noted that, under certain conditions (in rapidly frozen solutions or in a purinic acid), thymine, which is particularly resistant to ultraviolet irradiation in aqueous solution becomes particularly sensitive to such irradiation (Roersch et al. '58; Beukers et al. '59). No similar observation has been reported for 5-methylcytosine. Thus, depending on the experimental conditions and, probably, the nature of the reaction that occurs, either uracil or thymine are the most sensitive of the pyrimidine bases of the nucleic acids. In each case cytosine appears as the relatively resistant specimen. From such a broader point of view the sensitivity of the pyrimidine bases towards ultraviolet irradiation parallels to a large extent their sensitivity to ionizing radiations.

The preceding summary of available information is schematic and neglects many details. Such a simplification appeared nevertheless necessary for the sake of a general interpretation. A more detailed discussion will probably be possible at a later stage.

6.2. INTERPRETATION

We shall now discuss the principal structural features which may be considered to be responsible for the relative resistance of the preceding compounds to radiation damage and to account for the nature of the damage produced.

6.2.1. Resonance stabilization of the bases. - Both purines and pyrimidines belong to the class of resonating molecules possessing a considerable number of mobile or π electrons associated with the conjugated double bonds and the lone-pairs of the heteroatoms. These π electrons form a unique electrical cloud distributed over the whole molecular periphery, and the existence of this electronic delocalization bestows upon the molecules a complementary stabilization which is known as their resonance energy. It may reasonably be expected that the existence of this resonance energy will represent a factor of stabilization of the molecules against radiation damage and even that the degree of stabilization will be related to the value of the corresponding resonance energy.

Table 1 contains the resonance energies of the biologically essential purines and pyrimidines mentioned in Section 6.1. The resonance energies have been evaluated by the molecular orbital method of quantum chemistry (cf. Pullman and Pullman '52) and are given in units of β (the usual resonance integral of the method), whose approximate value is 20 kcal mole^{-1}. We have indicated both the total resonance energies of the compounds and the resonance energy per π electron in each of them. Because the different molecules contain different numbers of π electrons, it is obvious that it is the resonance energy per π electron which gives the proper measure of the relative stabilization of these molecules through electronic delocalization.

The examination of the data contained in Table 1 confirms this hypothesis to a very large extent. With the exception of a small inversion between guanine and hypoxanthine, the relative values of the resonance energies per π electron reproduce correctly the relative resistances of the compounds studied to radiation damage. Thus, the purines as a whole are characterized by a greater value of the resonance energy per π electron than are the pyrimidines. Moreover, among the purines, adenine which is the most stable of all the biochemically important bases, has by far the greatest value of resonance energy per π electron. Similarly, cytosine, which is more resistant to radiation damage than the other pyrimidine bases, has a relatively great value of this energy too, greater anyway than that of the

TABLE VI. 1.

RESONANCE ENERGIES OF BIOLOGICALLY ESSENTIAL PURINES AND PYRIMIDINES (in units $\beta \approx 20$ kcal $mole^{-1}$).

Compound	Resonance energy	Number of π electrons*	Resonance energy per π electron
Adenine[1,2]	3.894	12	0.32
Guanine[1,2]	3.838	14	0.27
Hypoxanthine[2]	3.385	12	0.28
Xanthine[2]	3.484	14	0.25
Uric acid[3]	3.374	16	0.21
Cytosine[4]	2.280	10	0.23
Uracil[4]	1.918	10	0.19
Thymine[4]	2.050	12	0.17
5-methylcytosine[4]	2.412	12	0.20
Orotic acid[4]	2.350	14	0.17
Barbituric acid[4]	1.743	12	0.15

* Including hyperconjugation, when present

[1] Pullman, Pullman and Berthier ('56).

[2] Pullman and Pullman ('58a).

[3] Ibid. ('58b).

[4] Ibid. ('59); the resonance energies of thymine and 5-methylcytosine indicated in this publication are slightly erroneous.

other pyrimidine bases. It may also be useful to stress that the purine and pyrimidine bases which enter into the constitution of nucleic acids have relatively greater resonance energy per π electron and are more stable than their metabolic degradation products (xanthine, uric acid, barbituric acid). Their immediate precursors in the de novo

TABLE VI. 2.

RESONANCE ENERGIES OF SOME INTERMEDIATES IN THE ORIGINAL SYNTHESIS OF NUCLEIC ACIDS (in units of β).

Compound (Ribotide of)	Resonance Energy[1]	Number of π electrons[2]	Resonance energy per π electron
Glycinamide	0.763	8	0.095
Formylglycinamide	1.150	10	0.115
5-aminoimidazole	1.998	8	0.250
5-amino-4-imidazole-carboxamide	2.573	12	0.214
5-formamido-4-imidazole-carboxamide	2.948	14	0.211

[1] Hyperconjugation included.

[2] Pullman and Pullman ('58c).

synthesis of nucleic acids (hypoxanethine, orotic acid) seem to be of relatively high stability.

The question may of course be raised whether the earlier intermediates in this original synthesis offer similar stability and whether there exists, from the point of view considered here, some particularly resistant or particularly weak elements in this long chain of synthesis. Table II contains the results of calculation of resonance energies for some important intermediates in the de novo synthesis of purines as established by the work of Buchanan et al. ('57) and of Greenberg and Jaenicke ('57). It can be seen that the species in the very initial steps in this synthesis have rather small resonance energies per π electron and that these energies increase greatly in the latter intermediates Nevertheless, the increase is not absolutely progressive and thus the formation of the imidazole ring is obviously connected with a peak of stabilization. Further ramification of this skeleton leads to a slight decrease of the resonance energy per π electron, until we reach a new peak with hypoxanthine. Thus the most stable intermediates in this de novo synthesis of nucleic acids are associated, from the point of view considered here, with the very steps of formation of the cyclic hetero-rings of imidazole and purine.

Another interesting problem which may be considered in connection with the data of Tables 1 and 2 is related to the particularly high value of the resonance energy per π electron in adenine. This molecule appears thus as representing the thermodynamically most stable biochemical purine or pyrimidine. The question may then be raised whether this particularly high stability may not be to some extent responsible for the apparently unique role of adenine in nature. Thus, it is adenine which enters into the constitution of a great number of fundamental coenzymes (DPN, TPN, FAD, coenzyme A) and other important transfer agents (5-adenosylmethionine) and which also is the carrier of the pyrophosphate chain in the energy rich ATP or ADP. It may not be unreasonable to imagine that, inasmuch as nature was induced to utilize a purine or pyrimidine base as a complementary constituent of these fundamental biochemicals - complementary because it does not represent the essential "functioning" center of these compounds, a phenomenon of natural selection played a role, leading to the choice of the energetically most stable base. This conclusion may be completed by the observation that the same phenomenon might perhaps have been operating partly in the choice of the bases which form the well-known pairs of DNA. Thus, as far as our present knowledge goes, the purine - pyrimidine pairs, which effectively exist in DNA, - the adenine-thymine and the guanine-cytosine pairs - seem to be more stable than the artificial combinations recently created experimentally between other related constituents (Doty et al. '59; Felsenfeld and Rich '57).

The last observation brings us to the final problem to be considered in this section. Hydrogen bonds are broken in DNA under the effect of all types of radiations and, again, it seems probable that the different hydrogen bonds present will not be ruptured equivalently. Recent quantum mechanical calculations by the authors have shown in this respect that the gain in resonance energy due to hydrogen-bond formation is greater for the guanine-cytosine pair than for the adenine-thymine pair, the numerical values being respectively 0.210β and 0.161β, which with the usual value of 20 kcal $mole^{-1}$ for β represents respectively 4.2 and 3.2 kcal $mole^{-1}$. Because the energies involved are small, the distinction will be of particular importance for the heat-induced rupture of these bonds and it is thus the hydrogen bonds between adenine and thymine which should be the easiest to disrupt. A confirmation of this state of affairs may be found in recent results of Marmur and Doty ('59; cf. Rosenkranz and Bendich '59) who have shown that the denaturation temperature of nucleic acids rich in the guanine-cytosine pair is higher than that of nucleic acids rich in the adenine-thymine pair.

6.2.2. Electronic structural features. - The sensitivity of biochemicals to ionizing radiations is, on the other hand, obviously related to their reactivity towards free radicals generated, especially in the water medium, by these radiations. Consequently, this sensitivity will be determined also, to a large extent, by the structural factors which govern, in general, the reactivity of organic systems - and in particular, as in the present case, of conjugated organic systems - towards this type of reagent.

The reaction consists apparently in the addition of a pair of free radicals to one of the "double" bonds of the purine or pyrimidine bases. The structural features which may be responsible for the relative ease of the reaction in different compounds will consequently be (a) the existence of a bond with a high mobile bond order (great double bond character) and (b) the presence of high free valences on the atoms forming this bond. (Concerning these indices and their role in chemical reactivity see, e.g., Pullman and Pullman '55, '58).

Figure 1 shows the values of the maximum bond order in the principal purines and pyrimidines considered here and the values of the free valencies at the terminal atoms of this bond (Pullman and Pullman '59a for pyrimidines; '58d for purines). Examination of this figure indicates the following points.

(a) The bond which has the greatest mobile bond order is the 5-6 carbon-carbon bond in the pyrimidines and the 7-8 nitrogen-carbon in the purines.

(b) The mobile order of the C_5-C_6 bond of the pyrimidines is, on the whole, much greater than the mobile order of the N_7-C_8 bond of purines. The pyrimidines should thus, from that point of view, be much more reactive towards the fixation of free radicals than the purines. Moreover, the mobile order of the C_5-C_6 bond of uracil and thymine is much greater than that of cytosine. The first two of these pyrimidine bases should thus be more reactive towards free radicals than the last one. From the same point of view, uracil should be slightly more reactive than thymine. As a matter of fact, this last conclusion is still substantiated by the values of the free valencies at the carbons of the susceptible bond. Thus, in particular, carbon 5 of uracil has an exceptionally great free valence and should consequently be particularly reactive toward free radicals. All these results permit one to predict the order of decreasing activity toward the addition of free radical:

uracil > thymine > cytosine > purines

This theoretical prediction is in good agreement with the experimental data quoted in the first part of this paper.

(c) The N_7-C_8 bonds of adenine and guanine have practically the same mobile bond order. Consequently, it is the value of the free valencies of the atoms which will determine their reactivity towards free radicals. Most probably it is the free valence of carbon 8 which will have a deciding influence on this reactivity, this carbon being the essential center of radical reactivity of the purines (Pullman '59). From that point of view, guanine should be more reactive

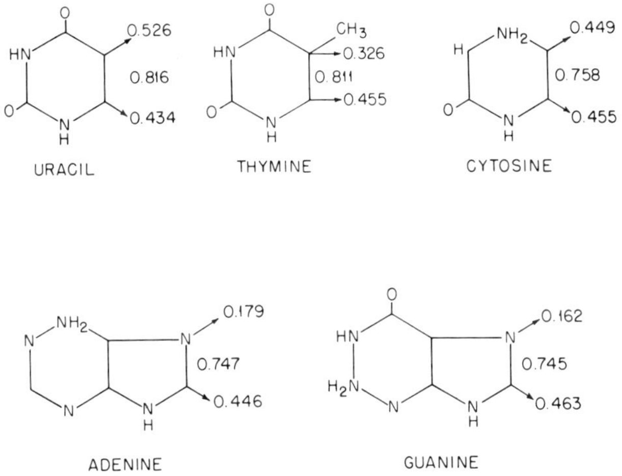

Fig. VI. 1. The bond with the maximum mobile order and the free valencies on the terminating atoms in purines and pyrimidines of the nucleic acids. (For the calculation of free valence, N_{max} = 1.732 for C atoms and 1.414 for N atoms).

than adenine, a result which again is in agreement with experiment.

On the whole, the conclusions which can be drawn from the distribution of the structural indices generally related to radical reactivity thus run parallel in the present case to those derived from the considerations based on resonance energy stabilization. The two factors contribute thus similarly to the relative stability or instability of the purine and pyrimidine bases of the nucleic acids toward ionizing radiations.

There is one aspect of the preceding results which merits emphasis. The site of action of ionizing radiations and the free radicals generated by them on the purine-pyrimidine pairs of DNA is largely located in the pyrimidines. In this connection it may be recalled that the site of action of the "radiomimetic" cytotoxic agents, - nitrogen mustards, ethyleneimines, epoxides etc. - which are electrophilic agents, is located predominantly, so far as these pairs are concerned, in the purines of the pairs, (Lawley '57a, b). Thus, in spite of some striking similarities in the effects of these two types of reagents on biochemicals, the attacked targets are different. Moreover, as already indicated, when the free radicals attack guanine they most probably do so at C_8, while the attack of nitrogen mustards on that molecule leads to the quaternization of N_7*.

The role of the electronic structural features regarding the effects produced by the ultraviolet radiations is more difficult to establish, in the first place, because of the great sensitivity of the results to experimental conditions (vide infra) and secondly because of the possible intervention of the excited states of the molecules. The results of molecular orbital calculations, such as those referred to in this paper are essentially valid for the ground state of molecules and are much less satisfactory when applied to the structure of the excited states. Nevertheless, some conclusions can be drawn to a first approximation. Definite chemical products resulting from the action of the ultraviolet radiations are, as far as it has been established, hydration products of the pyrimidines (the purines resisting the formation of such products), the hydration occurring again at the 5-6 bond of the pyrimidines. This is again an addition reaction, the ease of which will again be determined to a large extent

* Alexander and Stacey ('59) suggest that the attack of free radicals on guanine occurs on N_7, by analogy with what is known about the attack of alkylating agents. This conclusion, however, is unjustified because the types of reagent are of different kinds. On the other hand, it may be useful to quote the opinion of these authors, according to whom the alkylation of the purines of nucleic acids is a secondary reaction, a transposition which follows the esterification of the phosphates.

by the value of the mobile-bond order of the susceptible link. As the reaction consists in the addition of the elements of water in the form of H^+ and OH^- ions, a complementary factor in this case will be the polarity of this link. Inasmuch as the electronic distribution calculated for the ground state may be significant for the evolution of the reaction, the previously quoted results concerning the values of the bond orders are in agreement with the greater susceptibility of the pyrimidines over that of the purines and, roughly speaking, with the greater quantum yield of photolysis of uracil and thymine over that of cytosine. Moreover, the polarity of the link, which in the three pyrimidine bases corresponds to the existence of a formal negative charge on C_5 and a formal positive charge on C_6 (Pullman and Pullman '59a), imposes the addition of OH^- to C_6 and of H^+ to C_5 in agreement with experiment.

It is also known that the hydration photoproduct of cytosine is less stable than is that of uracil or of thymine. This state of affairs also can be easily interpreted. The tendency of the hydration product to revert to the original base corresponds to its tendency to revert to its completely conjugated form, so as to recover the part of resonance energy lost upon hydration. The tendency to reversibility will be the greater, the greater the gain in resonance energy upon dehydration. The data contained in Table 3 indicate that this energy gain will effectively be appreciably greater (about 4 kcal $mole^{-1}$) in the case of cytosine than in the case of uracil (or thymine)* . It seems possible that photoreactivation of microorganisms is related to this reversible photolysis of pyrimidines (Shugar and Wierzchowski '58).

---

*It may be useful to remind the reader that considerations of this type have been very useful for the interpretation of the metabolic transformations of carcinogenic aromatic hydrocarbons (Pullman and Pullman '54, '55).

In a more refined approximation, the structural features considered here and which are characteristic of the distribution of the electronic cloud in the isolated molecules, should be replaced by the corresponding "localization energies" related to the activation energies for corresponding types of reaction (cf. Pullman and Pullman '52, '58). Nevertheless, it seems very improbable that such a refinement would introduce any significant change in the present case, the differences between the different indices in different molecules being quite appreciable.

TABLE VI. 3.

GAIN OF RESONANCE ENERGY UPON DEHYDRATION OF HYDROXYHYDRODERIVATIVES OF PYRIMIDINES.

THE HYDROXYHYDRO-DERIVATIVE AND ITS RESONANCE ENERGY	THE BASE AND ITS RESONANCE ENERGY	GAIN OF RESONANCE ENERGY UPON DEHYDRATION
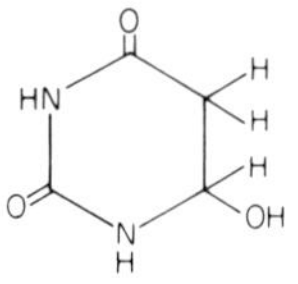 1.128 β	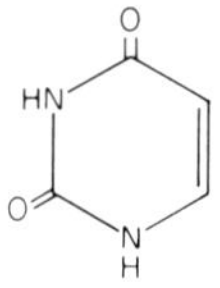 1.918 β	0.790 β
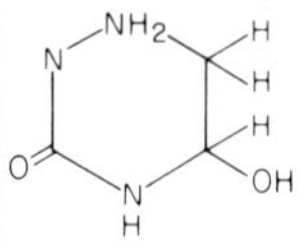 1.289 β	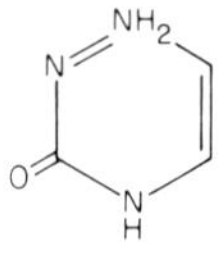 2.280 β	0.991 β

6.3. REMARKS

6.3.1. On the liberation of purine bases. - As stated in part 6.1. of this paper, the experimental results concerned with the effect of ionizing radiations which indicate the liberation of free purine bases, suggest that the glycosidic linkage of these bases is more labile than the same linkage of the pyrimidine bases. As quoted there, this conclusion may not be unequivocal, because the pyrimidine bases, which might also have been liberated, are more sensitive to destruction. Nevertheless, we believe that the preceding conclusion is correct. We have effectively shown in a recent publication ('59b) that the glycosidic linkage of the purine ribotides is more electron deficient than the glycosidic linkage of the pyrimidine ribotides. This electron deficiency, manifesting itself in the "bipositivity" of the glycosidic linkage involves electronic repulsion between the formal positive charges on the nitrogen and carbon atoms of the linkage, resulting in the weakening of the bond. The glycosidic linkage of the purine ribotides must thus be weaker than that of the pyrimidine ribotides and will consequently be more sensitive to the action of radiations.

6.3.2. On some correspondence between radiation induced and chemically induced damage. - The preceding remark draws our attention to a rather close correspondence existing between certain aspects of radiation-induced and chemically induced damage of nucleic acids. Thus, the glycosidic linkage of the purines appears also to be hydrolyzed more easily than the glycosidic linkage of the pyrimidines, in both enzymatic and acid hydrolysis*; there is even a marked parallelism between the relative susceptibility of different purine and pyrimidine ribotides to radiation-induced breakage of this linkage. In fact, the preferential breakage of the glycosidic linkage of the purines of the nucleic acids upon acid hydrolysis results in the production of a

* Cf. these authors '59b; also Daly, Alfrey and Mirsky '50; Tamm, Hodes and Chargoff '52; Hodes and Chargoff '56; Shapiro and Chargoff '57.

purinic acid. This preferential ease of hydrolysis has also been correlated with the amount of the electron deficiency on the susceptible bond (Pullman and Pullman '59b). As a related reaction, which also indicates the greater lability of the glycosidic linkage of the purines compared with that of the pyrimidines, one may quote the action of mercaptoacetic acid on DNA which results in the liberation of the purines (Jones and Latham '56).

In contrast, when the chemical reaction involves the breakage of the heterocyclic ring, the pyrimidines appears to be less resistant than the purines. The most striking example is probably the hydrazinolysis of nucleic acids which leads, through the destruction and elimination of the pyrimidines, to apyrimidic acid (Takemura '59; Takemura and Miyazaki '59). In this case, however, the relative susceptibility of different pyrimidines towards destruction does not parallel their susceptibility to radiation-induced breakdown. The reason is that hydrazinolysis is not concerned with the 5-6 bond of the pyrimidines but provokes the opening of other bonds of these molecules. However, in spite of this difference the general analogy between chemically and radiation-induced damage appears quite striking.

6.4. CONCLUSION

If we did not overlook any important bibliographical reference, this paper appears to be the first application of the results of quantum-mechanical calculations of energetic and structural characteristics of biochemicals to the study of their behavior towards the effects of ionizing radiations. The results obtained seem encouraging. Radiation doses necessary to provoke important biological effects are considerably smaller than those resulting in an extensive degradation of these compounds and the biological effects are probably localized in character. One of the principal advantages of the mode of approach presented here is the possibility of discerning, through theoretical examination of the electronic structure of the molecules, the sensitive elements in their constitution and of helping to understand the features responsible for this sensitivity. It may be hoped, consequently, that such a procedure may contribute to the elucidation of the relation between the chemical changes induced by radiations and the observable biological effects.

DISCUSSION

GRIFFING: I think the comment which I am about to make links the papers of Professor Kasha and Professor Pullman. Let me preface what I have to say with two defense statements. I know nothing about big molecules and to me benzene is a big molecule; secondly, I have profound respect for the great progress that organic chemists have made with the concept of resonance. If one asks how much we know about the distribution of electrons in molecules, it is irrelevant whether one considers the method as outlined by Professor Pullman or uses the best self-consistent field calculations — we really know nothing about the relative distribution of electrons even in isolated molecules. Sometimes we get pretty good energies for ground states, but we know nothing about the wave functions.

The calculated values we obtain for the dipole moment of a molecule (even in the best calculations) are usually not even the same order of magnitude as the experimental ones and often have an opposite sign. So one certainly has to be very careful about drawing any conclusions based on the calculated electron distributions regardless of the method. Let me say, even if in the isolated molecule these electron distributions were correct, as soon as one brings up an interacting molecule — a single interacting molecule — these distributions are profoundly changed. Let me take an example. In the interaction of two hydrogen molecules approaching linear collision, the Van der Waals minimum is quite small, representing an interaction energy of a few small calories. However, we have introduced into the calculated energy a classical repulsion between nuclei which must now be compensated for by a redistribution of the electrons between the two molecules. This classical energy in this case is 22 ev. If one looks at the simple cases where fairly good wave-functions have been obtained, the binding energy (i. e., interaction energy) is a delicate balance between the large nuclear repulsion and the additional coulombic attractions which arise because each electron now moves in a field that has a higher nuclear charge (ap-approaching twice the charge as the distance between the two atoms or molecules approaches zero). Therefore, I would like to emphasize that there are profound re-arrangements in a molecule when it is interacting with any kind of an environment.

Now, I would like to return to the comment I originally started to make on the difference between the electron distributions of triplet and of singlet states. Here is another distinction between the a priori self-consistent field solutions and the semi-empirical procedure used by Professor Pullman. The latter would give the same electron distribution, the triplet and singlet belonging to the same configuration. In the SCF calculations the charge distributions are not the same; furthermore, in order to obtain good wave-functions and good charge distributions, configuration interaction (this is the mixing-in of excited states to improve the wave function) must be included. There will be more singlets than triplets and their energy distributions will be quite different, meaning that their contributions to the wave-function of any given state will probably be quite different even in the isolated molecules. As soon as spin-free environment must be considered, many more states will be mixed, and it would indeed be astonishing if these many factors added up to give the same electronic distribution for singlet and triplet states.

If I may sum up, even the best quantum mechanical calculations, which give a good value for the energy of a molecule, give very poor wave-functions and interaction energies. Therefore, I would look skeptically at such small variations as Professor Pullman has shown. I think the correlations are beautiful but, their reliability, from my point of view I am afraid, is fortuitous.

The Pullman School have developed a large number of internally consistent numbers related to complex organic molecules and their properties. The numbers are based on a semi-empirical theory related to the concept of resonance in organic chemistry. It is interesting but not surprising to find that similar treatments of a few molecules important in radiation chemistry fit into this general scheme. No doubt as time permits, Professor Pullman can show that many other cases can be fitted into the general scheme.

This essentially seems to show that the resonance theory of organic chemistry does not contradict the observed experimental results quoted. However, before this can be considered as a quantum mechanical theory relating the two sets of numbers, it remains to be shown how these results are related to the solution of the Schrödinger equation of such a complicated system. This is the primary problem of theoretical organic chemistry even for adiabatic processes, at the present stage of development. Before a reliable structural quantum mechanical theory of radiation chemistry can be begun, progress must first be made in the ways of handling non-adiabatic processes. It seems to me, the role of simple theories and semi-empirical theories in the future depends upon an elucidation of just what kinds of understandable approximations can be made to make the transition to the complicated molecules and the highly non-adiabatic conditions of radiation chemistry. Perhaps even more experimental work must be done before any structural approach to the problem can profitably be made.

PULLMAN: Well, I don't think that Professor Griffing fully appreciates what I was attempting to show. What I really wanted to show you was the relative values of certain standard quantities for a number of molecules. For instance, suppose one considers the relative value of resonance energy for a pair of molecules. It is certain that we obtain a number by the use of approximate methods which is in the correct relative order. We have found that no more refined calculation can give a contrary result.

We have done enough work on it to be absolutely sure about it. It is purely a question of economy of effort. Once you have studied a large number of very complicated molecules then you just realize that if you use even simple methods you are able to explain a lot of phenomena which you might otherwise think require very elaborate calculations.

The second answer concerns the problem of polarization. I didn't speak about it here, but as a matter of fact, calculations have been made involving not only the electron distribution in the static molecule but also polarizability. Here again, as long as what you want to know are relative electron polarizabilities in different compounds, it is satisfactory in the great majority of cases to use a static model — and of course, you always have the final point of checking by experiment.

We find that the most stable double bond in pyrimidines is the 5, 6 double bond. This result simply means that polarization effects in big molecules are to a very large extent determined by the pre-existing electronic distribution in such molecules.

Approaching reagent can polarize the electronic distribution in a manner which depends on the pre-existing distribution. It is obvious that one cannot use simple calculations in an absolute way. They are not that reliable. They are entirely reliable if you use them in a cognizant way.

BURTON: I was confident when Professor Griffing made her remarks that Professor Pullman would be prepared for them because he, on his part, certainly anticipated that his paper would evoke such a comment.

Now I have another closely related comment for which he is probably equally prepared. I can understand how his calculations may have significance for thermal processes, but I don't understand why they have significance for excitation processes.

PULLMAN: Well, if radiation chemistry is involved, then one might simply look at the electron distribution in appropriate free radicals.

BURTON: My comment does not pertain to free radicals, but to sensitivity of molecules to the effect of high-energy radiation.

PULLMAN: Yes, but this calculation of relative stability by its very nature is related to sensitivity to radiation although the latter may be an indirect effect. There are two properties which one can consider. One is the resonance energy stabilization and I agree that it is not easy to see how this energy resonance stabilization is related directly to sensitivity to high-energy radiation.

The second point is the electron distribution. This is somewhat more subtle because it is related to the indirect effect of the radiation. The indirect effect involves reactivity toward free radicals, but the first step I agree is not clearly seen. Maybe the real stabilization involves some primary product of radiation. For instance, one can show that, if you form a free radical from a particular purine or pyrimidine base, the stability of such a free radical will be parallel to the stability of the parent molecule.

RUPERT: Wang recently concluded that the photolysis of cytosine is really a tautomerization, rather than addition of water across the 5, 6 double bond. I think the conclusions you wanted to draw would still be valid.

KASHA: I think Dr. Griffings' comments were well taken in part, but I think that without moderation they might be possibly confusing, and I would like to make one comment to serve as a sort of buffering action. The molecular orbital method and the Pullmans' use of it is not intended as a precise numerical computing technique. But the relative values of the orbital energies are frequently qualitatively quite significant, and may lead to important chemical conclusions. However, I myself do not know how invariant these qualitative results are with respect to the arbitrary empirical parameters chosen. The Pullmans' claim is that their parameters give them consistent results, whose order is fairly insensitive to parameter choices and to refinements in the calculation. I agree with Dr. Griffing that a certain amount of luck is involved, at least on the point that the Pullmans' calculations are really gaseous state calculations with environment and entropy effects neglected. Solvation energies, for example, are frequently much greater than the orbital-energy differences the Pullmans consider qualitatively significant. The semi-empirical feature of the method seems to make it useful in a relative way, if one has a related series of molecules in similar situations. But as far as environmental effects go, it must be nearly pure luck.

In the case of the Pullmans' recent paper on the inherent stability of GC vs. AT rich DNA, I cannot accept the stability argument as a proved example of the applications of the method, because there are only two ways this result could have come out: right, or wrong, since there was no comparison with a larger series of cases than just the two base pairs.

PULLMAN: Unfortunately there are only two pairs of purines and pyrimidines in nucleic acid.

KASHA: Well, perhaps in this case, one must demonstrate that no other theory could account for the relative stability. (Added in proof) The alternative might be to show that the Pullmans' charge-transfer stabilization fits a whole series of synthetic polynucleotide melting temperature results.

7

John L. Magee
University of Notre Dame

Elementary Processes in Action of Ionizing Radiation

The average energy loss of a high-energy charged particle is large enough so that several chemical processes, such as dissociations, can frequently result from the same initial event. In certain condensed molecular systems, such as aqueous solutions, track effects are known to occur because of the resulting initially inhomogeneous distribution of reactive intermediates. In this paper we are interested in the elementary processes which determine the initial distributions of chemically reactive intermediates in condensed molecular systems. We shall be particularly interested in the special case of low LET radiations where the initially affected regions are called "spurs".

Following an initial event two types of processes compete:

a) migration of energy to increase the size of the affected region, and

b) degradation of the initially formed electronic energy

*Supported in part under Atomic Energy Commission contract At (11-1)-38.

† The author is appreciative of discussions with Professors Milton Burton and W. H. Hamill and Dr. Koichi Funabashi.

to potential energy (molecular dissociation, metastable electronic states) and thermal energy.

As an example, after formation of a secondary electron with ca. 100 ev energy, the first electronic processes involve the stopping of this electron, which takes place in $\approx 10^{-15}$ sec. At the end of this period of time only relatively long-lived electronic states remain (i. e., long on this time scale). At this time the geometry of affected molecules is primarily determined by the range of the secondary electron. Some of the electronic excitation corresponds to ionization. The most rapid dissociation processes require 10^{-14} seconds and thus become important relatively later. On the time scale of interest, radiative transitions (requiring 10^{-9} sec or more) are unimportant.

In the next two sections we discuss in turn the migration of the electronic excitation and the mechanism of dissociation processes, which start to compete at later times. Then we consider factors which determine size and geometry of spurs.

The discussions of excitation migration and dissociation involve the use of specific models which have only limited applicability to real systems. However, they have the advantage that well-defined calculations can be made and they do have features which make their consideration useful.

7.1. MIGRATION OF ELECTRONIC EXCITATION

The problem of the migration of electronic excitation energy has been considered extensively both in experimental and theoretical investigations. Förster ('51) presents a convenient summary of this work. From the experimental viewpoint Burton, et al. (cf. Burton and Dreeskamp '59 and Lipsky and Burton '59) have been most concerned with the role of energy migration in radiation chemistry.

We are mostly concerned with theoretical aspects of the problem. Franck and Teller ('38) made the first explicit application of exciton theory to photo-chemical action. More recently, Simpson and Peterson ('57) and McRae and Kasha ('58) have considered the interpretation of molecular spectra in a manner pertinent to our interests.

Energy can always be transferred at some finite rate from an excited molecule to similar molecules nearby. However, this transfer process is in competition with degradation processes, such as radiationless transitions to the ground state and dissociations. Therefore, as a practical matter, some electronic states can be considered as completely localized.

On the time scale of interest in this paper radiative processes cannot compete and are ignored. The fraction of energy which ends up in states which can radiate is in any case small in all known systems.

In this section we consider briefly the nature of delocalized states, neglecting the competitive dissociative processes. For this purpose we consider a linear chain of coupled systems containing electronic excitation. Subsections 7.1.1 and 7.1.2 can be omitted by the reader who is not interested in details of the quantum mechanical treatment of the linear model of coupled systems. A summary of this consideration is given in sub-section 7.1.3.

7.1.1. Non-localized stationary states of a condensed system. -

The electronic states of interest are solutions of the Schrödinger equation

$$H\psi_j = E_j\psi_j \qquad \text{VII. 1.}$$

where H is the Hamiltonian operator and E_j an energy of a stationary state which has the wave function ψ_j.

We shall concern ourselves at first with states which are obtained by super-position of a one-molecule electronic excitation on the system. For such a case all ψ_j's can be given as linear combinations

$$\psi_j = \sum_{k=1}^{N} a_{jk}\,\eta_k \qquad \text{VII. 2.}$$

where the number of molecules in the system is N and η_k is the wave function of the system with the excitation located on the kth molecule. We shall limit detailed discussion to a linear model.

The problem is now to find allowed sets of the a_{jk} and E_j. The E_j's are the allowed energies, as altered by the interactions, and the $|a_{jk}|^2$ give the probabilities of finding the excitation on the kth molecule when the system has the energy E_j.

The set of equations which determine the a_{jk}'s is

$$E_j a_{jp} = \beta a_{jp-1} + \alpha a_{jp} + \beta a_{jp+1} \; ; \; p = 1, 2, 3, \ldots N \qquad \text{VII. 3.}$$

where β $(= \int \eta_p^* \, H\eta_{p+1} \, dt)$ is the resonance integral between neighboring molecules, α $(= \int \eta_p^* \, H\eta_p \, dt)$ is the so-called coulombic integral, and we are neglecting exchange integrals between molecules farther separated than next-neighbors.

The set of equations 3 leads to a secular equation of Nth order. Fortunately it can be solved for the general case and yields the result

$$E_j = \alpha + 2\beta \cos \frac{j\pi}{N+1} \; ; \; j = 1, 2, 3, \ldots N \qquad \text{VII. 4.}$$

The allowed energies of the linear system therefore, are in the band between $(\alpha - 2\beta)$ and $(\alpha + 2\beta)$, with the spread 4β. The wave function of the state which has the energy E_j is, for large N,

$$\psi_j = \left(\frac{2}{N}\right)^{1/2} \sum_{k=1}^{N} \sin \frac{kj\pi}{N+1} \, \eta_k \; ; \; j = 1, 2, 3, \ldots N \qquad \text{VII. 5.}$$

7.1.2. Time-dependent behavior of excited states. -

The time-dependent behavior of a system is to be obtained from the time-dependent wave equation

$$H\psi = \frac{h}{2\pi i} \frac{\partial}{\partial t} \psi \qquad \text{VII. 6.}$$

where $i \equiv \sqrt{-1}$ and h is Planck's constant. Thus the stationary states of equation 5 contain the time-dependent factor $\exp(-2\pi i E_j t/h)$, which is of no importance if the system is in a pure state. The states obtained by particle impacts, however, are not stationary states. At the initial instant of time t = 0 they correspond to location of the excitation on a particular member of the chain, say the pth. The state which exists then is, therefore

$$\psi(t=0) = \eta_p = \left(\frac{2}{N}\right)^{1/2} \sum_{j=1}^{N} \sin \frac{pj\pi}{N+1} \psi_j \qquad \text{VII. 7.}$$

At later times

$$\psi(t) = \left(\frac{2}{N}\right)^{1/2} \sum_{j=1}^{N} \sin \frac{pj\pi}{N+1} \psi_j \exp(-2\pi i E_j t/h) \qquad \text{VII. 8.}$$

It follows that the probability that the excitation will be located on the qth member of the chain at time t is

$$\underline{P}(p, q, t) = \left[J_{p-q} \left(\frac{4\pi|\beta|t}{h}\right) \right]^2 \qquad \text{VII. 9.}$$

where $J_{p-q}(4\pi|\beta|t/h)$ is the Bessel function of integral order p-q of the argument $4\pi|\beta|t/h$, which we shall hereafter call γ. The integral β has positive or negative value depending upon the nature of the interaction. The absolute value, $|\beta|$, is used so that the dimensionless measure of time, γ, has the same algebraic sign as t.

The most significant measure of the spread of excitation from the initial location is

$$\overline{(q-p)^2} = \sum_{q=1}^{N} (q-p)^2 \underline{P}(p, q, t) \qquad \text{VII. 10.}$$

If we limit the consideration to long chains it follows

that

$$\overline{x^2} = 2\sum_{n=o}^{\infty} n^2 J_n^2(\gamma) \qquad \text{VII. 11.}$$

where q - p = x and it has been found by the use of tabulated Bessel functions that $(\overline{x^2})^{1/2}$ is the following linear function of time,

$$(\overline{x^2})^{1/2} = 0.707\gamma \qquad \text{VII. 12.}$$

The excitation is spread over Δn units of the chain at time t. If we write

$$\Delta n = vt \qquad \text{VII. 13.}$$

where v is a measure of the spreading velocity, we find

$$v = 2.22 \times 10^{15} |\beta| \qquad \text{VII. 14.}$$

where $|\beta|$ is measured in electron volts and the constant has the appropriate dimensions.

7.1.3. Summary of results on linear model of coupled systems. - Figure 1 shows how the excitation spreads along a coupled linear chain as a function of time. In this figure x is the location of units with respect to the one which was initially excited, and γ is the dimensionless measure of time, $\gamma = 4\pi\beta t/h$ where β is the resonance integral between neighboring units in the chain.

The most significant measure of the spreading velocity is given by equations 13 and 14. The fact that the excitation spreads linearly in time is an indication of the particle-like character of excitons.

The most critical parameter for a system is the resonance integral β. For systems of present interest this quantity can be as large as 1 or 2 ev or may be many orders of magnitude smaller. Rapid energy spread is expected in systems with large β. A domain of molecules (Lipsky and Burton '59) is clearly expected to share electronic excitation as a single unit, and this has important consequences

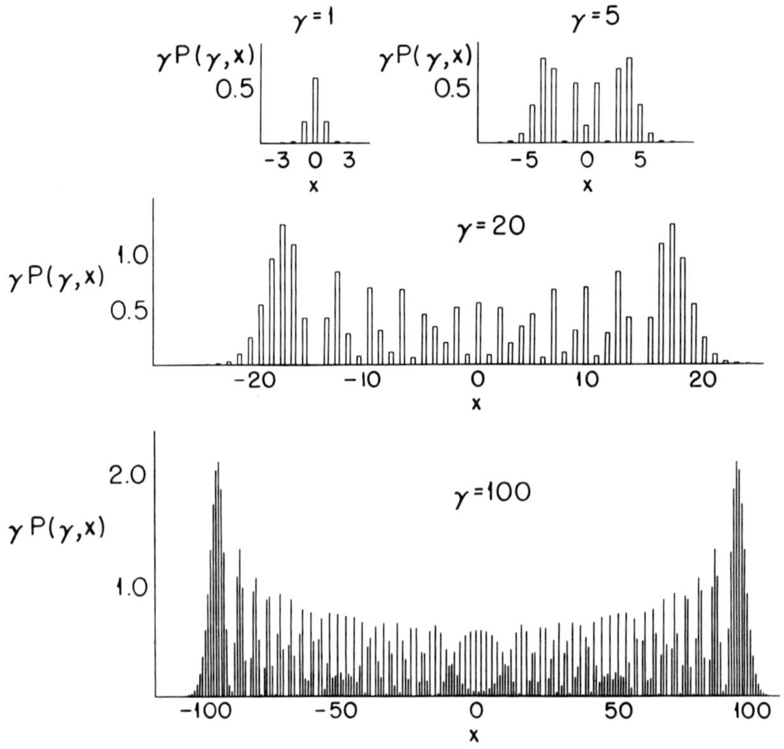

Fig. VII. 1. Values of the probability $P(\gamma, x)$ for finding the excitation on various members of a linear chain, where positions are designated by x, at various times γ. Four values of the time γ are shown. At $\gamma = 0$ the excitation is located on the chain member at $x = 0$.

in radiation chemistry. On the other hand, if β is as small as ~ 0.01 ev, delocalization can have little significance.

7.1.4. Strong-coupling, weak-coupling and the integral β. - It has been noted (Franck and Teller '38, Simpson and Peterson '57) that two special cases exist.

In the strong-coupling case an electronic state of the

entire coupled system is established and there are various vibrational states of the entire system in each of its electronic states. For this case the resonance integral β is the matrix element of the electronic interaction only.

In the weak-coupling case individual vibronic levels of the N components interact with each other. In this case the resonance integral β involves overlap integrals of the vibronic wave functions.

We are interested in several types of interactions which are characteristic for systems of interest.

a) The interaction of a positive ion with its neighbor leading to charge migration. This process can lead to a spreading of the positive charges presumably formed close together (i. e., in spurs) before the associated electrons return. For a system of small molecules we expect values of β of at most 0. 1 ev since in this interaction an electron must move from one molecule to its neighbor, and the wave functions do not overlap appreciably. Using equations 13 and 14 to estimate spreading distances, it is clear that the charge can move at most about four molecules before the electron returns. For such systems the positive charges are quite well localized. On the other hand for long chain molecules, with primary chemical binding along the chains, values of β can be larger than an electron volt. The formalism discussed above actually applies quite well to this case. Even for cases in which electron recapture is expected, the charge spreading can be very significant. For example, if $\beta = 2$ ev and $t = 5 \times 10^{-14}$ sec, $\Delta n = 220$ units.

For such cases the early spreading is so rapid that potential track effects can be reduced considerably. It should, however, be remembered that in systems of long chain molecules the chains are coiled up so that in our example the actual distance the excitation travels from its point of origin is on the average considerably less than 220 bond distances, perhaps closer to $\sqrt{220}$ bond distances.

b) The interaction of an "allowed" state with its neighbors leading to excitation migration. This is an example of dipolar coupling. For this type of interaction, values of β

can be various sizes leading both to strong- and weak-coupling cases. Again it is likely that chain structures can in general lead to a more rapid spreading than can systems composed of small molecules, except in the special case of domain structure of the latter.

c) The interaction of states which are not "allowed". For this type of interaction, values of β can only be small, leading to weak-coupling and slow spreading. This case is important because there are more excited electronic states of this type than allowed states, and mechanisms exist for their formation in spurs (Burton, Hamill and Magee '59).

7.2. DISSOCIATION PROCESSES

In the weak-coupling case excited states can really be considered as made up of excited molecules whose properties are very similar to the properties of the isolated molecules. If such excited states can decompose in a reasonably short time (10^{-12} - 10^{-13} second) they will do so before the locus of the excitation has changed significantly. Of course, if such a state does not dissociate or radiate, the excitation can migrate large distances. The phenomenon of radiation sensitization may be explained by such a mechanism.

It is clear that strong coupling alters the behavior of the excited state as observed in the isolated molecule. Fano ('59) has investigated this effect in a model of dipolar coupled oscillators which has applicability to molecular systems. In order to investigate the influence of coupling on dissociative processes we have considered a specific model of a linear chain of coupled diatomic molecules.* Figure 2 shows the potential energy relationships for an isolated molecule of the chain. We now consider the potential energy surfaces for a coupled system made up of one excited

* This calculation was performed with the collaboration of Dr. Koichi Funabashi. It will be reported in more detail in a forthcoming publication in the Journal of Chemical Physics.

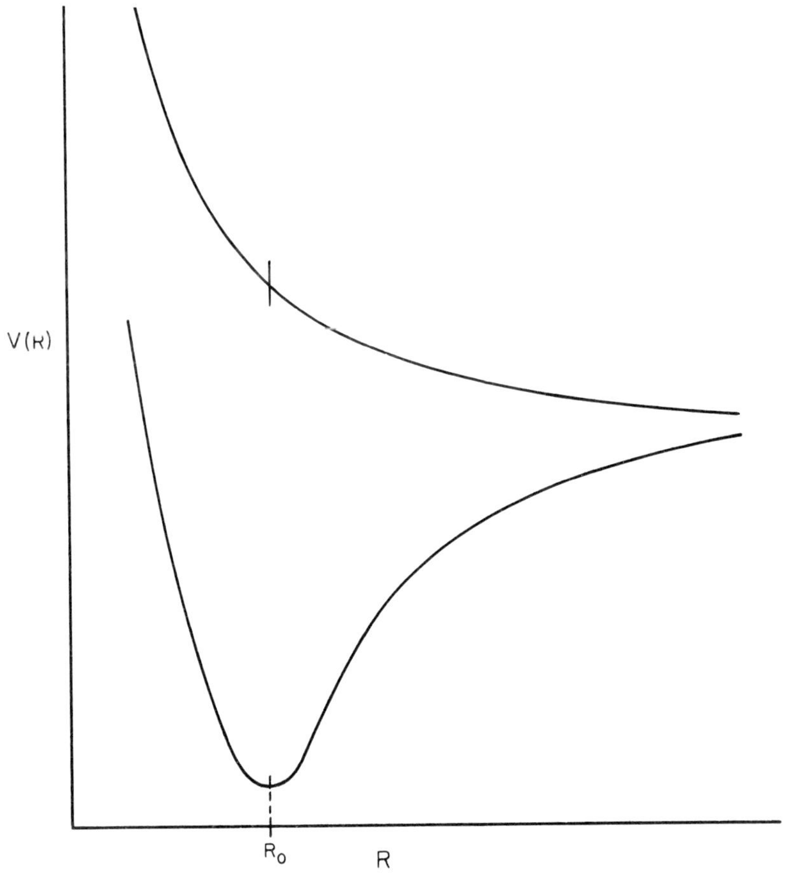

Fig. VII. 2. Potential energy curves for an isolated diatomic molecule which is of the type composing the linear chain model discussed in Section 3.

molecule and N - 1 ground-state molecules. In Figures 3 and 4 cross-sections through such potential surfaces are illustrated for the special cases of N = 2 and large N. In each of these figures the abscissa is the internuclear separation in one of the molecules with all other internuclear separations held at R_0. For such a system it is possible to dissociate only one molecule exothermically so that this parti-

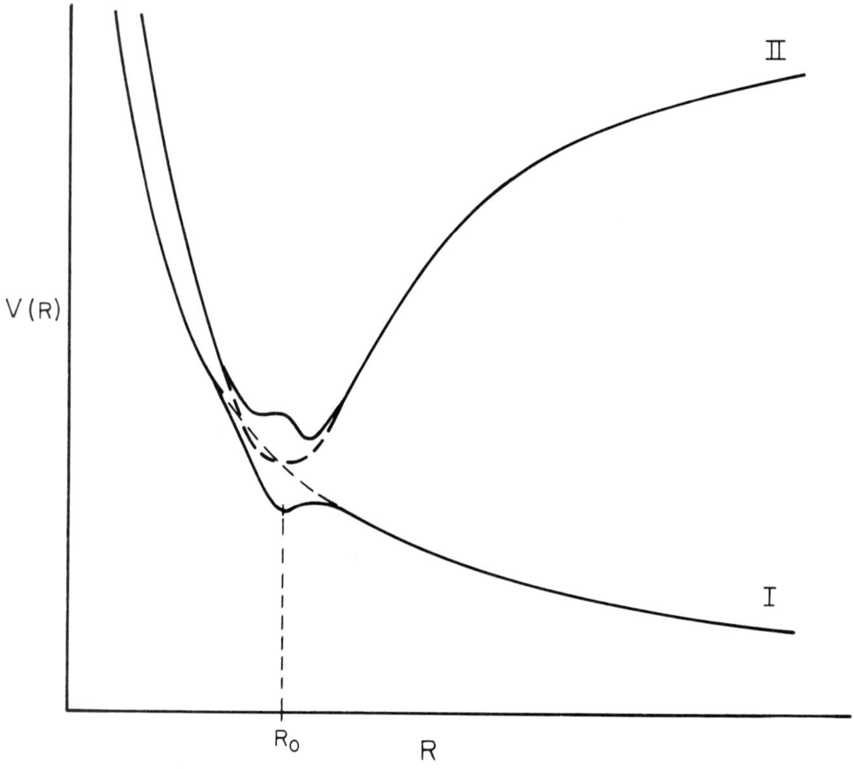

Fig. VII. 3. Potential energy curves for the two excited states of a two-molecule system. The abscissa is the internuclear separation of one of the molecules; the other is held at R_0.

cular cross-section gives significant information regarding all dissociative processes of interest.

For the system of two molecules there are two excited states. Each of them (Figure 3) has a minimum in its potential curve for the cut which we are examining. Points on curve I at the extreme right of Figure 3 correspond to dissociation of one of the molecules with the other molecule in the unexcited state at interatomic separation R_0; similarly, points on curve II at the extreme right correspond to dissociation of one molecule with the other molecule in an excited electronic state at interatomic separation R_0.

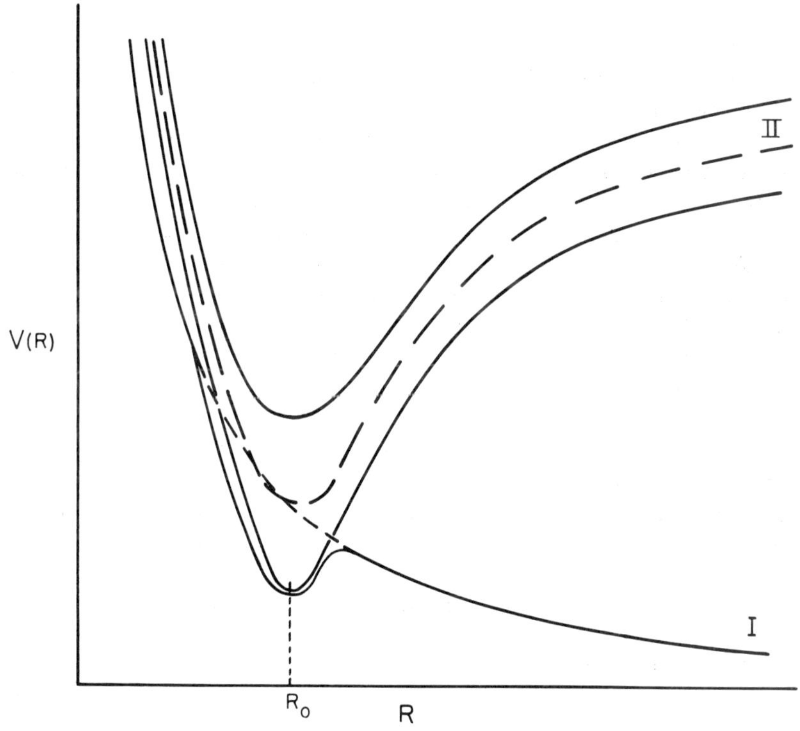

Fig. VII. 4. Potential energy curves for the N excited states of a long chain. The lower state is indicated by the solid curve and N-1 states are contained in the band. The dotted lines give the positions of the potential energy curves of an isolated molecule in the two relevant states.

For the system of N molecules there are N excited states. The potential curve for the lowest state separates from the other N-1 and is shown as the lowest solid curve of Figure 4. Potential curves of the other N-1 states are essentially parallel and are included between the upper two solid lines. At the extreme right of the figures on curve I no electronic excitation remains in the chain of N-1 molecules. At the extreme right of the figure on curve II the chain retains the electronic excitation. The curves were calculated for variation of the internuclear separation of a molecule which is far from the ends of the chain but the curves are actually not very sensitive to this position.

One sees from Figure 4 that it is extremely difficult for a system in strong-coupling to find a dissociation mechanism. Transitions to the excited state are generally expected to be 0-0 transitions and do not introduce vibrational excitation. Only the lowest state is at higher energy than its dissociation products and for it there is an activation energy for dissociation introduced by the resonance along the chain. Thus no initially excited chain dissociates immediately. Internal conversions can occur from higher states in the band with creation of vibrations, or energy can be acquired by collision processes. Such decompositions can occur only slowly. It appears more likely that decompositions of the chain itself can occur only if there are departures from strong-coupling. It is also likely that in most cases a chain in strong-coupling does not decompose at all, but eventually transfers energy to a localized state.

7.3. LOCALIZED ELECTRONIC EXCITATION

There are various mechanisms for localization of excitation. Any delocalized state tends to become localized as vibrations are excited or as rearrangements or dissociations occur. These are factors which either weaken the interaction between neighbors or destroy the equivalence of neighbors. A particular mechanism which applies specifically to radiation chemistry is the formation of triplet states. Molecules of interest have ground states which are almost invariably singlet and so excitation of a triplet type introduces interaction between neighbors of different multiplicity which is necessarily small. Only weak-coupling can result with almost complete localization of triplet excited states.

Triplet states are expected to form in spurs when parent ions recapture their electrons, as the following simple consideration demonstrates. To a good approximation there are no magnetic forces in a spur and so the overall multiplicity remains singlet. However, when several molecules are ionized, there are combinations of singlet and triplet states of the component molecules which are essentially degenerate and recombination occurs randomly yielding a mixture of singlets and triplets. For example, one ionization can yield only a singlet excited molecule, but a spur with two ion pairs has an equal probability to form two singlet or two triplet molecules. Large spurs yield 75 per cent triplets and 25 per cent singlets.

The fraction of triplet states formed on electron recapture for the general case is a simple combinational problem. This fraction has been calculated as a function of spur size and the results are summarized in Table 1.

TABLE VII. 1.

FRACTION OF TRIPLET STATES IN SPURS

Ions in spur	Fraction triplet	
1	0	0.000
2	2/4	0.500
3	9/15	0.600
4	36/56	0.640
5	140/210	0.667
6	540/792	0.682
7	2079/3003	0.693
8	8008/11,440	0.705
.		
.		
.		
∞	3/4	0.750

Localized states may also be formed by reactions between excited states. Specifically, two states which have strong dipole moments may react to form two states which have no dipole moment and thus tend to become localized. Of course, if such a transition can occur, the reverse transition can also occur, but such processes always tend to be dissipative and delay the spreading of excitation energy.

Another reaction between excited states is the formation of two triplet states from two singlet states. This process should occur for highly excited states only and is similar to the recombination of ionized states since it involves exchange of electrons.

It is likely that the most important mechanisms for localization involve (a) dissociations in delocalized states because of deviation from strong-coupling, or (b) transfer of excitation to a molecule which cannot share delocalized excitation because of its isolation (i. e., sensitized decomposition).

7.4. DISCUSSION

We have attempted to develop a picture of the very early processes which precede the chemical steps in radiation chemistry and now we shall attempt a brief summary.

Even for low LET radiations most of the energy is lost in spurs rather than in formation of isolated excited molecules. The overall geometry of a "spur" at the instant of its creation - when all the energy is still in electronic excitation and ionization - is determined by the path of the secondary electron. At this instant the affected molecules are quite close together; for a spur containing 100 ev of energy the diameter of the region is approximately 20Å. Electrons which are free at this time will be recaptured in their parent spurs for most systems of biological interest. Excited electronic states result from this recapture and since a large fraction of these are triplet states, they are confined to the vicinity of original formation. Radicals or reactive molecules formed in this process must diffuse to find reaction partners.

For the particular case of water, Samuel and Magee ('53) proposed a Gaussian initial distribution for H and OH radicals

$$C(r) = C_0 \exp\left[-(r/r_0)^2\right]$$

where $C(r)$ is the local concentration at the point r and r_0 is given by

$$r_0 = 5\ N_0^{1/3}\ \text{Å}$$

where N_0 is the total number of radicals formed in the spur. It was proposed in the original paper that in water all electrons are recaptured in their spurs and that all excited states are highly localized. Although details of this model require revision, it now appears that the gross features are essentially correct.

Much of the radiation chemistry of liquid systems which involve small molecules requires the interpretation that essentially only localized excitation is produced and spur sizes are quite small (Hamill '60). The spur size for water would appear to be about the minimum possible size.

In certain cases where there is crystalline structure or where positive holes may have unusually high velocities, electrons resultant from the ionization act may escape recapture in the original spurs. In such cases decompositions and localized excited states may be formed so widely dispersed as to be essentially distributed uniformly.

In all cases after neutralization there will remain a certain number of electronic states which are highly delocalized and effects will be transmitted to relatively large distances. For systems like water and aqueous solutions of small molecules, there is a negligible amount of energy involved in such states. For others, like benzene, a very significant amount of energy can be transmitted in such states.

A particularly favorable system for formation of delocalized excited electronic states is a long chain molecule. The primary chemical binding along the chain favors large values of the resonance integral β and thus a good velocity of spreading of the excitation. In such systems it is to be expected that excitation energy is ultimately trapped preferentially in significant functional groups. Delocalized excitation in domains of liquid systems of small molecules behaves in a very similar manner. Such excitation is transmitted to special locations where it is degraded (Lipsky and Burton '59).

Significant differences are always to be expected in the action of ionizing radiations and visible and ultraviolet radiation because of spur effects. Examination of the elementary processes involved leads to a wide range of possibilities and it is impossible at present to predict spur structure or the relative importance of specific spur effects. Some elementary processes are the same for all types of radiations and it is in the understanding of these that progress has been most rapid.

DISCUSSION

BURTON: I would like to know the time required for distribution of excitation in one hundred or so molecules of a coupled chain. When gamma is one hundred, what is the time?

MAGEE: The time depends upon the value of β, the coupling constant. If $\beta = 1$ ev, when $\gamma = 100$ the time is 3×10^{-14} seconds.

BURTON: This means that for a normal excitation, of around 5 ev, the energy is distributed in a chain of 100 molecules in a time which is very small in comparison with the single vibration period.

MAGEE: The value of β has nothing to do with the energy of transition. It is the coupling constant, and so they are completely independent. If $\beta = 1$ ev, the distribution time is about 10^{-14} seconds for a chain of 100 molecules, regardless of the transition energy.

KALLMANN: This is something like diffusion theory; something is missing.

MAGEE: This is not diffusion theory. It is exciton theory.

KALLMANN: Yes, but diffusion of excitons is the mechanism.

MAGEE: In this model there is no diffusion. The absolute value of X^2 is linear in time and if the process were diffusion X^4 would be proportional to time.

There are very definite phase relations all along the chain. The exciton has a Schrödinger equation for its motion, and the problem I have considered is very much the same as that of a wave packet for a free particle. If the diffusion equation were involved the result would be different.

POLLARD: This seems to be just like a pulse of electromagnetic radiation going down a loaded wave guide. You start with a pulse of energy and it has to keep its phase relationship both ways. There is a maximum where the wave reaches.

MAGEE: This relation comes out because there are phase relations all along. It is exciton motion. It is like a pseudo particle moving along, and it is not diffusive motion.

POLLARD: Would you get the equivalent of a reflection at a point which has weak-coupling?

MAGEE: I have taken a very long chain and there is no reflection.

POLLARD: But if you put the weak-coupling in the middle?

MAGEE: I have never worked that problem.

LIPSKY: I would like to make a few comments about Dr. Magee's paper. First of all, it does not seem very surprising to me that, in general, an aggregate of molecules will be less prone to photo-dissociate than will an isolated molecule. I reason as follows. In a molecular aggregate the electronic excitation resides on any one molecule a very short time — so short that the nuclei cannot move appreciably before the excitation has moved to another center. The result is that the nuclei spend most of their time in a potential field characteristic of the ground electronic state. My second point is this — and perhaps a spectroscopist can give me the answer — I would imagine that if you looked at the spectrum of one of the dimers that Dr. Kasha was talking about previously you would find the vibrational spacing in the excited state of the dimer very similar to the spacing in the dimer's ground state. The spacings would, in any event, not be as characteristic of an excited state as would be the case in an isolated molecule, namely the monomer.

ALLEN: Dr. Förster, do you have something to contribute to this point?

FÖRSTER: I can only say that I agree with Dr. Lipsky but I cannot say anything else to make it clearer.

LIPSKY: I would like to make one other point. This concerns something which has puzzled me for some time. I believe I spoke to Dr. Kasha about it once. In scintillation experiments one finds that cyclohexane can transfer energy to terphenyl with an efficiency of about one third that of benzene. This is not bad since benzene is a rather efficient solvent. Yet, as Dr. Kasha first pointed out to me, the first excited singlet state of cyclohexane should be repulsive. Then how can it transfer energy so efficiently? I would like to suggest now, after listening to Dr. Magee's talk, that cyclohexane exists in molecular aggregates in the liquid and that the first excited state is thereby considerably stabilized -- at least to the extent that now energy transfer can compete with dissociation.

BURTON: In the course of our work on mixtures we have rapidly arrived at the conclusion that cyclohexane transfers energy to benzene. Such transfer of energy can be a prelude to a transfer of energy to a scintillator solute which thereafter emits luminescence. It has been my habit to speak of excitation as being transferred from cyclohexane to benzene. However, I was seriously troubled by the knowledge that theorists find it rather difficult to think of a long-lived excited state of cyclohexane. Without going into detail, I just want to make the point that it now appears that ionization is transferred from cyclohexane to the benzene in a very short time. We know that the decay time for the cyclohexane when transfer of energy occurs to the scintillator directly is $<< 3 \times 10^{-10}$ sec. The transfer to the benzene must be even faster. The rate constant for the transfer of excitation from benzene to p-terphenyl directly is 9.3×10^{10} $1/\text{mole}^{-1}$ sec^{-1}. The transfer constant for the transfer of energy from cyclohexane to p-terphenyl directly is $>> 10^{12}$ $1/\text{mole}^{-1}$ sec^{-1}. It is because the latter transfer constant is so great that we are confronted with apparent anomaly. Cyclohexane does not transfer energy to a scintillator solute as effectively as does benzene. Again without going into detail the only satisfactory explanation is that benzene transfers excitation and cyclohexane transfers ionization.

KALLMANN: I agree with you in this respect that probably cyclohexane has a very short decay time and it is rather difficult to see how so much energy is transferred. I would say, however, that a way to get out of this difficulty is to assume that the transition probability for the lowest excited state of cyclohexane is much larger than that for benzene.

We have interpreted our experiments assuming that the decay time is very short but the transfer probability is high because the first absorption band may be much stronger; but we can really not prove it.

FÖRSTER: I only would like to say that benzene also is not very good for this purpose. Benzene has very weak oscillator strength, partially compensated by its longer life time, but you use it as a pure liquid where you have strong concentration quenching and, as far as I know, most experiments are with air-saturated solutions where you have also additional oxygen quenching. Therefore, it might not be very fair for cyclohexane to compare it with benzene which, from a theoretical standpoint at least, is also not very good.

BURTON: All of these comparison between benzene and cyclohexane are made both with air present and with air absent.

8

C. J. Hochanadel
Oak Ridge National Laboratory

Radiation Chemistry of Water

This presentation is an appraisal of the status of the radiation chemistry of water, with emphasis on the radiolysis of liquid water by ionizing radiations, although some attention is given to the radiolysis of water vapor and ice, and also to the photochemistry of water. It is not intended as a comprehensive review of effects of radiation on water and aqueous solutions, but rather as a general survey of the important principles that have been developed. For more detailed information the reader is referred to the original papers and to several recent reviews (e. g., Dewhurst et al. '54, Lefort '55, Miller '57, Hart '57, '58, and '59, Dainton '58 and '59, U. S. S. R. Symposium '57.)

A complete mechanism of radiation action accounts for the energy transmitted to water in terms of the various physical and chemical processes that lead to the final stable products. Such an account requires detailed knowledge of the exact nature, modes of formation, yields, spatial distribution, and subsequent reactions of all intermediates. Although some information is available from optical spectra, electron impact spectra, mass spectra, and electron spin resonance measurements on vapor and solid systems, there is little or no direct information concerning the intermediates in liquid water. Almost all our knowledge is inferred from chemical kinetic studies of various solutions as a function of scavenger concentration, pH, LET*, total radiation dose, dose rate, temperature, and

* LET is the linear energy transfer; i. e., $-\partial E/\partial x$, where x is the distance.

TABLE VIII.1.

POSSIBLE REACTIONS IN LIQUID WATER WITHIN 10^{-12} SECONDS FOLLOWING PASSAGE OF AN IONIZING PARTICLE

Numbers in parenthesis on the left give either the energy at the maximum in the optical absorption band, or the appearance potential.

Numbers in parenthesis on the right give the relative mass spectral abundance for impacting electrons of 100 ev energy.

			PRIMARY H_2O		
H_2O^*	(~ 7.4 ev)	→	$OH + H +$ ~ 2 ev k.e.		1
H_2O^*	(~ 9.2 ev)	→	$OH^* + H$		2
$H_2O^+ + e^-$	(~ 12.56 ev)		(non-bonding electron)	(100)	3
or $H{-}O{-}H \cdots {}^{+}OH_2$ + e^-		→	$H_3O^+ + OH + e^-$		4
$H_2O^{+*} + e^-$	(~ 18.7 ev)	→	$OH^+ + H +$ k.e. $+ e^-$	(23)	5
	(~ 19.5 ev)	→	$H^+ + OH + e^-$	(5)	6
	(~ 18.6 and ~ 26 ev)	→	$O^+ + H_2 + e^-$	(2 and 1.7)	7
H_2O (singlet) + e^-		→	H_2O (triplet) + e^-		8
e^- (5.6 ev) + H_2O		→	$H^- + OH$	(0.6)	9
e^- (7.5 ev) + H_2O		→	$O^- + 2H$	(1.5)	10

SECONDARY

H_2O^+		$+ e^-$	$\rightarrow$	$H_2O^* \rightarrow OH + H + k.e.$	11
H_3O^+		$+ e^-$	$\rightarrow$	$H_3O \rightarrow H_2O + H + k.e.$	12
H_2O		$+ e^-_{aq}$	$\rightarrow$	$H + OH^-_{aq}$ ($> 10^{-11}$ sec)	13
H_2O^+		$+ H_2O$	$\rightarrow$	$H_3O^+ + OH$	14
OH^+		$+ 2H_2O$	$\rightarrow$	$H_3O^+ + 2OH$	15
H^+		$+ H_2O$	$\rightarrow$	H_3O^+	16
H^-		$+ H_2O$	$\rightarrow$	$H_2 + OH^-$	17
O^-		$+ H_2O$	$\rightarrow$	$OH + OH^-$	18
	or	$+ H_3O^+$	$\rightleftarrows$	$OH + H_2O$	19
O^+		$+ H_2O$	$\rightarrow$	$H^+ + HO_2$	20
H_2O^*		$+ H_2O^*$ (or H_2O)	$\rightarrow$	$H_2O_2 + H_2$ (or 2H)	21

NET

$$H_2O \rightsquigarrow H, OH, (H_2, H_2O_2) \qquad 22$$

TABLE VIII. 2.

TIME SCALE FOR RADIATION CHEMISTRY OF WATER

Seconds	
10^{-18}	1 Mev electron traverses a molecule.
10^{-16}	5 ev electron traverses a molecule.
10^{-15}	"Vertical" excitation.
10^{-14}	Molecular Vibration. Possible dissociation. Electron thermalization and capture by spur in liquid.
10^{-13}	Collision time in liquid. Internal conversion.
10^{-12}	Conversion to H and OH probably complete. Radical jump time in diffusion ($D \sim 2$ to 8×10^{-5} cm^2 sec^{-1}).
10^{-11}	Dielectric relaxes. Electron trapped in liquid.
10^{-10}	Collision time in gas at 1 atmos.
10^{-9}	Radical combination nearly complete in cylindrical track (a).
10^{-8}	Radical combination nearly complete in spherical spurs (fast e^-). Lifetime for radiation of excited singlet.
10^{-4}	Radical diffuses inter-spur distance in 1 Mev electron track. Radiative lifetime of triplet state.

$$H_2O \xrightarrow{10^{-18} - 10^{-15}} H_2O^+,\ H_2O^* \xrightarrow{\sim 10^{-12}}$$

$$\underbrace{H, OH \xrightarrow{\sim 10^{-8} - 10^{-7}} H_2,\ H_2O_2,\ H_2O,\ H,\ OH}_{\text{Reaction of radicals with solute}}$$

Reaction of radicals with solute

(Cf. Section 8. 5. 3a)

isotopic content. The many studies of radiolysis of water and aqueous solutions can, for the most part, be explained by use of a simple radical diffusion model assuming H and OH radicals to be the predominant chemical intermediates produced along tracks of charged particles. Although the model was developed mostly from radiation-chemical studies, this presentation is given in reverse order. Possible primary and secondary processes are first discussed, followed by a description of initial spatial distributions of primary species and discussion of the diffusion kinetics model. Finally, experimental results are compared with semi-quantitative values calculated from the model. These include values of radical and "molecular" yields, and yields for total water decomposition, and the effects of scavenger concentration, LET, pH, isotopic content, temperature, and phase on these yields.

8.1. POSSIBLE PRIMARY AND SECONDARY PROCESSES

Various primary and secondary processes suggested by measurements of optical spectra (Watanabe and Zelikoff '53), electron impact spectra (Lassettre '59), and mass spectra (Smyth et al. '33) (Mann et al. '40) are listed in Table 1. An approximate time scale (Magee '53) for these processes and subsequent events is given in Table 2. Of the many possible reactions listed, only a few are considered to be important in liquid water. These are 1, 4 (or 3 and 14), 12 (or 11), and possibly 13. The remaining reactions, although generally of negligible importance, may be significant in special cases. At any rate, the net result of all the processes listed is decomposition of water to H and OH radicals, with the possibility of small yields of H_2 and H_2O_2. Most results at present can be correlated assuming decomposition only to H and OH.

The very efficient decomposition of water vapor (see section 8. 5. 4.) indicates that most of the energy is expended in exciting the lower electronic states. The lowest allowed electronic level extends over a broad continuum centered at ~ 7. 4 ev. The absence of any luminescence (Brown and Miller '55) indicates that excitation in this band is to a repulsive state which dissociates into normal $H(^2S)$ and $OH(^2\pi)$. The H atom would recoil with about 2 ev kinetic energy, allowing it to separate from its OH partner in the liquid by one or two molecular diameters. This broad continuum has also been attributed (Mulliken '35) to predissociation into H_2 $(^1\Sigma)$ and O (^1D), in which case the excited O atom would probably react immediately with water to form H_2O_2. Experimental results favor the first possibility. Excitation into the continuum centered at ~ 9.3 ev probably also leads to a repulsive state dissociating into H (^2S) and OH $(^2\Sigma)$. (If OH were present it might be detected by emission $(^2\Sigma)$ or absorption $(^2\pi)$ which occurs at 3064 Å in the vapor).

In the rarified atmosphere of the mass spectrometer, where ~ 10^{-6} seconds elapses between formation and detection of ions, processes 3, 5, 6, 7, 9, and 10 are indicated. Relative abundances for impacting electrons of 100 ev energy are listed in parenthesis at the right. In the vapor, ionization of a non-bonding electron on the oxygen requires 12.56 ev. Assuming ionization in liquid to have meaning, the ionization potential in the liquid is no doubt somewhat less than 12.56 ev. And, in the condensed system highly coupled by hydrogen bonding, the ionized molecule would immediately (~ 10^{-14} sec) abstract an H atom to form H_3O^+ and OH as in 4 (von Bünau et al. '59). The alternative is reaction 3 followed by 14. The mass spectrum also indicates a small proportion of positive ions in electronic states other than the ground state. These probably dissociate in the gas phase according to 5, 6, and 7 (Laidler '54). In the unlikely event that these ions are also formed in the liquid, they would probably react immediately according to 15, 16, and 20. Reaction 15, which probably would occur in two steps, offers a possibility for direct formation of H_2O_2. The ions observed in the mass spectrum at the lowest appearance potential are the hydride ion H^- and the ion O^-. Dissociative capture of subexcitation electrons to form H^- could lead to direct formation of H_2 by reaction 17. Platzman suggested (Platzman '60 unpublished) that this process might occur with a yield as large as ~ 0.5. Any O^- would react either as in 18, or more probably according to the thermal equilibrium indicated in 19. (See Section 8. 5. 1. for discussion of the acid-base properties of OH).

Electrons produced by ionization in the liquid are probably thermalized while still in the coulombic field of the associated positive ion, and react with either H_3O^+ or H_2O^+ according to 12 or 11. There is also a possibility that by the time the electron is thermalized it would have escaped the coulombic field of the positive ion, and in a period corresponding to the dielectric relaxation time (10^{-11} sec), would become trapped in the solvent. In a somewhat longer time, reaction 13 would occur to form H. (See Section 8. 2).

Niira ('52) presented theoretical evidence for formation of triplet states of water. The lowest electronically excited state of water may be a triplet state which could be formed directly by impact of slow electrons or by recombination of electrons with H_2O^+. Although there is no direct experimental evidence concerning either the existence of such species or their possible reactions, they would be interesting chemically because they are biradicals and have long radiative lifetimes ($\sim 10^{-4}$ sec in liquid).

Reaction 21 is also a possibility since primary events are closely spaced in spurs. Most experimental evidence, based on the ability of a solute to affect the yield of one molecular product (H_2 or H_2O_2) and not the other, indicates this reaction is unimportant.

The table does not list another possible primary event, i. e., ionization of an oxygen K electron leading to an Auger transition followed possibly by multiple bond breakage both in the original molecule and in adjacent molecules. There is no experimental evidence concerning this possibility.

8. 1. 1. Direct Experimental Evidence for Chemical Intermediates. - OH radicals have been detected both by absorption and by emission at 3064 Å in water vapor decomposed by various means (e. g., Oldenberg et al. '36, '39). It has also been reported (von Bünau et al. '59) that absorption bands attributable to OH and HO_2 are superimposed on the continuous Cerenkov background emitted by liquid water irradiated by ionizing radiation. Evidence for the existence of HO_2 also comes from mass spectrometric studies (Foner

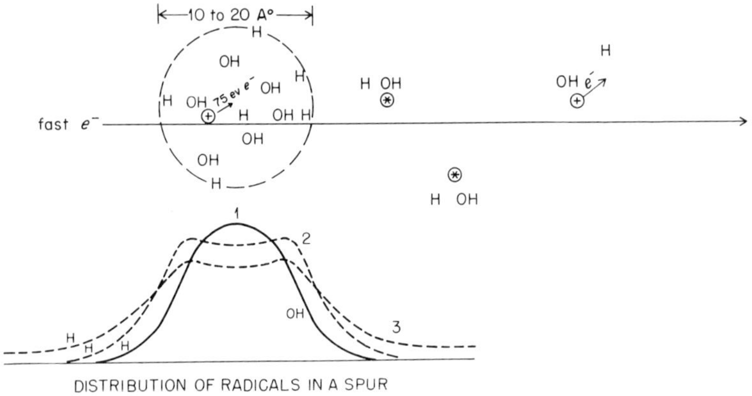

Fig. VIII. 1. Geometry of primary events produced by a fast charged particle traversing liquid water.

and Hudson '53). An absorption band at 2800 Å in irradiated ice (Ghormley and Stewart '56) may also possibly be attributed to OH. H atoms have been observed by paramagnetic resonance absorption in ice irradiated with electrons at 4.2° K (Piette et al. '59). Although there is some controversy, most evidence indicates that H atoms are not stable in pure ice at 77° K although they are stable in various frozen acid solutions at that temperature (Livingston et al. '54, '55, '59, Matheson and Smaller '55). Chemical evidence concerning the nature of the intermediates is discussed in Section 8.5.

8.2. GEOMETRY OF PRIMARY EVENTS

The geometry of primary events in liquid water is diagrammed roughly in Figure 1. The general picture (Platzman '53, Magee '53) is thought to be as follows. About 10^{-18} seconds is required for a 1 Mv electron to traverse and either excite or ionize a molecule. For a fast electron the primary ionizations are spaced, on the average, several thousand angstroms apart. For each ionization there are about two molecules excited to non-ionized electronic states. For slower particles, events along the primary track are more closely spaced, as shown in Table 3.

TABLE VIII. 3.

MEAN DISTANCE BETWEEN PRIMARY IONS

0. 5 Mev e^-	~ 5000 Å
1 Kev	~ 50 Å
100 ev	~ 5 Å
5 Mev α	8 Å
1 Mev α	2 Å

Most of the electrons ejected in the ionization acts have energy less than 5 ev. Some of them, however, have energies ranging up to a few thousand electron volts and can ionize and excite other molecules in a small volume element called a spur. The relative frequency of spurs containing various numbers of ions, as indicated both by cloud chamber measurements and by theory, is shown in Table 4.

TABLE VIII. 4.

DISTRIBUTION OF SPUR SIZES, according to Lea

Ion pairs per spur	1	2	3	4	> 4
Frequency	0. 43	0. 22	0. 12	0. 10	0. 13

The average energy of secondary electrons is estimated to be ~ 75 ev. It is believed that this energy is dissipated in a volume element ~ 20 Å in diameter, producing ~ 5 dissociated water molecules as a result of various excitation and ionization processes.

Details of this energy dissipation have been described about as follows. A 75 v electron makes about 10 collisions before its energy falls below the lowest excitation potential of water (~6.5 ev). This process requires about 10^{-15} sec, during which time the electron travels a total distance of about 30 Å. The several sub-excitation electrons produced in the spur loose their energy more slowly by exciting molecular vibrations and by dipolar interaction. They are thermalized (~ 0.025 ev) in ~ 10^{-13} sec, after travelling a total distance of ~ 1000 Å with frequent large angle deflections, and end up at a mean distance ~ 20 Å from the parent ion. At this distance the coulomb field of the positive ion is sufficiently strong (the dipoles have not had time to orient) to draw the electron back, thus producing a highly excited water molecule which dissociates into H and OH (Magee '53). The H atom receives enough kinetic energy to move a few molecular diameters through the relatively open water structure before being thermalized.

The spurs, of various size, are randomly spaced along the track of the primary particle. For very fast particles the individual spherical spurs are widely separated, while for sufficiently slow particles the spurs overlap and form a continuous cylindrical volume element 10 to 20 Å in diameter. For particles of intermediate velocity the track structure is a compromise between the two types. The distribution of radicals in a spherical spur is generally assumed to be Gaussian, as shown by distribution 1 in Figure 1. This distribution is probably correct for the OH radicals which are formed at the site of the primary event, regardless of the mode of formation. The H atoms, however, which may receive various amounts of kinetic energy depending upon the mode of formation, are probably formed initially in a broader distribution such as 2.

There is also the possibility, as suggested by Platzman ('53) that the mean distance of the electron from the positive ion upon thermalization may be > 50 Å, at which distance the electron is essentially out of the coulomb field of the parent ion and free to wander. The diffusing electron gradually polarizes the dielectric and eventually forms (or finds) a potential well and is trapped. This process may require a time as long as the dipolar relaxation time ($\sim 10^{-11}$ sec) during which the electron could have diffused a great distance. The electron is eventually captured by a water molecule in the region to give an H atom according to reaction 13. This latter process may require a time considerably longer than 10^{-11} sec. During all this time the diffusing electron could react with suitable solutes. The initial distribution of H atoms formed by this process is essentially uniform over the irradiated medium. Possibly both mechanisms contribute to H atom formation, giving a resultant H atom distribution something like 3. Most chemical evidence, as we see later, seems to favor the mechanism proposed by Magee, giving an initial distribution of H similar to 1 or 2.

It should be emphasized that there are three spatial correlations that are important in determining quantitative results (and in some materials qualitative results also). These are summarized under the following headings.

1. **The initial separation of the H and OH radicals in a pair.** This is a major factor in determining the amount of primary or "geminate" recombination (Burton et al. '58). The initial separation is governed by the amount of kinetic energy imparted to the H atom (mostly), which in turn is determined by the nature and energetics of the dissociation process; 2. **The average distance between radicals in a spur.** This is a major factor in determining the extent of radical combination during the process of spur expansion. 3. **The distance between spurs.** This is the most important factor leading to quantitative (and even qualitative) differences between radiations having different LET.

An initial H-OH separation of one molecular diameter may allow ~ 50% primary recombination, while a separation of two diameters may result in only ~ 20% recombination. Most radical pairs are produced in spurs in the proximity of several other pairs. This situation is true even for fast electrons where the typical event along the primary particle track is formation of a single pair (~40% of the spurs). For a spur containing 6 H and 6 OH radicals in a volume element 20 Å in diameter the radical concentration is ~ 5 M and the radicals are separated by about two molecular diameters. For this multi-pair system there is a high probability of combination of both like and unlike radicals before homogeneous distribution is attained by diffusion. Also, the close proximity of pairs in a spur, with subsequent inter-pair combinations, probably cuts down the amount of primary recombination. This effect would be largest at high LET where spur overlap brings pairs still closer together.

8.3. DIFFUSION KINETICS THEORY

Because most of the intermediates are produced in spurs, the initial concentration is non-homogeneous. The concentration is, therefore, time dependent, and subsequent reactions are diffusion controlled. Theories have been developed to describe the behavior of a spur as a function of time while undergoing concurrent outward diffusion and chemical reaction. The initial theory of Lea ('46, '47) has been developed in its present form mostly by Magee and coworkers (Magee '53, Samuel and Magee '53, and subsequent papers). The models used start with H and OH radicals in some initial distribution in a spur (Figure 1), and in thermal equilibrium with the surroundings. The basic equation

$$\frac{\partial(H)}{\partial T} = D_H \nabla^2(H) - k_{H,H}(H)^2 - k_{H,OH}(H)(OH) - k_{H,S}(H)(S)$$

describes the concentration of H atoms as a function of time, at any distance from the center of the spur, as the concentration changes by outward diffusion (term 1), combination with other H atoms (term 2), combination with OH radicals (term 3) and reactions with solutes (scavenger) (term 4). The equation can be generalized still further by adding terms to account for reactions with other intermediates and other solutes. Similar equations describe the behavior of OH radicals or other intermediates. The initial concentration distribution is usually assumed to be Gaussian and to remain so during diffusion. The initial concentration in a spur is specified in terms of time (t_c) required for expansion from a point to the assumed initial spur diameter.

For radiations of low LET, where spurs are isolated, the concentration is assumed to have spherical symmetry; for large LET, > ~ 3 ev $Å^{-1}$, (complete spur overlap) cylindrical symmetry; and for intermediate LET, a combination of spherical symmetry followed by cylindrical symmetry. The important quantities which can be calculated and compared with experiment are: (1) the fraction of radicals combining to give the "molecular yield" (term 2) and the fraction reacting with scavenger to give the "radical yield" (term 4); (2) the effect of LET on these yields; and (3) the effect of scavenger concentration on these yields; i. e., the extent of increase in term 4 at the expense of term 2 for radiations of various LET. Other experimental variables that have been studied, but which alter the model in ways which are as yet not completely understood, are the effects of isotopic substitution, pH, temperature, and phase, on these yields.

The success of this theory in making quantitative predictions depends upon the accuracy of the model, which in turn depends upon the amount and accuracy of information available for formulating and testing the model. Development of a detailed, quantitative treatment requires information on the nature of the intermediates, their initial spatial distribution, diffusion coefficients, and rate constants for reaction with like intermediates, with other intermediates, and with various solutes. Calculations must also take into account the distribution of spur sizes, and random spacing of spurs along the primary track. Actual calculations are limited to some extent by lack of physical and chemical information, and also by mathematical tractability. However, using a simplified model, and assuming reasonable values for the parameters involved, remarkable agreement has been achieved between calculated and experimental values for the quantities listed above. With the accumulation of more physical and chemical information, and the use of electronic computers (Flanders and Fricke '58, Kuppermann '59, Dyne and Kennedy '58, '60) to allow more versatile mathematical calculations, the quantitative situation is rapidly improving. For example, calculations have been made on the sensitivity of radical and molecular yields to changes in various parameters for both spherical symmetry and cylindrical symmetry (Dyne and Kennedy '58, '60). Summaries of the development of diffusion theory have been given by Kuppermann ('59) and Platzman ('58).

8.4. RESULTS IN PURE WATER

Most chemical results can be explained assuming decomposition of water into H and OH radicals, part of which combine during the process of spur expansion ("molecular yield"), and part of which "escape" and are scavenged by small concentrations (10^{-6} M to 10^{-3} M) of solute ("radical yield"), (Allen '48). The overall initial sequence is shown at the bottom of Table 2. The effect of solute concentration is discussed in Section 8. 5. 3a.

Irradiation of pure water with gamma rays or fast electrons (low LET) produces steady state concentrations of H_2, H_2O_2, and O_2. In absence of added solute the diffusing H and OH react with H_2 and H_2O_2 produced in spurs, according to reactions 1, 2, and 3 (Table 5). Reactions 1 and 3 lead to steady state concentrations of H_2 and H_2O_2; reaction 2 and subsequent reactions of HO_2 lead to O_2 formation (Allen et al. '52, Hochanadel '52, '56).

Ghormley ('56) found that for 1. 5 Mv electron irradiation, the steady state H_2O_2 concentration increases approximately as the square root of dose rate, reaching a value ~ 9×10^{-6} M at the upper limit of ~ 10^8 rads min^{-1}. For an electron current of ~ 0. 1μa (~ 2×10^7 rads min^{-1}), the shortest lived intermediate that affected this concentration had a lifetime ~ 10^{-3} sec. Increasing dose rate favors interaction of intermediates from spurs of adjacent tracks, thereby increasing the extent of radical-radical reactions leading to decomposition products over the radical-decomposition product back reactions. With an efficient scavenger for free radicals present (Fe^{+2} or Ce^{+4}), Sutton and Rotblat ('57) observed a dose-rate dependence only above 10^{10} rads min^{-1}.

The steady state concentration increases with LET up

TABLE VIII. 5.

TYPICAL SCAVENGER REACTIONS IN RADIOLYSIS OF LIQUID WATER

H_2	$+ \ OH$	$\rightarrow$	H_2O	$+ \ H$		1
H_2O_2	$+ \ OH$	$\rightarrow$	H_2O	$+ \ HO_2$		2
H_2O_2	$+ \ H$	$\rightarrow$	H_2O	$+ \ OH$		3
O_2	$+ \ H$	$\rightarrow$	HO_2			4
Br^-	$+ \ OH$	$\rightarrow$	Br	$+ \ OH^-$		5
Fe^{+2}	$+ \ OH$	$\rightarrow$	Fe^{+3}	$+ \ OH^-$		6
Fe^{+2}	$+ \ HO_2$	$\rightarrow$	Fe^{+3}	$+ \ HO_2^-$		7
Fe^{+2}	$+ \ H_2O_2$	$\rightarrow$	Fe^{+3}	$+ \ OH^-$	$+ \ OH$	8
$Fe^{+2}\cdot HOH$	$+ \ H$	$\rightarrow$	$Fe^{+3}OH^-$	$+ \ H_2$		9
Fe^{+3}	$+ \ H$	$\rightarrow$	Fe^{+2}	$+ \ H^+$		10
Fe^{+3}	$+ \ HO_2$	$\rightarrow$	Fe^{+2}	$+ \ H^+$	$+ \ O_2$	11
Ce^{+4}	$+ \ H$	$\rightarrow$	Ce^{+3}	$+ \ H^+$		12
Ce^{+4}	$+ \ HO_2$	$\rightarrow$	Ce^{+3}	$+ \ H^+$	$+ \ O_2$	13
Ce^{+4}	$+ \ H_2O_2$	$\rightarrow$	Ce^{+3}	$+ \ H^+$	$+ \ HO_2$	14
Ce^{+3}	$+ \ OH$	$\rightarrow$	Ce^{+4}	$+ \ OH^-$		15
$HCOOH$	$+ \ H$	$\rightarrow$	H_2	$+ \ COOH$		16
$HCOOH$	$+ \ OH$	$\rightarrow$	H_2O	$+ \ COOH$		17
C_2H_5OH	$+ \ H$	$\rightarrow$	H_2	$+ \ C_2H_4OH$		18
C_2H_5OH	$+ \ OH$	$\rightarrow$	H_2O	$+ \ C_2H_4OH$		19
	H_2^+	$\rightleftharpoons$	$H + H^+$	$pK_a \leqq 2.7$		20
	OH	$\rightleftharpoons$	$O^- + H^+$	$pK_a \sim 10$ (possibly higher)		21
	HO_2	$\rightleftharpoons$	$O_2^- + H^+$	$pK_a \sim 2$ to 3		22
	H_2O_2	$\rightleftharpoons$	$HO_2^- + H^+$	$pK_a \ 11.75$		23

to ~ 10 ev Å, beyond which there is no evidence that a steady state is reached at all (Allen and Schwarz '58, Hart et al. '56). The steady state concentration, therefore, increases as the fraction of radicals combining in spurs to form decomposition products increases, and the fraction excaping and thereby becoming available to promote back reaction decreases.

8.5. RESULTS IN SOLUTIONS*

With solute present to scavenge the diffusing H and OH, the back reaction can be prevented, and yields of H_2 and H_2O_2 (molecular yield) evaluated. Also, from the nature and amount of change in the solute or solutes, yields of H and OH (radical yield) can also be evaluated.

Solutions which provide the most useful information on radiolysis of water are those which: (a) can be studied under a variety of conditions, e.g., pH, (b) provide a simple pattern of products which can be accurately analyzed, (c) quantitatively scavenge the intermediates at low solute concentration (small activation energy), (d) allow postulation of a simple mechanism on the basis of which yields of intermediates can be related to yields of measured products. Reactions of some typical scavengers with the intermediates are listed in Table 5. The reactions as written are stoichiometric, but not necessarily mechanistic.

8.5.1. Chemical nature of the intermediates. - All of the intermediates show both oxidizing and reducing properties. The nature and rate of the reaction may depend upon conditions governing the structure of the intermediate or the reacting solute or both. For example, all the intermediates have acid-base properties and, at sufficiently high pH, assuming sufficient time for conversion, they would react as the basic species (e.g., O^-, O_2^-, HO_2^-). Estimated pK values are listed in Table 5. The "H atom"

* A measured yield is given as G(Product). Derived values of yields of intermediates are given as G_H, G_{OH}, G_{H_2}, $G_{H_2O_2}$. Yields referred to in this discussion are initial yields; i.e., the yield at low total dose where the relative amount of product is so small that it does not compete appreciably with added solutes for intermediates.

or its equivalent displays a variety of reducing and oxidizing properties determined by the mode of formation, pH, etc. For example, the H atom produced in reaction 1, Table 5, (sometimes termed H') and that produced in radiolysis of water display markedly different relative reactivity toward O_2 and H_2O_2 (Barr and Allen '59). For the H produced from water decomposition $k_{H + O_2}/k_{H + H_2O_2} \sim 1.5$ to 1.8 (Allen and Schwarz '58, Hochanadel '56) whereas for H' from reaction 1, $k_{H' + O_2}/k_{H' + H_2O_2} \sim 1000$ (Hochanadel, unpublished). A detailed treatment of radiolysis of water based on two kinds of H has been developed by Allen and Schwarz ('58). It has been suggested that the various species are acid and basic forms (Allen and Schwarz '58) represented by:

$$e^-_{aq.} \overset{H+}{\rightleftharpoons} H \overset{H+}{\rightleftharpoons} H_2^+.$$

The exact nature of the species under particular conditions is not known. However, the fact that reaction 1 is relatively slow, and that it is about three times faster in H_2O than in D_2O at 25° C (Bunn et al. '59) indicates this reaction occurs by bond breakage rather than by electron transfer and, therefore, a true H atom is produced. The relative stability of H_2O_2 in acid solution suggests that this species is the acid form since it reacts much more rapidly with O_2 than with H_2O_2. Other evidence for "H atoms" of varying reactivity has been obtained by Armstrong et al. ('59) and by Baxendale and Hughes ('58). It is not entirely clear at present whether more than one type H atom is produced in the decomposition of water.

Evidence for acid base properties of OH radical was obtained by Hart et al. ('52).

Either H or OH radicals can be generated separately by other means, e.g., thermally or photochemically, and, by studying their reactions individually, evidence can be obtained to substantiate or eliminate postulated radiolytic mechanisms. For example, photolysis of H_2O_2 solution produces OH radicals. By studying the photolysis of dilute H_2O_2 with H_2 added, it was found that the ratio of rate constants for reaction with OH (reactions 1 and 2, Table 5) was the same as in the radiolysis of water (Hochanadel '52,

'55, '56). With both H_2 and O_2 added, the competition of O_2 and H_2O_2 for H atoms (reactions 4 and 3) produced in reaction 1 can also be studied. In this case the ratio of rate constants was much greater than the ratio of rate constants for reaction with H atoms produced in radiolysis of water (Hochanadel '55, and unpublished results) in agreement with the postulate of Barr and Allen ('59) that "H atoms" produced by these two methods are different. Hydroxyl radicals can also be produced by reaction of ferrous ion with H_2O_2 (Fenton's reaction), and the relative reactivities of H_2 and D_2, e. g., with OH have been studied by this means (Bunn et al. '59). Either OH or H can also be generated by the photooxidation or photoreduction of water, sensitized by metal ions (e. g., Dainton '54). The ratio of rate constants, $k_{Tl^+ + OH}/k_{Ce^{+3} + OH}$, was found to be the same (~ 40) for OH radicals generated either by this means or generated radiolytically (Sworski '57).

Another subject which probably should have been mentioned earlier, is the effect of subexcitation electrons, or subexcitation energy. Platzman ('55) suggested that subexcitation electrons (i. e., electrons with energy less than the lowest electronic excitation energy of water) may produce effects in a minor component far out of proportion to the concentration of that component. An effect of this type may have been demonstrated (Hart '59) in the radiolysis of formic acid solutions.

8.5.2. Reaction kinetics. - For a system in which several solutes compete for the same intermediate, the fraction of intermediate, R, reacting with a particular solute is given by:

$$f_i = \frac{k_{Ri}(S_i)}{\sum_j k_{Rj}(S_j)}$$

where k_{Ri} is the rate constant for reaction of intermediate with that solute, and (S_i) is the concentration of that solute. There are no measurements of absolute rate constants for reactions of intermediates, either with other intermediates or with solute, in the radiolysis of liquid water. Calculations based upon diffusion theory and the dependence of molecular yields on scavenger concentration indicate, however, that rate constants for some radical-solute reactions (reactions 5, for example) are comparable to those for radical-radical reaction (~ 10^{10} liter $mole^{-1}$ sec^{-1}), and presumably occur on every collision (Burton and Kurien '59, Senvar and Hart '58). Many relative rate constants have been evaluated. Measured relative rates for reaction with H atom, for example, vary by a factor ~ 10^5 (e.g., Riesz and Hart '59).

8.5.3. Evaluation of radical and molecular yields. - Reactions of the stable decomposition products of water with the intermediates (1-4) lead to steady state concentrations as described above.

With Br^- added to pure water, reaction 5 competes with 1, and only a trace of Br^- (~ 10^{-6} M KBr) inhibits reaction 1 sufficiently to allow measurement of the initial yield of hydrogen (Allen et al. '52, Hochanadel '52). This is presumably hydrogen produced by H atom combination during spur expansion to give the "molecular" hydrogen yield, G_{H_2}. For gamma irradiation, the initial yield is constant at KBr concentrations in the range ~ 10^{-6} to ~ 1 M.

With excess H_2 and O_2 initially added, all of the radicals are converted to HO_2 radicals which disproportionate to H_2O_2 and O_2, (Hochanadel '52), giving the stoichiometry:

$$G(H_2O_2) = G_{H_2O_2} + \frac{1}{2}(G_H + G_{OH}) = G_{H_2} + G_H$$

$$G(-H_2) = G_{OH} - G_{H_2}$$

$$G(-O_2) = \frac{1}{2}(G_H + G_{OH})$$

Formic acid and oxygen provide a scavenger system similar to the hydrogen-oxygen system (Hart '52). In this case, however, both formic acid and oxygen compete for H atoms (16 and 4). The formate radicals react with O_2 to give CO_2 and HO_2, and the HO_2 radicals disproportionate to H_2O_2 and O_2. If concentrations are adjusted so that reaction

4 occurs to the exclusion of 16, then

$$G(H_2) = G_{H_2}$$

$$G(H_2O_2) = G_{H_2O_2} + \frac{1}{2}(G_H + G_{OH})$$

$$G(CO_2) = G_{OH}$$

$$G(-O_2) = \frac{1}{2}(G_H + G_{OH})$$

This system has been extensively studied. The behavior of ethanol-oxygen is similar to that of formic acid-oxygen solutions (Jayson et al. '57). Either formic acid or ethanol provides a better scavenger than hydrogen for OH radicals, both because of greater reactivity and greater range of concentrations available. Reactions of these molecules with H or OH are abstraction reactions. Other organic solutes such as ethylene and benzene undergo addition reactions with H, and in the case of ethylene, with OH also.

Kinetics for the oxidation of ferrous sulfate by radiation have also been extensively studied, and the system has provided much information on the radiolysis of water (e.g., Rothschild and Allen '58). It also provides the best chemical dosimeter, and is generally used as such. Reactions 6 - 11 are all possible under some conditions, but conditions (particularly concentration and pH) can be adjusted to favor particular reactions. For example, in acid solution containing Fe^{+2} and O_2, reactions 4, 6, 7, and 8 predominate, giving

$$G(Fe^{+3}) = 2G_{H_2O_2} + 3G_H + G_{OH}$$

$$G(H_2) = G_{H_2}$$

$$G(-O_2) = G_H$$

As Fe^{+3} is produced, it competes with Fe^{+2}, mostly for HO_2 radicals according to 11. This reaction is favored by higher pH, presumably because of dissociation of HO_2, and the ease of electron transfer between O_2^- and Fe^{+3}. In oxygen-free Fe^{+2} solution, results indicate that Fe^{+2} is oxidized by all the intermediates according to 6, 8, and 9, giving

$$G(Fe^{+3}) = 2G_{H_2O_2} + G_H + G_{OH}$$

$$\text{and } G(H_2) = G_{H_2} + G_H$$

Oxidation by H atoms may occur by electron transfer with H_2^+ (Rigg et al. '52), or by H atom abstraction from water of hydration as written in 9 (Uri '52). The fact that this reaction, and also reactions 6 and 8, are two to three times slower in D_2O than in H_2O suggests that they all involve bond breaking in water of hydration (Baxendale and Hughes '58; Dainton '59; K. Coatsworth, E. Collinson and F. S. Dainton '60, in press). The system is generally studied at pH < 2.5 and, therefore, lack of pH dependence does not necessarily rule out the H_2^+ mechanism as sometimes suggested. The Fe^{+3} produced competes for H atoms according to reaction 10. Again, the reaction is favored at higher pH, indicating that the hydrolyzed species, $FeOH^{+2}$ is more reactive towards H than is $FeSO_4^+$ or other sulfate complexes.

In ceric sulfate solution, Ce^{+4} is reduced by H, HO_2, and H_2O_2 according to 12, 13 and 14, and Ce^{+3} is oxidized by OH (Allen '54), giving

$$G(Ce^{+3}) = 2G_{H_2O_2} + G_H - G_{OH}$$

$$G(H_2) = G_{H_2}$$

$$G(O_2) = G_{H_2O_2}$$

From yields measured for several systems, yields for all intermediates can be evaluated. For example, Figure 2 shows yields in 0.4 M H_2SO_4 solution for: (1) Fe^{+3} production in solution containing $Fe^{+2} + O_2$; (2) Fe^{+3} production in solu-

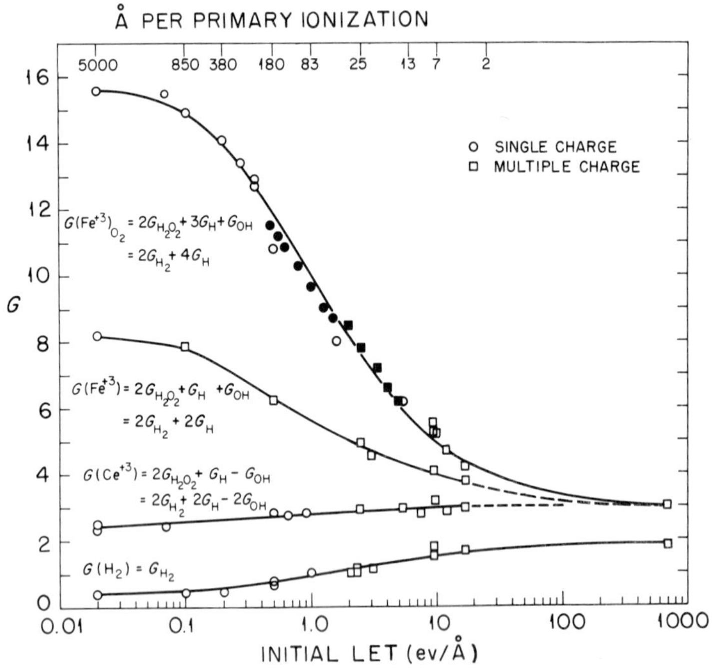

Fig. VIII. 2. Average yields for Fe^{+2} oxidation (in presence and absence of O_2), Ce^{+4} reduction, and H_2 formation as a function of initial LET of the impinging particle. Values for both multiply charged and singly charged particles are shown. The average distance between primary ions is shown at the top of the Figure. Values were taken from many sources, e. g. , the filled circles and squares are by Schuler and Allen ('56) using deuterons and helium ions. See Table 6 for LET values for the various radiations.

tion containing only Fe^{+2}; (3) Ce^{+3} production in solution containing Ce^{+4} (with or without O_2); (4) H_2 formation in KBr solution or in solution 1 or 3. From stoichiometry in these systems: 1 - 2 = $2G_H$; 2 - 3 = $2G_{OH}$; and from G_{H_2} and the stoichiometric equation

$$G(-H_2O) = 2G_{H_2} + G_H = 2G_{H_2O_2} + G_{OH},$$

$G_{H_2O_2}$ can also be evaluated. Radical and molecular yields evaluated using many different types of scavenger systems show excellent agreement. Values for 0.4 M H_2SO_4 solution are plotted in Figure 3. LET values for the various radiations are listed in Table 6.

8.5.3a. Effect of concentration. - For dilute solutions, nearly all the change in solute results from energy absorbed in the water (indirect action) and, therefore, the total measureable change is independent of solute concentration. The proportion of decomposed water detectable as H and OH or as H_2 and H_2O_2, however, depends to some extent on solute concentration. Diffusion calculations indicate that radical combination (molecular yield) is essentially complete in spherical spurs in ~ 10^{-8} to 10^{-7} sec. Within this time the spurs have expanded sufficiently that the remaining radicals are in essentially homogeneous distribution and are scavenged (radical yield) by low concentration of solute. For example, for gamma irradiation, Br^- at a concentration ~ 10^{-6} M scavenges the diffusing OH radicals and protects the molecular H_2 from back reaction (reaction 1, Table 5) (Hochanadel '52). The estimated rate constant for reaction of Br^- with OH is ~ 10^{10} liter mole^{-1} sec^{-1} (Burton and Kurien '59) giving $k_R(S)$ ~ 10^4 sec^{-1}. To compete appreciably with formation of molecular products (H_2 and H_2O_2), the half-life of radical-solute reaction must be less than ~ 10^{-7} sec, or $k_R(S) > 10^7$ sec^{-1}. Therefore, to a first approximation, yields are independent of scavenger concentration in the range for which ~ 10^7 sec^{-1} > $k_R(S)$ > ~ 10^4 sec^{-1}. For lower scavenger concentration some back reaction will occur, and for greater concentration, the molecular yield will be diminished and the radical yield enhanced. The situation is suggested at the bottom of Table 2.

The scavenging of radicals that would have combined in spurs may not, however, be reflected in appreciable change in measured product yields. For example, since both OH and H_2O_2 oxidize Fe^{+2}, interference with H_2O_2 formation in spurs by Fe^2 would not be reflected as an increase in

TABLE VIII. 6.

INITIAL LET OF VARIOUS RADIATIONS FOR WHICH YIELDS HAVE BEEN MEASURED; see Figures 2 and 3.

Radiation	Initial LET, ev per Å
Co γ-rays	0.02
1 or 2 Mev e^-	0.02
S^{35} β^- (av. 46 kev)	0.07
250 kvp X-rays	0.10
10 kv X-rays	0.20
8 kv X-rays	0.28
H^3 β^- (av. 5.5 kev)	0.36
20 Mev d^+	0.45
8.4 Mev p^+	0.55
5.2 Mev d^+	1.3
2.0 Mev p^+	1.7
38.0 Mev He^{++}	2.2
delta rays	~ 0.8 to 3.0
2.7 Mev $(H^3)^+$	3.0
0.9 Mev p^+	3.0
12.0 Mev He^{++}	5.0
0.3 Mev p^+	5.4
Max. for p^+ or d^+ (~ 0.1 or 0.2 Mev)	~ 8.
5.3 Mev α (Po^{210})	8.8
Li^6 (n, 2.05 Mev α) 2.73 Mev H^3	~ 10.
3.4 Mev α	12.
β (n, 1.50 Mev α) 0.85 Mev Li^7	~ 17.
Max. for H (~ 1 Mev)	~ 26.
65 Mev fission recoil (mass 138)	~ 700.

$G(Fe^{+3})$, except insofar as replacing an OH radical by Fe^{+3} would affect the interplay of radicals within the spur. The effect in this case is negligible. On the other hand, oxygen diminishes the molecular hydrogen yield by scavenging H atoms in spurs, giving enhanced $G(Fe^{+3})$ through the HO_2 radicals produced (e.g., Trumbore and Hart '59).

For radiations of large LET, for which the diffusing region is cylindrical, the molecular reaction is nearly complete in 10^{-9} to 10^{-8} sec and, therefore, higher scavenger concentration is required to inhibit back reaction and to diminish molecular yields. For example, even in presence of 10^{-3} M Fe^{+2}, the yield of HO_2 from reaction 2 in the expanding tracks produced by Po α-particles is ~ 0.25 (Donaldson and Miller '56), and presumably decreases with increasing Fe^{+2} concentration. For cobalt gamma-rays the yield is ~ 0.02. The HO_2 is determined by O_2 formed on reduction of Cu^{+2}. Evidence for reaction 3 in expanding spurs was deduced from effects of O_2 concentration and pH on H_2O_2 formation in oxygenated KBr solutions irradiated with deuterons or helium ions (Schwarz et al. '59). From these studies the rate constant for reaction 3 was estimated to be ~ 100 times greater than that for reactions 1 or 2.

The dependence of molecular yield on scavenger concentration was first studied by Sworski ('54) who found that for gamma radiation $G_{H_2O_2}$ was depressed approximately in proportion to $(KBr)^{1/3}$. Expressed in general form,

$$G_{R_2} = G_{R_2}^0 - A(S)^n$$

where G_{R_2} is the molecular yield at scavenger concentration (S), and $G_{R_2}^0$ is the yield extrapolated to zero concentration. The equation has been found to hold for depression of both $G_{H_2O_2}$ and G_{H_2} by several solutes in the concentration range up to ~ 0.1 M. At higher concentration the scavenger efficiency drops off, and a better empirical fit is obtained using scavenger activity rather than concentration (Mahlman '58). Burton and Kurien ('59) found the general equation to hold for depression of $G_{H_2O_2}$ by Br^- using 24 Mev x-rays, Cobalt-γ-rays, 50 kvp x-rays, and 3.4 Mev

α-particles, the exponent n being 0.38, 0.33, 0.26 and 0.15 respectively. At present there is no theoretical justification for this expression, and it is probably merely curve fitting by an insensitive function. Were it to have theoretical significance, the cube-root dependence implies that a distance is important.

Comparison of the dependence of molecular yields on scavenger concentration with diffusion theory was made by Schwarz ('55). The test is made by plotting the fraction of radicals combining (1-S) against log q (where $q = k_S t_o (S)$ and $t \sim 1.25 \times 10^{-10}$ sec), or more simply G_{R_2} or $G_{R_2}/G_{R_2}^0$ against log (S). Schwarz found that for depression of both G_{H_2} and $G_{H_2O_2}$ by a variety of solutes (S) in the concentration range 10^{-5} to several molar, the curves could be superimposed by displacing them laterally an amount depending upon k_S, indicating that all the solutes behave in similar fashion. At low concentration the scavenger is somewhat more efficient and at high concentration somewhat less efficient than diffusion theory predicts. In the calculations it is assumed that decomposition occurs by a single mode giving a Gaussian distribution of radicals. However, as pointed out earlier, there may be several modes of decomposition, each giving a different spatial distribution of radicals (or even different species) which would be scavenged with what appeared to be different efficiencies.

8.5.3b. Effect of LET. - As shown in Figure 2 yields for various reactions have been measured for many different radiations covering almost the entire spectrum of LET, and including both singly and multiply charged particles. The average distance between primary ionizations (or spurs) is shown at the top of the Figure. Yields are plotted against initial LET, usually of a completely absorbed particle. A more complete treatment taking LET distributions into account, would allow evaluation of yields for single values of LET (Burch '59). Derived values of radical and molecular yields in 0.4 M H_2SO_4 solution are shown in Figure 3 as a function of LET. Two different yields for water decomposition are also plotted; $G(-H_2O)_{net}$ gives a value based only on stoichiometry of measured products; $(G\text{-}H_2O)_{gross}$ includes an estimate of H-OH combination in spurs, thereby giving an estimate of "initial" decomposition. This is discussed more fully in section 8. 5. 4. From the figure several things are obvious: (a) at low LET, radical yields are much higher than molecular yields, and G_H is greater than G_{OH} (correspondingly $G_{H_2O_2} > G_{H_2}$); (b) with increasing LET, radical yields decrease and molecular yields increase until, at sufficiently high LET, the radical yields approach zero, indicating nearly complete radical combination during track expansion, and also, in the LET range above ~ 5 ev Å, G_{OH} exceeds G_H, and G_{H_2} exceeds $G_{H_2O_2}$; (c) although $G(-H_2O)_{net}$ decreases with LET, $G(-H_2O)_{gross}$ increases with LET.

The fraction of total radicals combining (based on $G(-H_2O)_{gross}$) varies from ~ 0. 5 for low LET (see Table 7) to ~ 1. 0 at very high LET. Measured values for radiations of various LET are in semi-quantitative agreement with calculated values based on diffusion theory. The fact that G_H exceeds G_{OH} at low LET indicates that H is produced in slightly broader distribution than OH as indicated in Figure 1. It should be emphasized, however, that diffusion theory requires the spur size to be of the order 20 Å diameter in order to explain ~ 50% radical combination during spur expansion. For spherical spurs the fraction of radicals combining is very sensitive to the initial spur diameter, and broadening the distribution of H results in a considerable decrease in G_{H_2} and a slight increase in $G_{H_2O_2}$. For cylindrical tracks, however, yields are much less sensitive to track diameter, and broadening the initial H atom distribution probably actually results in slight increases in both G_{H_2} and $G_{H_2O_2}$ (Dyne and Kennedy '60). This result can explain the fact that the ratio $G_{H_2O_2}/G_{H_2}$ diminishes with increase in LET.

As we see in the following sections, decreasing pH, changing to D_2O, or increasing temperature, all result in effectively broader H atom distribution.

TABLE VIII. 7.

EFFECTS OF pH, D_2O, AND TEMPERATURE ON DECOMPOSITION OF WATER BY GAMMA RAYS.

	Neutral		0.4 M H_2SO_4		
	H_2O, 25°C [a]	D_2O, 25°C [c]	H_2O, 25°C [d]	H_2O, 65°C [e]	D_2O, 25°C [f]
G_{H_2}	0.45	~ 0.28	0.45	0.43 [f]	0.38
$G_{H_2O_2}$	0.71 [b]	--	0.79	0.75	0.79
G_H	2.75	~ 3.24	3.63	3.79	3.94
G_{OH}	2.23	--	2.95	3.16	3.12
$G(-H_2O)_{net}$	3.65	~ 3.80	4.53	4.66	4.70
$G(-H_2O)_{gross}$	5.0	--	6.0	6.1	6.1
% radicals combined	50	--	46	43	42

a (Hochanadel '52).
b (See Allen and Schwarz '58).
c (Hardwick '59).
d (See e.g., Sworski '54 - values corrected to zero scavenger concentration).
e (Ghormley and Hochanadel, unpublished).
f (Mahlman and Boyle '58 - values corrected to zero scavenger concentration).

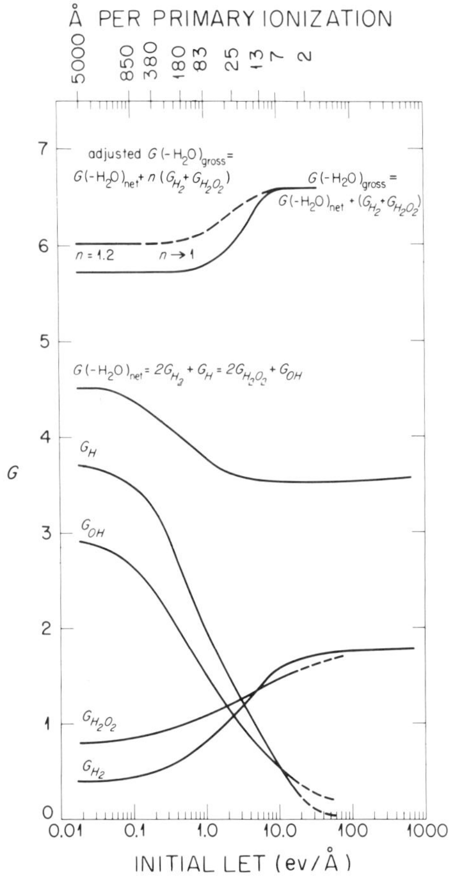

Fig. VIII. 3. Yields of the intermediates H, OH, H_2, and H_2O_2 in the decomposition of 0.4 M H_2SO_4 as a function of initial LET of the incident particle. The curves are smoothed curves derived from values shown in Figure 2. The yields $G(-H_2O)_{net}$ and $G(-H_2O)_{gross}$ are discussed in Section 8.5.4.

Although it is not shown in Figure 3, there is some experimental and theoretical evidence for small maxima in the curves, resulting from the complex inter-play of radicals within spurs or tracks (Dyne and Kennedy '60).

8.5.3c. Effect of pH. - Yields in acid solution are considerably larger than in neutral solution. This result is shown in Table 7 for gamma radiation. Also, for irradiation of neutral solutions by Co γ-rays, 18 Mev deuterons, 32 Mev helium ions, or 11 Mev helium ions curves similar to those in Figure 3 are obtained, except that yields for neutral solutions are displaced to lower values (Schwarz et al. '59). The effect is attributed to the hydrogen ion since added sodium ion or sulfate ion have no effect. The effect is much too large to be attributed to direct action (including subexcitation effect). The results indicate a broader initial distribution of H atoms in acid solution. It has been suggested that this may result from reaction of H_3O^+ with diffusing H atoms (or the equivalent, which may be an electron) thereby altering the correlation between H and OH, giving an effectively broader distribution of H and larger total radical yield.

8.5.3d. Isotope effect. - Radical yields and the yield for water decomposition are slightly larger for heavy water than than for light water as shown in Table 7 (Mahlman and Boyle '58, Armstrong et al. '59, Hardwick '59). Any effect on primary decomposition should give greater decomposition of H_2O than D_2O. The results indicate a broader initial distribution of D than of H. It has been suggested (Armstrong et al. '59) that, as a result of the greater relaxation time for D_2O (~ 25% greater at 25° C), electrons are trapped and react with water (reaction 13, Table 1) farther away from the positive ion (Platzman mechanism) in D_2O than in H_2O. However, H atoms produced by this mechanism are already in essentially homogeneous distribution, and broadening the distribution should have little effect. Also, the contribution of H atoms by this mechanism is probably small.

8.5.3e. Effect of temperature. - For decomposition of 0.4 M H_2SO_4 solutions by gamma rays, increasing temperature in the range 2° C to 65° C increases radical yields and decreases molecular yields about 0.1% per degree (Ghormley and Hochanadel, unpublished). Again, this effect is not completely understood, but in terms of the model increased temperature (which decreases the relaxation time) gives increased initial spur size.

8.5.4. Efficiency of water decomposition. - In a very significant experiment, Firestone ('57) measured the decomposition of water vapor by tritium beta particles using D_2 as scavenger which quantitatively reacts with H atoms to give HD in amount equivalent to the number of water molecules decomposed. The measured yield G(HD) = G($-H_2O$) = 11.7 indicates a very efficient utilization of energy. This is shown by the following simple calculation: the W value for water vapor is ~ 30 ev per ion pair. Since the lowest ionization potential is 12.56 ev, the 3.3 ions formed per 100 ev total energy expended in the vapor require at least 41.5 ev. The remaining 58.5 ev must decompose 8.4 molecules, corresponding to ~ 7 ev per molecule. This indicates that most excitation is to the lowest electronic level (max. at ~ 7.5 ev).

The efficiency of decomposition of liquid water is considerably less than that of the vapor. Yields for water decomposition (in 0.4 M H_2SO_4 solution) for radiations of various LET are shown in Figure 3. Two different values are plotted; $G(-H_2O)_{net}$ is calculated from stoichiometry of measured products according to

$$G(-H_2O)_{net} = 2G_{H_2} + G_H = 2G_{H_2O_2} + G_{OH}.$$

In this yield, re-formation of water is not considered, and therefore, the yield drops with increasing LET since radical combination increases with LET. The yield $G(-H_2O)_{gross}$ includes an estimate of the amount of water re-formed by H-OH combination. To a first approximation, the amount re-formed is equal to $G_{H_2} + G_{H_2O_2}$ and, therefore,

$$G(-H_2O)_{gross} = G(-H_2O)_{net} + G_{H_2} + G_{H_2O_2}.$$

This is the amount expected statistically assuming H and OH present in large number in equivalent initial spatial distribution and assuming equal rates for diffusion and radical combination (see Dyne and Kennedy '58, '60). The track of the heavy particle (large LET) most nearly approximates this condition since spur structure is lost and the H and OH are produced in large number in a single cylindrical volume element. This condition is not true for fast electrons (low LET) because the spurs, which contain an average of only a few radical pairs, are widely separated. For these conditions

$$G(-H_2O)_{gross} = G(-H_2O)_{net} + n\,(G_{H_2} + G_{H_2O_2}),$$

where n is 2 for two pairs of radicals, 1.5 for 3 pairs, 1.2 for 6 pairs, etc. Assuming n = 1.2 at low LET and 1 at high LET, the adjusted value of $G(-H_2O)_{gross}$ varies from ~ 6.0 at low LET to ~ 6.6 at high LET (neglecting the value for fission recoil particles). Corresponding values for neutral solution are somewhat lower, ranging from ~ 5.0 to ~ 6.1. The estimated $G(-H_2O)_{gross}$ does not include any allowance for primary recombination. The close coupling in the liquid may reduce the efficiency of water decomposition both by deactivating excited molecules thereby preventing decomposition, and by causing primary recombination of the radical pair before it can escape from the cage. The photolysis of liquid water containing formic acid as scavenger for H and OH, gave a quantum yield for water decomposition of 0.44 at 1860 Å and at 1935 Å (Weeks and Matheson '58). The extent of primary recombination, as indicated previously, is probably greatest at low LET (isolated spurs). Primary recombination can probably not, however, account entirely for the low efficiency. Although the corresponding energy levels in liquid water are somewhat lower than for the vapor, it may be that energy is wasted in the liquid in exciting higher states. It is also possible that other special conditions in spurs, (and especially in cylindrical tracks) such as local high temperature, high charge, or close proximity of electronically excited species, might promote more or less efficient decomposition, and also account in part for the higher efficiency of decomposition at higher LET. The reduced efficienty of organic scintillators at high LET presumably results from similar special conditions.

8.5.5. Frozen solutions. - The difficulty in preparing homogeneous frozen solutions of suitable scavengers, and the occurrence of reactions during warming and melting, makes results on frozen solutions both scanty and difficult to interpret. Certain general features are clear, however. The efficiency of water decomposition decreases considerably on changing from liquid to ice, and decreases still further with decrease in temperature. For example, with 0.1 M ethanol as scavenger, the yield of H_2 [$G(H_2) = G_{H_2} + G_H$ in the liquid] under gamma irradiation decreases from 4.05 in the liquid at 2° or 20° C, to 3.1 in ice at -10° C, 0.88 at -78° C, and 0.40 at -200° C (Allan et al. '59). The yield of "molecular" hydrogen, G_{H_2}, also decreases in like manner from 0.45 in the liquid at 20° C, to ~ 0.25 in ice at -10° C, to 0.1 in the range ~ -100° C to -220° C (Ghormley and Stewart '56). From these values, assuming the same stoichiometry in ice and liquid water, $G(-H_2O)_{net}$ drops from 4.50 in neutral liquid water at 2° or 20° C, to 3.4 in ice at -10° C, 1.0 at -78° C, and 0.5 at -200° C. In radiolysis of liquid water, there is relatively little change in radical and molecular yields, and in yield for total water decomposition, over the temperature range 2° C to 65° C (or even at 250° C), (Ghormley and Hochanadel, unpublished, Hochanadel '56), indicating that the extent of hydrogen bonding is of little importance. Freezing probably does not affect the primary dissociation of water, but the rigid structure does affect subsequent chemical reactions. The amount of primary recombination is undoubtedly greatly enhanced in ice, and the mobility of free radicals greatly reduced, the extent depending upon the temperature. On the other hand, in the more highly organized solid system, diffusion of electrons and positive holes may be rapid, and energy transfer by exciton migration may be important.

It is well known that irradiation produces free radicals that are stably trapped in the solid at low temperatures and react in the solid at higher temperatures. For example, H atoms are stably trapped in pure ice at -269° C but disappear before reaching -196° C (Piette et al. '59). In frozen acids they are stable at -196° C and disappear on warming to ~ -180° C (Livingston et al. '55). Hydrogen atom scavengers such as H_2O_2 and HNO_3 diminish both the amount of trapped H and the amount of H_2 observed after melting (R. L. Livingston and A. J. Weinberger, to be published). Also, structure seems to be an important factor; the yield of trapped H atoms depends upon the acid, and in some cases is greater for slowly frozen acids which appear glassy, than for quick-frozen acids which are crystalline. In the temperature range -180° C to -160° C, both a spin resonance peak (Matheson and Smaller '55) and an optical absorption band at 2800 Å disappear (Ghormley and Stewart '56). These are tentatively attributed to OH radicals. Also, added H_2O_2 disappears in this temperature range. In a still higher temperature range, at ~ -100° C, other electrons spin resonance lines disappear, back reaction between added H_2 and H_2O_2 becomes rapid, and addition of O_2 leads to considerably increased H_2O_2 production.

DISCUSSION

KALLMANN: I would like to know why some of your yield curves go down with stopping power and some go up.

HOCHANADEL: In the low stopping-power region intermediates are formed in spurs which are isolated from each other and contain on the average a half-dozen dissociated molecules. The spur expands and for this case calculations indicate that about half of these radicals will combine with each other and about half of them will essentially escape to infinity. The latter can be removed by a very small concentration of scavenger. In the high stopping-power region intermediates are formed in spurs which coalesce into a cylindrical distribution and in this situation most of the radicals combine during the process of expansion. It is expected, therefore, that radical-radical reactions become more important as stopping-power is increased and their yields should increase. At the same time the number of escaping radicals which react with scavengers should decrease.

KALLMANN: When investigating similar problems in inorganic materials one finds that with alpha particles or with slow electrons most of the excitation is concentrated in a small area because of the potential fields of the space charge. Here you produce so many positive ions that one has a potential of more than 100 V inside the channel of excitation so that even the electrons cannot diffuse out. As a consequence, one has a large electron concentration. In your case, of course, positive ions can diffuse but not in a time of 10^{-8} seconds; so I would think that electron diffusion in your case is also hampered by space charges. I would like to know whether any investigation of this point has been carried out.

MAGEE: (communicated): This is a fundamental consideration that went into development of the Samuel-Magee model ('53).

KUPPERMANN: I would like to make some remarks

about the overall picture of radiation chemistry of water. Essentially the situation is that a model has been proposed and it has been found that within the mathematical approximations necessary to derive conclusions, the qualitative results are reasonable Now the question is how much detail obtained from this model agrees with experiment. This is not to say that the model is necessarily correct. There may be other possibilities which would in principle explain experiments better than the diffusion model. On the other hand, much work has been carried out with this model, and there is no other model which at present can be compared with experiment to the same extent.

First of all I would like to emphasize that results obtained from elaborate calculations which take into count such things as distribution of spur sizes, different reaction rate constants for the various radicals, etc., are significantly different from results obtained using the simplest approximations. Thus many phenomena, which can really be explained on the diffusion model, are frequently said to be in disagreement with it.

In order for the H_2 and H_2O_2 yields to be approximately equal in this model or even roughly of the same magnitude, the initial distribution of H atoms and OH atoms cannot be very different. Actually the calculations show that if the distribution of the OH's corresponds to a diameter of about 20 Å, the distribution of the H's could not possibly have a diameter of more than 30 Å and still give the same order of magnitude of yield. This is one point.

Secondly, if you assume that the D atoms in D_2O are distributed a little further out than the H atoms in H_2O a significant isotope effect is obtained. It is possible that the diffusion model actually furnishes an adequate explanation of the isotope effect.

Finally, I would like to add that variations of yield with LET show certain crossings. The yield of H_2O_2 is larger than H_2 for low LET and becomes smaller for high LET. Even such details apparently are explainable with the diffusion theory if the calculations are simply done with reaction schemes closer to the real situation.

One cannot obtain such detail with the simplified model in which two radicals and one scavenger are used. On the other hand, if two radicals and their actual reactions are used it seems that the model does explain even fine points. It would be interesting to see whether any other kind of explanation would also furnish this fine degree of agreement which the radical diffusion model seems to produce.

ALLEN: There is one point that Dr. Hochanadel mentioned with which I certainly do not agree; i. e., he does not think that H atoms are scavenged by hydrogen ions. I think they quite probably are, since we know that the yield of water decomposed is increased greatly by acid as would be the case if the acid were acting as a scavenger. Perhaps these hydrogen atoms are really electrons, or perhaps the hydrogen ions scavenge hydrogen atoms to form H_2^+; I don't know. I don't want to be dogmatic about this. There are some other data that indicate, rather definitely I think, that one can have competition between solutes and hydrogen ions for these reducing species formed in water, so I really feel that putting in the possibility of hydrogen ions as scavenger rather increases the general applicability of this model and tends to clarify the situation.

FANO: I would like to comment on the comparison of gas and liquid. If we have a mechanism which, so to speak, in condensed phase replaces lower excitation with higher excitation, it allows more opportunity for degradation of excitation into heat. Therefore, in the condensed state the yields are expected to fall as they do. Such expectation is just on general grounds. If you have the situation which I discussed earlier, that in condensed phases excitations that would normally be 10 or 12 volts become say 15, then this is leaving much more margin for internal conversion and therefore, would lower the yield. That is apparently one possible effect independently of whether we consider spurs, etc.; i. e., other things being equal.

I would also like to make another comment. I wonder, if at the present time, you have a fair body of information that would already make it possible to start going in the reverse and form a theory by inductive processes, trying to see what each experiment means, forgetting the details of any model.

ALLEN: I think that we do have a lot of information on this and we are trying to find out what the experiment means as best we can. We welcome assistance in this effort from any source we can get. So, if you wish to help us try to figure out what these data mean, we will welcome anything you can offer.

FANO: All I wanted was to try to recommend a little more emphasis on just taking the bare facts outside the context of a detailed model.

ALLEN: I prefer to present the matter somewhat reversed from the way Dr. Hochanadel did, taking it more from the inductive rather than the deductive standpoint.

BURTON: Dr. Fano's comment regarding the degradation of energy in condensed systems is very pertinent to the decomposition of water. The most important measurable quantity is G for decomposition. It is a well-known fact that this quantity is smaller for liquid water than for gaseous water.

It is a rather interesting fact that the same phenomenon seems to permeate all systems. We have found that decomposition in the liquid phase is smaller than in the gaseous phase for the few organic systems we have investigated. Dr. Fano's comment offers the first suggestion of a possible explanation.

REID: I am not quite sure whether I get this deuterium versus hydrogen question right, but it seems to me that it is impossible to attempt an explanation without taking into account the fact that there are substantially stronger hydrogen bonds than deuterium bonds in the system. Presumably any bond breaking is most likely to be along the direction of a hydrogen bond, in an isolated molecule, and there are more of them in D_2O than in H_2O. In a molecule which is part of a hydrogen bonded polymer you simply go from a water molecule to an H_3O^+ when you move your hydrogen, and nothing much has happened so in a way this connects with Magee's idea that if you have a lot of molecules coupled together you cause less decomposition than if you have isolated molecules; and D_2O has more isolated molecules than H_2O.

MAGEE: I meant to say that if molecules are coupled together the locus of decomposition can be spread. It is a matter of competition of the various processes. As Dr. Reid says, the effect would tend to make spur sizes in H_2O larger than in D_2O. However, we are confronted with the fact that free radical yields in H_2O solution are lower than those in D_2O solution (see page 178). This result is contrary to what might be expected if the effect suggested by Dr. Reid were dominant.

9

William H. Hamill
University of Notre Dame

Direct and Indirect Effects on Chemical Systems

9.1. INTRODUCTION

The experimental approach to the radiation chemistry of liquids has often been characterized by consideration of gross phenomena and, from a more critical point of view, as a variation of photochemistry. Progress is now being made in stationary state kinetics of free radicals, of diffusion controlled processes (track effects) and of energy transfer processes. Substantially no progress has been made toward elucidating ionic processes. This is in contrast to experimental developments in gaseous systems where ion decomposition patterns, electron capture, charge transfer and neutralization effects have long been known. In particular, recent reports of gaseous ion-molecule processes emphasize the necessity of searching for chemical effects peculiar to ions. The techniques appropriate to gas phase work cannot, in the main, be employed for liquids. We must accept indirect evidence for ionic processes in irradiated liquids.

Among the questions to which experimental answers would be highly desirable in the present context are these: What is the ionization potential of a molecule or radical in a given medium and is this, in fact, a meaningful question? What is the cracking pattern associated with ionic processes in the liquid state? What is the electron affinity of a molecule or radical in a given medium? What are the cross-sections for these processes and how do they depend upon energy? If the preceding quantities cannot be measured absolutely, can they be measured relatively? Can dipolar solvation effects for ions be demonstrated? How many molecules does an electron encounter during its lifetime? What is the 100 ev yield of ion-pairs? Which reaction products result from ionization, which from electron attachment, which from neutralization? Do ion-molecule reactions occur in liquids and are they predictable from measurements in the gases.

This report considers tentatively chemical evidence for irradiated liquid systems in terms of dissociative electron attachment, charge exchange and ion-molecule reactions. Extrapolating information about ionic processes from the gaseous state to the liquid state is arbitrary and uncertain. As a working hypothesis it is worth trying, however, and the results obtained by mass spectrometry will be used on this basis. Fortunately, relative values of ionization and appearance potentials suffice for present purposes, and these are expected to be the same in the liquid as in the gas. Absolute values of appearance potentials would differ between the two phases with the probable exception of electron attachment having a zero energy threshold. Qualitatively, decomposition by electron impact should be less in the liquid state because the medium favors recombination, as illustrated by the photolysis of ethyl iodide (Bunbury et al. '56).

It must be considered that the available data are far from complete and apparent contradictions may arise. No single experiment is likely to be critical at this stage and we must be willing to accept patterns of behavior. Accordingly, many systems have been examined, particularly those which might be expected to provide chemical evidence of dissociative electron attachment or of charge exchange.

Dissociative resonance electron attachment in irradiated systems is more efficient for low than for high energy thresholds since the electron passes quickly through a high resonance energy range but remains at low energy until attachment or neutralization occurs (Magee and Burton '51). Organic halides should be useful reagents since the reaction

$$RX + e \rightarrow R + X^- \qquad (1)$$

often has a very low resonance attachment threshold. In the presence of a suitable reagent SH (either solvent or solute), the extent of reaction 1 can be measured from the amounts of either RH or HX

$$R + SH \rightarrow RH + S \tag{2}$$

$$X + SH \rightarrow HX + S \tag{3}$$

where HX is an acid. Interference from H-atoms in irradiated hydrogenous systems may result by reaction 4

$$RX + H \rightarrow R + HX \tag{4}$$

and from energy transfer by reaction 5

$$RX + M^{*} \rightarrow R + X + M \tag{5}$$

where M^* is an excited molecule. Reactions 1 and 4 can be differentiated chemically. Direct decomposition of RX by radiation will not be serious at moderate concentrations. Further evidence of electron attachment may be found from qualitative or quantitative modification of the yield of a product which is characteristic of the major component of the solution. This expectation is based upon a considerable difference in the exothermicity of reactions between cation and electron and between cation and anion. The electron affinity for halogen atoms, which determines this difference, ranges between 3. 2 and 3. 8 ev. Also, the chronology of the two processes may be quite different, with possible chemical consequences arising from dipolar solvation or diffusion effects.

If charge exchange occurs from a major component AB to a minor component CD

$$AB^{+} + CD \rightarrow AB + CD^{+} \tag{6}$$

we expect a decreased yield of product characteristic of AB (e. g. , AH) and an enhanced yield of product characteristic of CD (e. g. , CH). Also, this should occur only when the ionization potentials, I, are in the sense $I_{AB} > I_{CD}$.

Complications arise because we cannot be sure, a priori, about the relative rates of reaction 6 and the unimolecular decomposition 7

$$AB^+ \rightarrow A^+ + B \qquad (7)$$

The relatively abundant ionic fragments commonly have a lower ionization potential than their molecular precursors so that the exchange 8

$$A^+ + CD \rightarrow A + CD^+ \qquad (8)$$

would usually be impossible.

If ion-molecule reactions occur in the gas, they must also be considered in the liquid since, in addition to producing unique products, they may suppress expected results and induce unexpected results. To illustrate, if reaction occurs between AB^+ and AB, the potential energy of the system usually decreases by several ev and the exchange reaction 6 would become impossible in most instances, even though $I_{AB} > I_{CD}$. As additional examples, reaction of AB^+ + CD or A^+ + CD may result in formation of products characteristic of CD even though $I_{AB} < I_{CD}$. Such effects may provide limited evidence of ion-molecule reactions.

9.2. EXPERIMENTAL

Five-ml samples of liquids, prepared on a vacuum line, received ca. 5×10^{-2} ev of gamma radiation at 2×10^{16} ev ml^{-1} min^{-1} and approximately 20°. Gaseous products were analyzed mass spectrometrically. When HI was to be measured, samples contained 10^{-2} $\underline{M}$ I_2 as well as 0.1 ml H_2O to prevent loss of HI by reaction with free radicals or unsaturates. For HCl, water alone was added because HCl reacts less readily with free radicals than does HI. Also, added I_2 would react readily with Cl atoms as well as with free radicals. Methyl radicals were measured as CH_4 in separate runs by adding 10^{-2} $\underline{M}$ HI because Bunbury et al. ('56) found the reaction $C_2H_5 + HI \rightarrow C_2H_6 + I$ to be highly efficient.

9.3. RESULTS

L. J. Forrestal examined gamma-irradiated binary mixtures of cyclohexane with methyl iodide, chloroform, and benzyl acetate. The characteristic product of cyclohexane is hydrogen, both molecular and atomic; the total $G(H_2)$ = 5.8. H-atoms are readily converted to HI by adding a small amount of iodine, giving G(H) = 2.1. The residual hydrogen is molecular and $G(H_2)$ = 3.7. Its yield is sharply decreased by these additives, as shown in Figure 1. The data of Burton et al. ('58) for cyclohexane-benzene have also been included. The choice of variables, $G(H_2)$ vs. log(electron fraction), is convenient for expanding the scale at small concentrations of additive. No particular significance is attributed to the linear relationship. The yield of methane from solutions containing methyl iodide, shown in Figure 2, is very nearly the complement of that for hydrogen with $G(CH_4) + G(H_2) = 6.5$, except at the lowest concentrations of methyl iodide. G(HCl) for chloroform is qualitatively similar to $G(CH_4)$ for methyl iodide. The striking features of the results in Figures 1, 2 are the substantially similar effects exerted by various solutes upon $G(H_2)$ from cyclohexane and a characteristic "break" in the plot G(product) vs. (electron percent) at ca. 10% additive.

W. Van Dusen examined binary mixtures of benzene with methyl iodide, ethyl iodide, chloroform, carbon tetrachloride, methyl chloride and phenylacetyl chloride for which results appear in Figures 3 and 4. The yield of product characteristic of the second component exhibits a "break" (transition from exponential to linear) in the neighborhood of 10% additive. Also, there is a marked similarity for $G(CH_4)$ from methyl iodide in benzene and in cyclohexane. G(HCl) from solutions with added methyl chloride is much smaller, however, at corresponding concentrations of halides. In all solutions, one must allow for G(H-atom) $\simeq$ 1 for benzene.

9.4. DISCUSSION

Forrestal's work, as well as that of Meshitsuka and Burton ('59), is consistent with the interpretation that

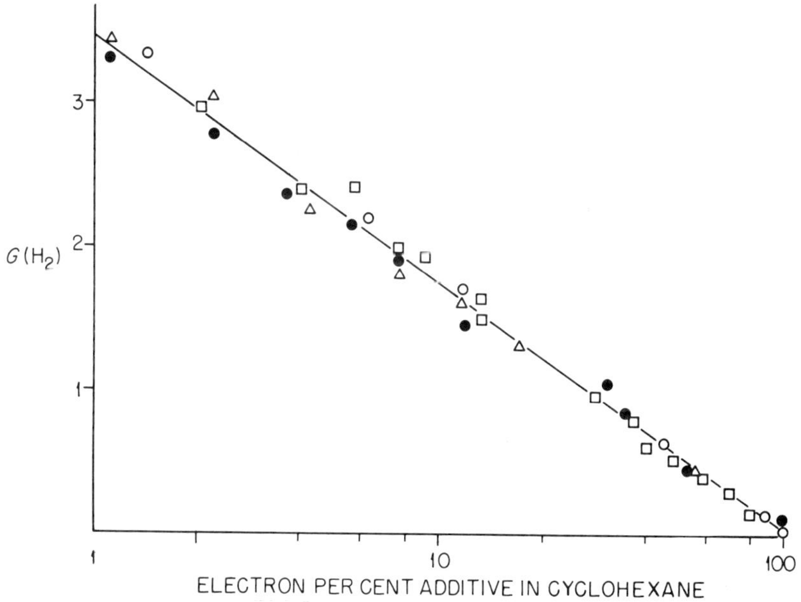

Fig. IX. 1. Yield of hydrogen from solutions of cyclohexane with various additives.

- ● methyl iodide
- ○ chloroform
- △ benzyl acetate
- □ benzene

G(HI) = 2.1, in cyclohexane containing ca. 10^{-2} $\underline{M}$ I_2, represents G(H atoms). This view is supported by an appreciable yield of methane from added methyl iodide and 10^{-2} $\underline{M}$ I_2 which would be expected by reaction 9

$$CH_3I + H \rightarrow (CH_3 + HI) \rightarrow CH_4 + I \qquad (9)$$

where parentheses denote a diffusion-controlled process. Schuler ('57) found a diminished, limiting $G(H_2)$ for cyclohexane below 0.5 mole % I_2 or CH_3I. Moreover, his data

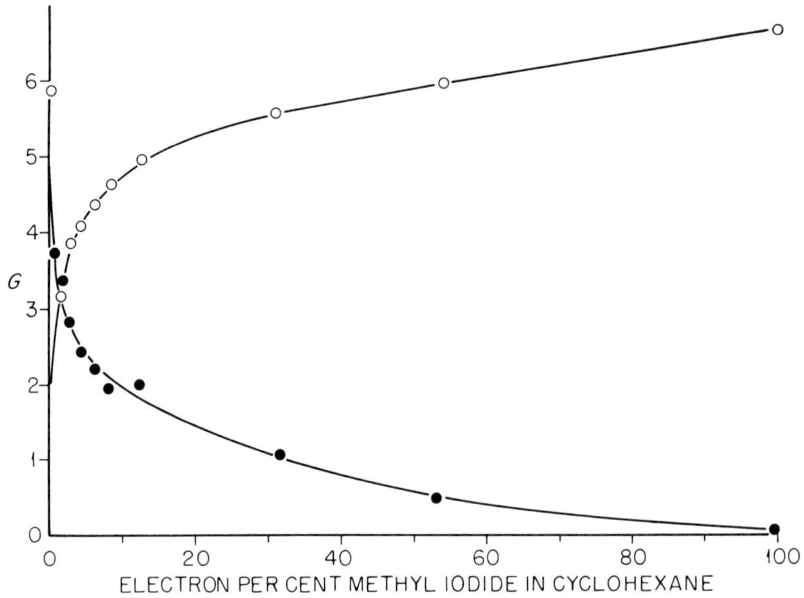

Fig. IX. 2. Yields from cyclohexane + methyl iodide with 10^{-2} M HI;
● $G(H_2)$
○ $G(CH_4)$

agree with the competitive mechanism (10, 11)

$$H + I_2 \rightarrow HI + I \quad k_{10} \qquad (10)$$

$$H + RH \rightarrow H_2 + R \quad k_{11} \qquad (11)$$

If $\Delta G(H_2)$ is the decrease in $G(H_2)$ at any value of $[I_2]$ and $\Delta G(H_2)_{max}$ is the constant value which $\Delta G(H_2)$ attains at $[I_2] \sim 0.01$ M (i.e., at the plateau), it follows from stationary kinetics that

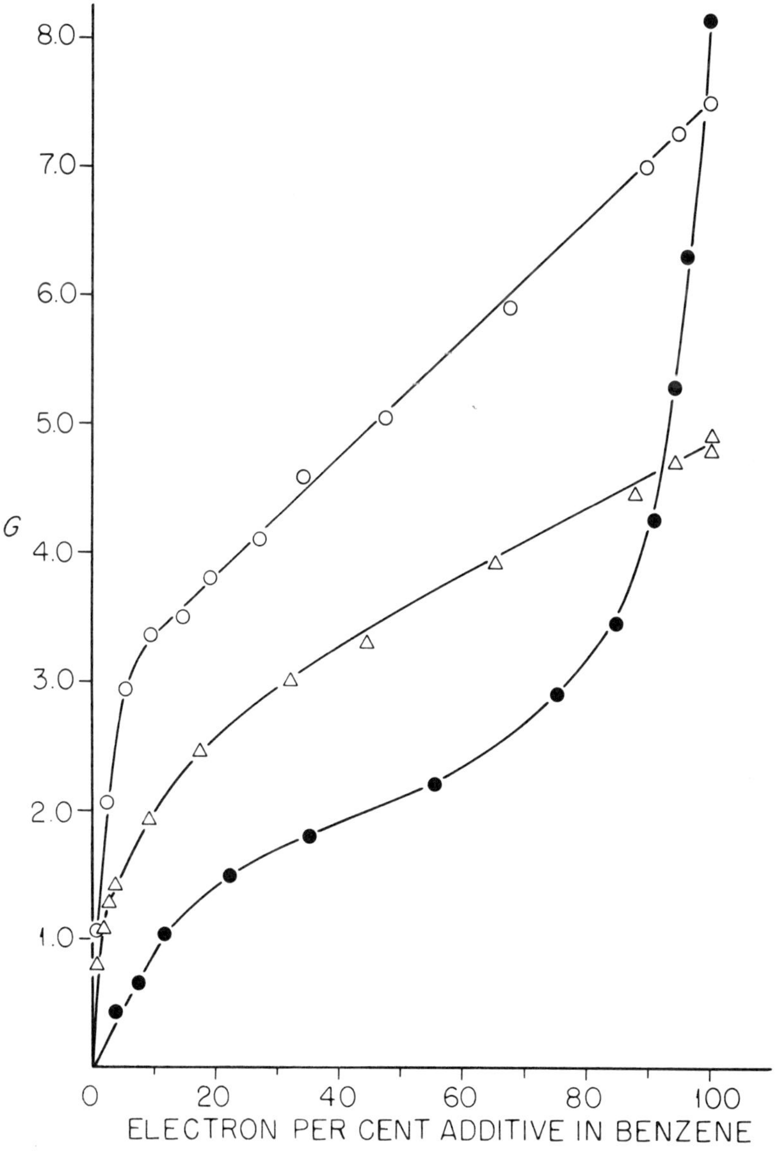

Fig. IX. 3. Yields from various mixtures.

● G(HCl) from benzene + chloroform

○ $G(CH_4)$ from benzene + methyl iodide with 10^{-2} M HI;

△ $G(C_2H_4) + G(C_2H_6)$ from benzene + ethyl iodide.

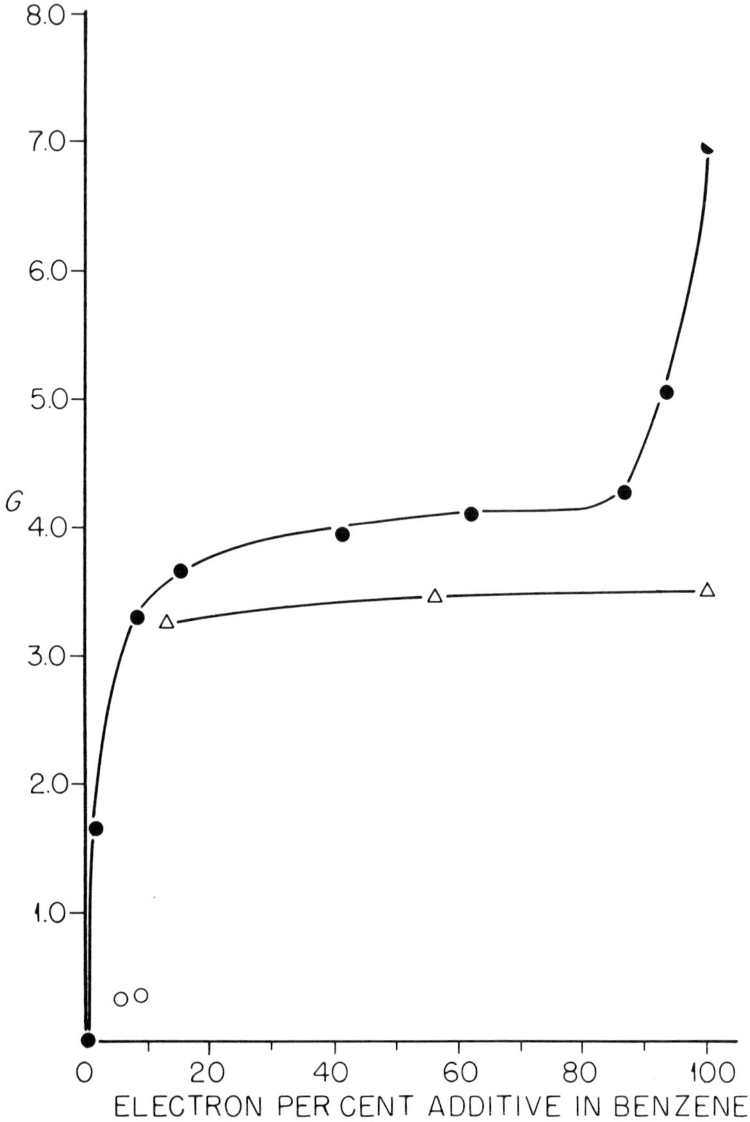

Fig. IX. 4. Yields from various mixtures.

● G(HCl) from benzene + carbon tetrachloride

○ $G(CH_4)$ from benzene + methyl chloride with 10^{-2} M $(C_2H_5)_3CH$

△ G(CO) from benzene + phenylacetyl chloride

$$\frac{\Delta G(H_2)}{\Delta G(H_2)_{max}} = \frac{k_{10}(H)(I_2)}{k_{10}(H)(I_2) + k_{11}(H)(RH)}$$

where $\Delta G(H_2)_{max} = 2.1$.

Both Schuler's results and earlier work on cyclohexane by Williams and Hamill ('53) were erroneously explained in terms of electron attachment because HI was not known to be a product in solutions of cyclohexane containing iodine or methyl iodide. Since reduction of I_2 by H yields HI, while reduction by electrons does not, and since the observed $\Delta G(H_2)$ is now found to be matched by an equal G(HI), it follows that there is no chemical evidence for electron attachment by I_2 at these small concentrations. The careful measurements of Biondi and Fox ('58) show an electron attachment cross section for I_2 of 4×10^{-16} cm^2 for thermal electrons, so that an inefficient reaction can scarcely be invoked to account for the lack of an effect.

The limited solubility of iodine precludes its use at higher concentrations Methyl iodide is even more suitable than iodine since it is as good a scavenger for H-atoms; its reaction with H-atoms gives HI and CH_3; its reaction with electrons gives CH_3 but no HI; its participation is specifically identifiable by converting CH_3 to CH_4; it is a good reagent for thermal electrons since Dibeler and Reese ('55) found an appearance potential of zero for the resonance electron attachment process 12

$$CH_3I + e \rightarrow CH_3 + I^- \qquad (12)$$

In cyclohexane solutions of methyl iodide (also 10^{-2} M in I_2) $G(H_2)$ decreases from 5.8 to 3.7 at a first plateau well below 1% CH_3I. This $\Delta G(H_2)$ is matched by an equal combined yield of $G(CH_4) + G(HI)$ and is assigned to reactions 4 and 9. $G(H_2)$ decreases more gradually with increasing amounts of solute to a second break at ca. 10% CH_3I while $G(H) = G(CH_4) + G(HI)$ remains constant. Replacing 10^{-2} M I_2 by 10^{-2} M HI does not affect $G(H_2)$ but we see in Figure 2 that $G(H_2) + G(CH_4)$ is now constant. Cyclohexane is protected at the expense of methyl iodide and the effect produced by 0.5 - 10% CH_3I is attributed to reaction 12.

For this higher concentration interval it might be proposed that the solute is interfering in track reactions between free radicals. Of these, only H atoms would be expected to react efficiently with solute. Since G(H) was shown to be constant over this range of concentration, the proposal is untenable A more nearly plausible alternative is transfer of energy from excited C_6H_{12} to CH_3I followed by dissociation of the latter. This postulate would leave us no plausible chemical effect in Figure 2 which might be assigned to electron attachment, while the facts do not clearly require this assumption. There is, however, a further basis for attributing $\Delta G(CH_4)$ for 0.5 - 10% CH_3I to reaction 12, since it is now necessary to attribute G(H atoms) = 2.1 to dissociation of an excited molecule, $C_6H_{12}^*$, which does not have an ionic precursor because G(H) is unaffected by 10% CH_3I. This leads to $G(C_6H_{12}^*) \geqq 2.1$ and G(electrons) ~ 3, which are both plausible. If the excited state is repulsive, dissociation would occur in < 10^{-13} sec. and would not be affected by small concentrations of solute unless migration of energy occurred. The energy of an excited singlet may, but does not necessarily, migrate appreciably within this time; if the state is a highly excited triplet, it is necessarily localized (Magee '60).

The preceding interpretation has the merit of being able to predict similar processes in other systems. Of those examined, benzene-methyl iodide corresponds most closely and the dependence of $G(CH_4)$ upon added methyl iodide is remarkably similar to the preceding case. The same is true of $G(C_2H_6) + G(C_2H_4)$ for benzene-ethyl iodide; although the appearance potential for electron attachment has not been reported for ethyl iodide it is probably zero. The appearance potential for the resonance electron attachment reaction 13

$$CCl_4 + e \rightarrow CCl_3 + Cl^- \qquad (13)$$

is 0.20 ev according to Hickham and Berg ('58). The results in Figure 4 for G(HCl) from benzene-carbon tetrachloride in the range 1 - 10% CCl_4 are therefore also consistent with the preceding explanation. On the other hand, methyl chloride in benzene yields much less HCl than does carbon tetrachloride at the same concentration, as might be expected from an appearance potential of 10 v for Cl^- from CH_3Cl by ion pair formation (Dibeler and Reese '55). No comparable conclusions are possible for chloroform since the appearance potential for Cl^- has not been reported. Neither is it possible to account on this simple basis for the very different results from solutions of this additive in benzene and in cyclohexane.

From one measurement at 7.6 mole % CH_3Br in cyclohexane (Schuler '57), it appears that methyl bromide resembles methyl iodide as might be expected from an appearance potential of zero for Br^- (Dibeler and Reese '55). The abrupt increase for G(radicals) in styrene observed by C Chapiro ('50) for several percent added carbon tetrachloride resembles the present results for benzene-carbon tetrachloride and can also be explained by reaction 13.

The results for isobutane-cyclohexane suggest the charge exchange reaction 14

$$i\text{-}C_4H_{10}^+ + C_6H_{12} \rightarrow i\text{-}C_4H_{10} + C_6H_{12}^+ \qquad (14)$$

which is consistent with their relative ionization potentials. In order to account for the similar effect of 10% C_6H_6 on carbon tetrachloride in Figure 4 it would be necessary to postulate that CCl_4^+ is stable in the liquid phase since $I_{CCl_3} < I_{C_6H_6}$ (measurement by L. W. Theard) and CCl_4^+ does not appear in the mass spectrum. The case is therefore doubtful. It can readily be deduced, however, that CCl_4^+ may exist in the liquid state since we know that

$$CCl_4 + e(11.50 \text{ ev}) = CCl_3^+ + Cl + 2e \qquad (15)$$

and Farmer et al. ('56) found

$$CCl_3 + e(8.78 \text{ ev}) = CCl_3^+ + 2e \qquad (16)$$

$$CCl_4 = CCl_3 + Cl; \quad D(CCl_3\text{-}Cl) = 68 \text{ kcal mole}^{-1} \qquad (17)$$

Combining reactions 16 and 17 we see that the fragments in 15 have no relative kinetic energy while ion-induced dipole forces between them require a minimum in the potential energy curve. This argument is valid only at the onset of ionization. For 70 v electrons D. Kubose finds experimentally that the maximum kinetic energy of CCl_3^+ is only ca. 0.5 v. In the liquid state, CCl_4^+ may be expected to survive long enough to undergo charge exchange with benzene by 18

$$CCl_4^+ + C_6H_6 \rightarrow CCl_4 + C_6H_6^+ \qquad (18)$$

The preceding proposal cannot be reconciled with the results of Cooper and Stafford ('58) that G(HCl) from carbon tetrachloride is quite constant with added n-hexadecane up to 4.40 mole %. Charge exchange would be allowed from CCl_4^+, but not from CCl_3^+. Before considering this case further, let us examine another which resembles it.

The effect of added benzene on G(HCl) in chloroform can be accounted for by either of the processes

$$CHCl_3^+ + C_6H_6 \rightarrow CHCl_3 + C_6H_6^+ \qquad (19)$$

$$CHCl_2^+ + C_6H_6 \rightarrow CHCl_2 + C_6H_6^+ \qquad (20)$$

The ion $CHCl_2^+$ predominates in the mass spectrum of $CHCl_3$ and the ionization potential of the free radical $CHCl_2$ is 9.30 v according to Lossing et al. ('58). Since cyclohexane does not greatly depress G(CHl) in chloroform, we must conclude that $CHCl_3$ is not abundant in the liquid state. On the other hand, cyclohexane slightly diminishes G(HCl) from n-propyl chloride, a result which is consistent with the ionization potentials and the limited abundance of $C_3H_7Cl^+$ in the mass spectrum.

The difficulties of accounting consistently for the effects observed in mixtures containing carbon tetrachloride or chloroform in terms of charge exchange require an alternative approach. Accordingly, mixtures of benzene with alkyl halides were examined by L. W. Theard at high pressures in the mass spectrometer. The following reactions were observed in high yield:

$$C_6H_6 + CCl_3^+ \rightarrow C_7H_5Cl_2^+ + HCl \qquad (21)$$

$$C_6H_6 + CHCl_2^+ \rightarrow C_7H_6Cl^+ + HCl \qquad (22)$$

$$C_6H_6 + CH_3^+ \rightarrow C_7H_7^+ + H_2 \qquad (23)$$

Using CD_3I as a source of CD_3^+, the ions $C_7H_4D_3^+$, $C_7H_5D_2^+$ and $C_7H_6D^+$ appear, which suggests that the ions are tropylium rather than benzyl. The neutral products above are probable but unproven. If the tropylium ion were produced in a liquid system it would survive as stable un-neutralized tropylium halide (Doering and Knox '59). Thus far we have been unable to detect it. Since Cooper and Stafford ('58) reported $C_6H_5CCl_3$ from irradiated benzene-carbon tetrachloride (as well as HCl, $CHCl_3$, C_2Cl_6, C_6H_5Cl and C_6H_5-C_6H_5) we postulate that in the liquid phase the course of reaction is modified to reactions of the type

$$C_6H_6 + CCl_3^+ \rightarrow C_6H_5CCl_2^+ + HCl \qquad (24)$$

and that charge neutralization proceeds by reactions of the type

$$C_6H_5CCl_2^+ + Cl^- \rightarrow C_6H_5CCl_3 \qquad (25)$$

for which ΔH = 4 ev - (solvation energy). The neutral fragments associated with the original ion pair are CCl_3 and Cl, which account for the other known products.

The failure of benzene to decrease $G(CH_4)$ in methyl iodide, or $G(C_2H_4) + G(C_2H_6)$ in ethyl iodide, is only apparently inconsistent with their ionization poetntials. We must also consider the known ion-molecule reactions of the type

$$RI^+ + RI \rightarrow R_2I^+ + I \qquad (26)$$

observed by Pottie et al. ('59). The reactions

$$R_2I^+ + C_6H_6 \rightarrow C_6H_6^+ + RI + R \qquad (27)$$

are energetically impossible because the potential energy of RI^+ has been diminished by at least D(R-I) = 2 v.

The effect of benzene on $G(H_2)$ from cyclohexane observed by Burton et al. ('58) can be accounted for by the allowed reaction

$$C_6H_{12}^+ + C_6H_6 \rightarrow C_6H_{12} + C_6H_6^+ \tag{28}$$

Similarly, Manion and Burton ('52) found effects consistent with the energetically allowed exchange

$$C_6H_{12}^+ + C_6H_{10} \rightarrow C_6H_{12} + C_6H_{10}^+ \tag{29}$$

but none for

$$C_6H_{10}^+ + C_6H_6 \rightarrow C_6H_{10} + C_6H_6^+ \tag{30}$$

which is energetically forbidden.

The interpretation that 50% electron attachment requires about 5% of an efficient reagent is equivalent to assigning a domain of twenty molecules (or less) to the electron. This view is not consistent with the theory of Platzman ('53) or even of Samuel and Magee ('53). The present results are qualitatively closer to the latter and could be made to agree quantitatively by assigning a reaction cross-section about 10^{-2} times that for molecular collision. However, Buchel'nikova ('58) observed a cross section 1.3×10^{-16} cm^2 for reaction 13 and values ranging to 10^{-18} cm^2 for other electron attachments. In the present work there is no evidence of appreciably different reaction efficiencies for various solutes, although the number used is rather limited.

Energy transfer does not appear to be able to account for the effects being considered. West and Paul ('32) found the same quantum yield for the photolysis of ethyl iodide at 2537 Å in the pure liquid and at 0.03 M in benzene. Van Dusen has found similar effects for dilute solutions of methyl iodide and of carbon tetrachloride in benzene. That is, energy transfer occurs at much lower concentrations than the effects which we have found with high-energy radiation.

A striking feature of the yield - concentration dependence is the symmetry of the effects attributed to electron attachment and to charge exchange. A common mechanism was assigned to cyclohexane-methyl iodide and to benzene-methyl iodide although the ionization potentials are in the order $I_{C_6H_{12}} > I_{CH_3I} > I_{C_6H_6}$. Actually, it is possible that charge exchange operates in one case and electron attachment in the other. This inability to distinguish between two distinct experimental effects simply on the basis of the concentration dependence (which is most clearly seen for benzene-carbon tetrachloride) suggests mechanisms more nearly related to each other than electron attachment and charge transfer would imply. In particular, when we consider that the domains of the effects being considered approximate 15 molecules, it may be preferable to describe the phenomena in terms of charge transfer complexes. Types of mechanisms which are suggested by the preceding results are

$$AB^* + RX \rightarrow AB^+ + R + X^-$$

$$AB + RX^* \rightarrow AB^+ + R + X^-$$

$$AB^* + RX \rightarrow AB + R^+ + X^-$$

where AB* and RX* result from primary energy absorption. Such mechanisms would still reflect strongly variations in ionization potential and electron affinity. On the other hand, they are sufficiently flexible to include so-called ionic polymerizations which are otherwise rather anomalous.

DISCUSSION

DOLE: I think this is a very interesting paper and I would like to ask the speaker if he by chance is familiar with a paper by Chang, Yang and Wagner, where they studied the radiolysis of terminal olefins and have presented their results in terms of an ion-molecule reaction?

HAMILL: I have read the paper, but I do not recall the details. It may very well be an ion-molecule reaction. I do not recall in detail either whether a complete analogue to such a reaction has been observed in the mass spectrometer. Perhaps you can tell us that.

I might say that it is necessary to involve ion-molecule reactions in one or two instances. Some of you must have noticed that in the mixture of benzene-ethyl iodide there is no sharp decrease in the yield of products characteristic of the radiolysis of ethyl iodide, namely, ethylene and ethane, although the ionization potential of ethyl iodide exceeds that of benzene. To be consistent there should have been a sharp decrease similar to carbon tetrachloride. However, we have examined in detail the ion-molecule reactions of ethyl iodide and of methyl iodide and there is an ion-molecule reaction involving the parent molecular ion and the molecule in both cases giving a product of type R_2I_4.

This ion would certainly be unable to exchange charge with benzene and, therefore, can account for the failure of benzene to decrease the yield.

GORDON: Why is chloroform ineffective, in contrast to carbon tetrachloride? Chloroform does potentiate a number of photo-oxidations.

HAMILL: I would be inclined to interpret the results with photo-oxidation in terms of a bimolecular reaction between the aromatic amine and the organic halide to give the cation, the halide anion and free radical. Until we have analyzed the product we cannot know this, but the free radical cation could very well go on to give the onium salt. We do know that it is first-order in light intensity and the effective wave length, as Dr. Kasha said yesterday, is around

3600 Å.

It could not reasonably be a one-step ionization with an energy of about 4 ev. Since there is only one quantum involved, it would appear, therefore, to go through an excited state. If it is a cooperative effect, this could be in the nature of a charge transfer complex so it doesn't necessarily proceed according to the simple mechanism which I have written.

These are merely the end products. But in any event the energy required depends not only upon the nature of the halide. As our observation shows, some are effective, some are not. It must also depend on the excited species and, therefore, I would expect the cut-off point in our electromotive series, if I may call it that, to vary with the excited species.

BURTON: If I understood correctly, you were a bit troubled about the matter of ionization transfer from cyclohexane to benzene.

HAMILL: It is a remarkable coincidence that such different solutes as methyl iodide and benzene give results whose concentration dependences are indistiguishable. Although one could be electron attachment, the other could not, but could be charge exchange.

The only way I can see to reconcile these effects is to interpret the results from radiolysis in terms of something like a charge transfer complex. The concentration dependence is, therefore, going to tend to be the same for the two solutes.

The phenomena may differ in details. In some instances the charge transfer complex may form irreversibly, for example, chloride ion. In another case, it might involve the transfer of the positive charge, e. g., from cyclohexane to benzene. In any event these do appear to be cooperative phenomena since the region of 50 percent efficiency lies typically around 6 to 8 mole percent solute and therefore, some 15 molecules would appear to be the domain in which these changes are occuring.

BURTON: First, I would like to make a simple factual

comment. The dependency of a "protective effect" in cyclohexane by methyl iodide is clearly distinguishable from the dependency on benzene concentration. The indistinguishability mentioned by Professor Hamill is a result of the type of plot he is using. It is a familiar fact that plots as a function of logarithms must elminate the zero region and bring the remainder of the plotted information closer together.

Second, in the work that we did on mixtures of cyclohexane and benzene in our luminescence studies we arrived at the conclusion that whatever energy is transferred in cyclohexane to benzene is transferred completely when the benzene concentration is as much as 1 mole percent.

Nothing like 10 percent is required. For other reasons on which I do not think it appropriate to expand here, it seems to be ionization that is transferred.

I am very much impressed by the data which indicate that the electron apparently does not go any farther than the first coordination sphere. This situation is rather strange. I think, however, that it is not the electron that is determinative in this case. It is the positive charge that goes from the cyclohexane to the benzene; this makes it look as if the cyclohexane has a definite domain in which the positive charge can exist rather widely, so that the benzene at a very low concentration can make contact with any molecule in that domain and pick up the positive charge. This is not an electron transfer process in the sense in which you have employed the term. It seems to me that without a picture of this kind it would be hard to correlate the results with benzene and the results that you have found with the halides.

HAMILL: As I understand, you find the approach to saturation in the phenomenon you observe at about one percent benzene in cyclohexane?

BURTON: What we did was to study decay times in luminescence (Burton and Dreeskamp '59b). We examined solutions in which the solvent could be cyclohexane, benzene, or a mixture of the two and in which a typical scintillator solute (like p-terphenyl or PPO) was present in relatively small concentration. In pure (aerated) cyclohexane solution, the decay time of luminescence, τ, is 2.2×10^{-9} sec. In pure benzene, of the same solute concentration, τ is 3.5×10^{-9} sec. In the neighborhood of 0-10 mole percent benzene (in mixed solvent) τ has a broad maximum near 4.7×10^{-9} sec. It is quite difficult to distinguish the sharp rise of τ to a maximum value as a trace of benzene is added to the cyclohexane solvent. We attribute the high value of τ to the essential isolation of the benzene molecule; we attribute the decrease in τ at higher benzene concentration to the existence of domains by which excitation is effectively transferred to the scintillator. We attribute the sharp rise in τ when benzene is added to cyclohexane to very effective charge transfer from cyclohexane to the benzene. At such very small concentrations every benzene molecule is effectively in (or in near) contact with some domain of cyclohexane in which the ionization moves very freely. When the benzene strikes any molecule in that domain, it can pick up the positive charge. In my opinion, this phenomenon is something subtly different from an electron transfer in the sense that your work implies.

HUTCHINSON: Would this be 0.01%, say? I am just trying to get the order of magnitude of the smallest concentration of benzene with which you can work.

BURTON: We have worked with about 0.1% benzene. The exact figures appear in the published paper.

MAGEE: I think that the total G value associated with this process is perhaps much smaller than the G value associated with the process that Professor Hamill was talking about, and it may be that only a small amount of absorbed energy is involved in the process you are describing, whereas the majority of it was involved in the process that he was describing. Would you agree with this?

BURTON: The technique that we use is an indicator technique so that when I tell you anything about the time that a species lasts, this refers to the time that a particular species lasts independently of what happens to it. You see, if it enters into the chemical processes this is a secondary consideration. The point is that these decay times are characteristic of the species in the particular set of circumstances.

MAGEE: Yes, I know, but I don't think there is necessarily any kind of a paradox here. You deal only with a small amount of energy.

BURTON: I didn't suggest that there is a paradox. I suggested that there is an explanation.

MAGEE: Then I would like to make one other comment. I have noticed that in this meeting when we talk about effects of ionizing radiation we always get around to talking about the very lowest excited state and what happens to the very lowest excited state. People seem to base interpretations of what happens only on these very lowest excited states. Of course, it is clear that these states are reached in very complicated processes which frequently involve very highly excited states. I think it is an important matter that initially-formed highly excited states convert rapidly into these low states, and this is the problem I have been wondering about for many years.

I would be the first to say that a picture of the electron wandering around and being moderated as a free electron is really too naive. When Samuel and I were first talking about it we didn't like the language because we thought it was perhaps a little too naive, but it is extremely difficult to describe these phenomena in more elegant or more rigorous ways. It is a very, very key problem in radiation chemistry. I would say that in these systems it is quite likely the lowest excited states that are determinative. Some of them no doubt involve charge-transfer complexes. Thus, in any case one expects electrons to be transferred from one molecule to another and the concept of ionization followed by electron capture in the sense that it was used ten years ago is not an adequate description of the phenomena.

BURTON: (communicated May 1, 1960): Cramer, Kropp and I have now established that the excited state of benzene which transfers energy in luminescence is not the state which decomposes.

KALLMANN: I want first to answer the question whether excitation by high energy leads to energy transfer from the lowest excited state. We have investigated this question for three cases of energy transfer: fluorescence, quenching, and phosphorescence, which last may also be brought about by means of energy transfer. I would say that within 10 percent the fluorescence under ultraviolet light absorbed in the lowest excited state is equivalent with respect to concentration effects to that under excitation by high energy. We get curves which coincide almost completely under both types of excitation so that we believe that fluorescence, phosphorescence, and quenching are connected mainly to the lowest state.

I have a question for the speaker. You have shown curves which have large values at a certain concentration and then go down considerably with a very small change of concentration. Then they are almost horizontal and finally go down again at the other end. Similar curves were produced by Chapiro for free radicals of carbon tetrachloride. At large benzene concentrations, the number of free radicals was rather small. If he adds about 5 percent of carbon tetrachloride, the number of free radicals increases by a considerable factor. He thinks this proves that at the high benzene end the free radicals are formed by energy transfer from excited benzene to carbon tetrachloride, because he gets similar transfer if he uses ultraviolet light to excite the benzene.

On the other hand, the carbon tetrachloride is not affected by ultraviolet light because it is not absorbed. Now I want to ask whether you have any explanation for these very sharp changes at the two ends of your curves?

HAMILL: I must not have made myself very clear. In the first place, because I was pushed for time, I did not give proper recognition to other work.

However, I think I may take it for granted in this audience that most of you are acquainted with the work of Bouby and Chapiro as well as the work of Manion, Burton and others who found somewhat similar effects.

Regarding the statement about energy transfer, I will take one case where I am better acquainted with the facts, benzene-ethyl iodide, worked on years ago by William West. He found that light absorbed in benzene was efficiently transferred to ethyl iodide, at, I believe, .05 M, the quantum yield of product being the same as in pure ethyl iodide. Somewhat later West and a collaborator examined another system. The major component was some paraffin, (e.g., normal hexane), with small concentrations of napthalene and of ethyl iodide. The light was absorbed entirely by the napthalene at about 10^{-3} M and the energy was efficiently transferred to ethyl iodide at concentration below 10^{-2} M.

This is fairly clear evidence of energy transfer through an inert medium, and I call your attention to the fact that it takes place at concentrations far smaller than those with which we are concerned here.

I did not mean to exclude the possibility of events due to excitation or excitation transfer. In cyclohexane the G(H atoms) = 2.1, which I stated must have non-ionic precursors, is presumably due to decomposition of an excited state. Also I called attention to the fact that G(HI), which is a measure of these hydrogen atoms, gave no indication of decreasing appreciably as the concentration of methyl iodide increases to beyond 20 percent.

I certainly do not imply that energy transfer and excitation of neutral states are not important. There is evidence both in cyclohexane and in benzene that such states are present, but the main point I would make is that some of these features (in the Figures) must be attributed to ionic processes, or radiation chemistry has no meaning.

SPEAKER UNIDENTIFIED: This may be a very naive comment, but I have been looking at radiation chemistry of liquid mixtures. In reference to your paper the thought occurred to me that there might be some effect of non-homogeneous distribution of one solvent in others. Are these solutions ideal solutions or is the entropy of mixing different from the normal entropy of mixing? For example, when you mix a neutral molecule with one that has a dipole moment, I suspect that the free energy of mixing is far from ideal and you might have nearest neighbors that are not completely random. Such an effect, if it exists, can affect the radiation chemistry.

HAMILL: I would think that the solutions are not ideal. The mono-iodo alkanes are dipolar. The solutions are certainly non-ideal in the glassy state; there is some very clear evidence of this. One forms both non-solvated and solvated iodine in dilute solutions of ethyl iodide in glassed, irradiated hydrocarbons. Two iodine absorption peaks appear. After melting and freezing there is one peak for solvated iodine. However, these are extreme conditions. I don't see how the major aspects of the features I described can be attributed to non-ideality at room temperature.

10

M. Ebert
Hammersmith Hospital, London

Direct and Indirect Initial Effects on Biological Systems

It seems to be an established fact that in Egypt, and also in India, people knew how to make use of the indirect action in radiobiology some 3000 years ago. Since they had no radiation-producing machines, they used sunlight. The experimental material was the human skin, and the instigator of these experiments was medicine. Skin conditions believed to be vitiligo were customarily treated by painting the skin with a concoction and exposing it to sunlight. This concoction has been found to contain furocoumarins, which are known to convey the photodynamic action in bacteria (Pathak '58). Light alone has no effect, and the coumarins alone have no effect on the killing of bacteria, but the combination of the two agents is most effective. Energy absorbed by one molecule is conveyed to a biologically important molecule, which is changed and thus lethal damage is inflicted on the cell.

I do not intend to review the literature of the last 3000 years comprehensively. I shall only present a few observations published quite recently, which allow some conclusions to be drawn on the importance and the mode of action of the direct and indirect effects in radiobiology. The views expressed will not be popular with everybody in the field, and this I hope will be confirmed in the discussion.

10.1. INDIRECT EFFECTS

Indirect and direct effects have been studied quite carefully for the two extreme conditions: very dilute solutions in which only the indirect effect contributes to the chemical change, and pure substances in which the chemical change can be attributed to the direct effect only. Radiation chemistry has hardly touched conditions that prevail in concentrated solutions or colloidal suspensions, which certainly occur in living cells. Let us first consider the mechanisms underlying the indirect effects of radiations in water. Radiation is absorbed by the H_2O molecule and chemical change observed in another molecule. The transfer of energy required for this chemical change occurs in simple cases via free radical intermediates. These intermediates have been studied rather carefully in the case of H_2O and sometimes the view is expressed that the radiation chemistry of water and of aqueous solutions can be regarded as a closed chapter. The relative rate constants for reactions of the H and OH free radicals with a variety of organic and inorganic molecules are known in acid solutions, and there is even fair agreement in the literature about them. I am listing a few of them in Tables 1 and 2 (Swallow '60). It can be seen that the majority of the measured values do not differ greatly from one another for small organic molecules. On the basis of these values fair predictions could be made for systems in which these substances are competing for the radiation-produced OH and H radicals.

The H and OH radicals react with organic material, with a high probability. The first step of this reaction can in most cases be described chemically as H atom abstraction from the molecule RH.

$$RH + H \text{ or } OH \rightarrow R + H_2 \text{ or } H_2O \qquad (1)$$

The free radical R formed by the action of a free radical derived from the radiolysis of H_2O is identical chemically with an R which would result from the direct ionization of the intact molecule. Thus the H and OH radicals disappear to leave a new free radical R, whose reactions are difficult to predict. They depend quite critically, as is pointed out by M. Kasha in Chapter VI, on the structure and excitation state of R. The position of the unpaired

TABLE X. 1.

RELATIVE RATES OF REACTIONS OF HYDROGEN ATOMS, IN SOLUTION IN 0.1 N H_2SO_4 AT 20° C.

Reactant	Reaction	K_1*
CH_3COOH	$H + RH \rightarrow H_2 + R$	0. 10
CH_3COCH_3	$H + RH \rightarrow H_2 + R$	0. 34
$CH_3COOC_2H_5$	$H + RH \rightarrow H_2 + R$	0. 56
H COOH	$H + RH \rightarrow H_2 + R$	1. 0
CH_3OH	$H + RH \rightarrow H_2 + R$	1. 5
HCHO	$H + RH \rightarrow H_2 + R$	4. 7
D COOH	$H + RD \rightarrow HD + R$	6. 8
C_2H_5OH	$H + RH \rightarrow H_2 + R$	14
Sucrose	$H + RH \rightarrow H_2 + R$	35
Cu^{++}	$H + Cu^{++} \rightarrow H^+ + Cu^+$	56
O_2	$H + O_2 \rightarrow HO_2$	540
Benzoquinone	$H + Q \rightarrow HQ$	3000
$C_2H_5OC_2H_5$	$H + RH \rightarrow H_2 + R$	High
$CH_2(NH_2)$ COOH	$H + RH \rightarrow H_2 + R$	Low
$(C_2H_5)_3N$	$H + RH \rightarrow H_2 + R$	Low

* K_1 is the rate constant for reaction divided by that for H + H COOH → H_2O + COOH

electron within the molecule will determine possible reactions of R.

It is known that for a given molecule certain positions are more probable than others for the unpaired electron. This is particularly true for macromolecules as has been demonstrated by Gordy ('59) for irradiated proteins. Ionization was certainly produced at random in the dry protein molecule, but the unpaired electrons always occur on the cystine or glycine units of the molecule. This fact suggests very strongly that either the vacancy, or the impulse which leads to the vacancy, can be transmitted by the molecule over one or several peptide units until it comes to rest in a favored position.

What is known about the fate of R, the free radical produced in reaction 1? The R, by virtue of having an unpaired electron, will readily react with other molecules carrying an unpaired electron, such as NO. These reactions have been discussed by Howard-Flanders ('58) and need not be elaborated here. I would, however, like to mention an experimental demonstration of such a reaction by Sparrman et al. ('59) in dry seeds of Agrostis stolonifera. They observed the number of magnetic centers produced by radiation in the seeds and followed their rate of disappearance after irradiation. The seeds were stored in N_2, O_2 or NO. The rate of disappearance of the magnetic centers R was fastest in NO and very slow in N_2 and O_2.

Some reactions of R in dilute aqueous solutions with O_2 or HO_2 lead to peroxides or hydroperoxides as has been discussed and shown by Scholes et al. ('59). In the absence of O_2 or other radical scavengers two R are likely to dimerize or to undergo some form of disproportionation reaction, thus ending the reaction sequence with the formation of stable and unreactive molecules.

The fate of R in concentrated solutions is by no means as clearly described as it is in dilute solutions. I feel that the area of concentrated solutions is still in need of a great deal of radiation chemical investigation. The kind of results encountered in this field is illustrated by the work of Proskurnin and Kolotyrkin ('58). They followed $G(NO_2^-)$ in irradiated alkali nitrate at pH 12 for varying concentrations

TABLE X. 2.

RELATIVE RATES OF REACTIONS OF HYDROXYL RADICALS IN SOLUTION IN 0.1 N H_2SO_4 AT ROOM TEMPERATURE; calculated assuming the ratio of the rate constants for the reaction of OH with Fe^{++} and HCOOH to be 6. 2, as in 0. 8 N H_2SO_4.

Reactant	K_2*	Reactant	K_2*
$(CH_2COOH)_2$	0. 19	$CH_3CH(OH)COOH$	11
$CH_2(NH_2)COOH$	0. 22	$CH_2OHCOOH$	12
$CH_3CH(NH_2)COOH$	0. 74	CH_3CHO	14
H_2	0. 93		
HCOOH	1. 0	CH_3OH	14
Pyridine	1. 1	C_3H_7OH	16
$(C_2H_5)_2NH$	2. 5	$CH_2SHCOOH$	17
$(CH_3)_3COH$	4. 0	$(CH_3)_2CHOH$	19
C_4H_9COOH	4. 5	$CH_2OH(CH_2)_2COOH$	19
$(C_2H_5)_3N$	5. 0	HCHO	19
		$(CH_3)_2CH\ CH_2COOH$	20
CH_3COOCH_3	5. 3	C_2H_5OH	24
C_3H_7COOH	5. 3	$(CH_2)_5O$	28
Fe^{++}	6. 2		
$CH_3COOC_2H_5$	6. 8	$CCl_3CH(OH)_2$	29
		$CH_2 = CHCOOCH_3$	31

TABLE X.2. (continued)

Reactant	K_2*	Reactant	K_2*
$((CH_3)_2COH)_2$	7.7	$(CH_2)_4O$	38
$CH_3COOCH(CH_3)_2$	7.7	$C_2H_5OC_2H_5$	40
$((CH_2)_2COOH)_2$	7.7		

* K_2 is the rate constant for reaction of OH with reactant divided by that for $OH + HCOOH \rightarrow H_2O + COOH$.

TABLE X.3.

YIELDS ON IRRADIATION OF CALCIUM NITRATE IN SOLIDS AND LIQUID PHASE,

according to Proskurnin and Kolotyrkin ('58).

Material	Phase	$G(NO_2)$
$Ca(NO_3)_2 \cdot 4H_2O$	solid	0.77 ± 0.02
$Ca(NO_3)_2 \cdot 4H_2O$	melt	0.07 ± 0.01
$Ca(NO_3)_2 \cdot 4H_2O$ + glucose	melt	0.26 ± 0.01

of NO_3^-. They varied the concentration from 10^{-7} M to 10 M. $G(NO_2^-)$ rises from about 1 at 10^{-7} M to 4.5 at 5×10^{-4} M to remain constant up to about 10^{-2} M and rises again to about 9 for 10 M solution. The reaction for similar solutions has been studied previously by Mahlman ('56) and described satisfactorily. If one extrapolates $G(NO_2^-)$ from 10 M solutions to solid crystals of alkali nitrates containing water of crystallization, the expected value is from 50 to 200 times too high. The phase state introduces additional confusion, as has been found with melts of these nitrates. Table 3 gives the results of investigations on $Ca(NO_3)_2 \cdot 4H_2O$ irradiated in the solid and the liquid phases. This unexplained observation points to our lack of knowledge in radiation-chemical systems involving high concentrations of solute in H_2O. Charlesby has commented on experiments done with polyvinyl alcohol, where even the qualitative result of irradiation in the absence of H_2O, i.e., degradation, is different from that in the presence of small amounts of H_2O, i.e., crosslinking. These very concentrated solutions, which certainly occur in biological systems, will require detailed energy and exciton transfer concepts as well as free-radical reactions to describe a valid mechanism which will account for the observed results.

However, some fairly safe generalizations can be made about conditions in dilute aqueous solutions. The initially produced free radicals from H_2O are particularly reactive by virtue of some excess energy imparted to them in the process of their formation and by virtue of the high energy balance on forming H_2 or H_2O as reaction products. Energy of activation for any reaction is practically built into the primary radical. The important reaction products of this first step are R, the secondary free radicals which are much less reactive than the primary radicals and have therefore a considerably longer lifetime. This means, of course, that their reactions are considerably more selective. The reaction products of the secondary radicals are certain to show even more sluggishness than the secondary radicals, etc.

Some post-irradiation treatments which modify latent biological radiation damage, such as photoreactivation or some physiological factors involved in certain growth conditions, are at the moment being interpreted by postulating restoration or repair mechanisms operating at the damaged biological structure. This interpretation implies a chemically reactive configuration at the site of damage, or else a reactive intermediary. Formally, such behavior is not distinguishable from the process of interfering with radiation-produced toxic entities in the cell which retain a fraction of the reactivity of their predecessors, the free radicals H and OH. The decision which of the two alternatives is important depends on the evaluation of the importance of the direct versus indirect actions in biological material. Rupert's results suggest the repair of damage and not toxins on TP-isolated DNA.

It is possible that in the majority of cases the changes observed in the important molecule will be the same whether they are the result of direct or indirect action. It is more hopeful to look for different efficiencies for the two mechanisms in producing the observable change. However, not even this hope proves to be borne out always, as Butler ('59) has pointed out. Several different samples of DNA were irradiated in concentrations varying from 1% to 100%, and the drop in viscosity of the solutions taken as a measure of damage. The efficiency of one ionization in causing damage, whether produced in the water or the DNA molecule, was unchanged over this range of concentrations. The two conclusions to be drawn for this system are that direct and indirect actions have the same efficiency and that both lead to the same damage. It was estimated that in the concentrated solutions a radical could travel about 100Å without undergoing recombination reactions.

10.2. DIRECT CONTRIBUTIONS

When information was sought on the contribution of either the indirect or direct action to biological damage, there was one obvious way of tackling this question experimentally: measure the radiosensitivity of a molecule in solution, i.e., the sensitivity due to the indirect action; measure the sensitivity in the dry state, i.e., that due to the direct action; and compare these results with the sensitivity in the intact cell. I shall be quoting results which follow part of this general plan of attack on this important problem. Unfortunately it now seems that the choice of the correct model system is difficult to make and has so far not been successful. But much work has been done and I would like to quote some experiments which followed the mentioned general plan of attack on this problem.

A long time ago it was shown that dilute solutions of purified enzymes were very sensitive to inactivation by ionizing radiation. These were conditions in which the indirect effect only was operative. Enzymes in the intact cell were usually far more radioresistant. The reason was obvious from the work of Dale, Barron and others, who demonstrated the protective ability of added small concentrations of solutes. Pollard ('59) and his team (cf. Pollard and Barrett '59) and others have carried out a great number of measurements of inactivation doses of enzymes and biomolecules in the dry state, and have compared some of these with the sensitivity of the same molecule in the living cell. This comparison does not prove to be the answer to the general question of the relative importance of the direct and indirect effects. The reason is that each biomolecule behaves differently and the results do not lend themselves to generalizations. A likely explanation for the unforeseen behaviour of the biomolecules is furnished by their location, inside the cell, in regions of different concentrations of organic matter, or in regions in which the concentrations of radical scavengers, or perhaps H donors, are different. An example of three different enzymes in the same cell system has been given by Hutchinson et al. ('57). They irradiated yeast cells (which resist drying very well) both in the wet and in the dry state, and assayed the activity of invertase, alcohol dehydrogenase and Coenzyme A. They dried the cells by evacuating for up to 36 hours at a pressure between 10-100 μ Hg. At these pressures, H_2O is obviously removed from the cells, but not completely, and, in fact, the authors observed that the inactivation dose of CoA, even after exhaustive evacuation, was still three times higher than in purified enzyme preparations. The influence of small amounts of water, whether chemically bound or present otherwise, is, as pointed out previously, and for radiation chemical reasons, not predictable. The inactivation dose, defined as the dose to inactivate to 37 % of the original activity of the enzyme, was taken as the measure of sensitivity. The ratio of inactivation dose in the dry state to that in the wet state gives an indication of the probable contribution of the indirect action. It was found that this ratio was 1:100 for CoA, 1:20 for ADH and 1:2 for invertase, showing a varying degree of indirect action contribution. An interesting observation was also made on the indirect inactivation of CoA. The authors found that only 70% of CoA showed this high sensitivity, whereas the remaining 30% was less sensitive. These data provide an example for the comparison of three different enzymes in one cell, and show that the relative contribution of indirect action to the inactivation of enzymes can be different by a factor of about 100 within one cell.

A similar experiment carried out on one enzyme in different cells completes the caution not to predict anything in radiation biology. In a recent publication Hutchinson and Norcross ('60) investigated the radiosensitivity of CoA in four different types of cell: Bacterium E. coli, fresh peas, beef liver and beef heart. CoA was found to be most sensitive in E. coli, giving an exponential survival curve down to 10% survival. In beef heart and beef liver cells the enzyme was considerably less sensitive, with an inactivation dose within a factor of about 2 of that observed in the purified dry enzyme. In pea tissue about 30% of the CoA content was very sensitive and the remaining 70% of the enzyme was in the radioresistant state. However, the spread of the experimental result makes it difficult to draw definite conclusions, except that the inactivation of one enzyme in a variety of cells may show greatly different contributions of indirect action. This may well reflect differences in the structure carrying the enzyme and furthermore it is to be expected that the availability of an enzyme to indirect action is determined by the physiological state of the cell.

To go further, I want to quote results obtained by

Wilson ('59) on isolated DNAase. This enzyme was inactivated by irradiation in the isolated and dry state, in which only direct action contributes to inactivation. The sensitivity depended critically on the pH of the solution in which the enzyme was dried, which merely says that the molecular structure is important for the utilization of absorbed energy. But Wilson also found that the enzyme-substrate complex was less sensitive to radiation than the enzyme in the pH range from 4. 5 to 6. 5. This observation makes a comparison of the sensitivities of isolated enzymes with those in the intact cell still more difficult.

Hutchinson ('57) drew an interesting conclusion two years ago from some of his observations. He stated that there is no doubt about the contribution of the indirect effect in the inactivation of biologically important molecules, but he calculated the distance from which H and OH are likely to diffuse to contribute to inactivation. His estimate amounts to about 30 Å, a figure considerably smaller than that given by Zirkle et al. ('53). The volume of water in the shell surrounding the observed molecule would then be a determining factor for the "indirect" contribution. Since this volume changes with the shape of the biomolecule, i. e., whether coiled or uncoiled, spherical or fibre-like, a different contribution from the indirect effect for the same molecule in different configurations has to be expected.

10.3. CELL PROCESSES

The work described so far deals with individual enzymes, and as such the results are of interest and importance. When it comes to considering other more radiosensitive processes in the cell, however, the experiments quoted can be regarded as providing model systems only. The doses quoted were of the order of Mrads, whereas the viability of cells is impaired with doses 10^3 - 10^5 times smaller. Some work I want to refer to next has been carried out on cells and the survival measured under conditions similar to those employed in the enzyme work - similar in respect to water being the variable, whose influence on radiosensitivity was studied. The availability of water for the diffusion and production of free radicals was controlled by freezing the cells. That the diffusion of H and OH is affected by the freezing of H_2O had been demonstrated radiation chemically by Ghormley and Stewart ('56), who used H_2O_2 production in irradiated H_2O as their criterion.

Stapleton and Edington ('56) used the colony-forming ability of E. coli as a criterion and studied the change in sensitivity with temperature below 0° C. They found a continuous decrease in sensitivity from 0° C to -196° C. Their observations are very suggestive that the indirect effect was mediated by those radicals formed from H_2O, whose behaviour had been described by Ghormley and Stewart ('56).

Wood ('59) showed that the phase change of H_2O from liquid to solid had a great effect on the radiosensitivity of haploid yeast. When yeast cells were irradiated at -10° C in supercooled H_2O, i. e., in the liquid phase, their sensitivity was nearly the same as that observed from 0° C to +10° C in the solid state. The drop in sensitivity by a factor of about 2 with decreasing temperature in the frozen samples occurred between 0° C and -10° C and did not increase markedly down to -100° C, in conditions in which freezing occurred slowly. This was ascribed to a dehydration of the cells in analogy with the work of Meryman ('57) on mammalian liver cells. The latter had shown that slow freezing of liver slices resulted in the removal of a very great part of the intracellular water. Rapid freezing, on the other hand, did not allow the intracellular water to leave the cells following osmotic gradients, but froze it inside the cells. When rapid freezing was applied, the sensitivity was similar to that in the liquid phase. The interpretation is that, on irradiation, free radicals form in the ice crystals inside the cell and react with cell constituents on warming. Dehydration of the cells by slow freezing solidifies the water outside the cell membrane, and the radical intermediates formed in the ice have no chance of reacting with cell constituents.

The last two experiments, that on E. coli and that on haploid yeast, show that freezing of the liquid water in cells reduced their sensitivity (in terms of cell viability) by a factor of at least two. Assuming that the main physical chemical variable in these experiments was the phase change of water, with a reduced mobility of the free radicals H and OH, it can be concluded that at least one half of the radiation damage is mediated by radical intermediates from liquid water.

It is possible that that figure is an underestimate in the light of experiments on the transforming factor Pneumococcus, by Latarjet et al. ('59), who looked at the sensitivity of TP when irradiated in a frozen aqueous medium. They found that the sensitivity decreased with increasing yeast extract concentration, but that catalase added further protection. They argue convincingly that this is evidence for protection against the indirect effect, by interference with H and OH free radicals formed in the ice. These radicals are mobile enough to form H_2O_2. The catalase then protects TP against the radiation-produced peroxides by encouraging coupled oxidation of yeast extract constituents. This simple explanation is taken as evidence for protection against indirect effect and makes protection against the direct effect an unnecessarily complicated explanation invoking energy transfer from molecule to molecule. In these experiments, in the frozen state catalase provided an additional increase of about 50% in the inactivation dose. Protection against free radicals or some chemical intermediate was obviously possible by catalase, where an increase in organic matter (yeast extract) had no further protective ability.

10.4. OXYGEN EFFECT

There is a series of experiments in which radiosensitivity has been measured in micro-organisms exposed to various alcohols, glycerine and glucose. These agents were chosen for various reasons. An important property was their dehydrating effect on living cells either by direct exchange with water or by creating osmotic conditions favorable to dehydration (Rosenberg '58, Dewey, personal communication, Marcovich '58). A marked protection is afforded by these agents under aerobic conditions but not under anaerobic conditions. In the case of Serratia marcescens and Proteus vulgaris, this protection seems to be independent of oxygen concentration from 1% to 100%. Even relatively small concentrations of glycerol have a marked effect. Haploid yeast cells will support life for some time in 7 M glycerol; this is a 50% mixture of glycerol and H_2O, in which the oxygen effect is reduced to very low levels. A dose reduction factor of 4 is achieved under aerobic conditions, and of 5 if anoxia is added. These glycerol data are paralleled by the results of Stapleton and Edington ('56) mentioned earlier, in which the freezing of E. coli resulted in protection under anaerobic conditions. But the abrupt change in sensitivity with change in the phase state of water occurred in aerobic suspensions.

Another observation which may be relevant to the above findings is that of E. L. Powers et al. ('59) on dry spores of Bacillus megaterium. A small but increasing oxygen effect persists over a temperature range from -148° C to +40° C. It vanishes below -150° C. It is to be noted that the oxygen effect in these dry spores at temperatures up to 40° C is not greater than 1.25; its decrease with decreasing temperature might be related to changes occurring in the residual water of the spores.

All these results have an obvious bearing on the interpretation of the mechanism of the oxygen effect. They make it likely that the oxygen effect during irradiation is linked to the presence of liquid H_2O and thus connected with the primary radicals produced from H_2O. Hutchinson ('57) has made a calculation suggesting that the free radicals H or OH diffuse a mean distance of about 30Å in the cell before reacting with an organic molecule, and that they are unlikely to meet O_2 molecules which are spaced 700Å apart at O_2 concentrations which increase the radiosensitivity by a substantial amount. It appears to me that this calculation was made assuming oxygen to be distributed homogeneously throughout the cell - an assumption which surely is far from justified. Apart from O_2 gradients set up by local utilization of O_2, its solubility varies greatly for different cell constituents. A relatively high concentration of oxygen is to be expected in areas with low utilization, high solubility and large interfaces. It is likely that the cell nucleus conforms to these conditions. Nuclear damage is assumed to be the avenue leading to ultimate loss of viability of cells and it is this criterion which has been measured in connection with most experiments aimed at elucidating the oxygen effect.

These arguments are contrary to the suggestions of T. Alper ('56), who visualizes the action of O_2 to be the chemical fixation of radiation damage in the important molecule in competition with a repair mechanism, which may involve an H donor, as suggested by P. Howard-Flanders ('58).

Dr. Howard-Flanders has kindly allowed me to quote some unpublished data, which I hope he will describe more fully in the discussion. He has shown that T2 bacteriophage irradiated in 0.01 M thiourea shows no oxygen effect. When the phage is irradiated as a complex in E. coli B immediately after injection of its DNA, the presence of oxygen nearly doubles its sensitivity compared with its anaerobic condition. This oxygen sensitivity vanishes when phage replication inside the host cell has progressed for some time and is near completion, i. e., when the phage DNA is covered with a protein coat.

The increased sensitivity after absorption suggests that the phage DNA, freed from its protein coat, has become accessible to some peroxy radical which originated in reactions involving the primary radicals from water. The peroxy radical, being more reactive than molecular oxygen, cannot diffuse through the protein coat of the free phage. From this it appears to me that the peroxy radical can regain its reputation as an agent in the mechanism of the oxygen effect, a reputation which it had lost several years ago. This interpretation is no doubt different from that of Dr. Howard-Flanders, but it has the virtue of extreme simplicity.

10.5. SUMMARY

I would like to come to an end by outlining briefly, and in general terms, where we stand at the moment.

The direct action model assumes energy to be absorbed by the molecule which determines the biologically observable damage. There is no migration of energy involved. This model is very powerful in explaining experimental results on the inactivation of viruses and other small biological units. It is, however, inadequate for the explanation of very radiosensitive effects. These include in mammalian cells loss of reproductive integrity, mitotic arrest, and the loss of the biochemical function of nuclear phosphorylation. The original narrow definition of the direct action model has been modified by allowing for energy transfer within the molecules of one cellular structure and by allowing the excited target molecule to react with cellular constituents, oxygen or nitric oxide.

The indirect action model assumes that inactivation of the biological unit is brought about by energy absorbed in the material surrounding the target molecule. This leads to chemical intermediates, which convey the damage to the target molecule. It is undoubtedly true that free radicals are produced within a cell and that these react with cell constituents including target molecules. Although the oxygen effect loses much of its magnitude when water is removed by drying or freezing to low temperatures, it still persists. This result is best explained by the direct action model which allows excited target molecules to choose between finalizing and repairing their damage. The indirect action model which comprised originally a large volume from which damaging chemical intermediates were drawn, has been modified by giving an upper limit of 30Å to the mean diffusion path of free radicals.

We seem to witness a gradual softening of the boundaries of the direct and indirect effects in radiobiology. This really means that the two opposing hypotheses are merging in the description of the true events.

DISCUSSION

HOWARD-FLANDERS: (communicated): I have a comment to make on Dr. Ebert's discussion of our experiments on radiation action on the T2 bacterophage system. Indeed, his speculative reference to the possible role of a peroxy radical intermediate is an invitation to consider this and other mechanisms of radiation action.

The experiments on T2 bacteriophage referred to showed there to be no oxygen effect with the free phage, but a two-fold oxygen effect with the phage-bacterium complex formed by the adsorption of the phage to the host cell, E. coli B. The complex shows this oxygen effect immediately after adsorption has taken place, even if metabolic inhibitors are present. This oxygen effect can be observed with complexes at various stages of their development but is greatly reduced towards the end of the latent period at the time when intracellular phage particles are being formed.

Further light was thrown on the problem of the oxygen effect by Doermann and by Marcovich who carried out experiments on phage suspended in nutrient broth. The phage appeared to be fully protected against indirect radiation effects, as adding to the concentration of nutrient broth did not further protect the phage. They found that the addition of cysteine or cysteamine gave a considerable degree of additional protection, which was referred to as "hyper-protection". The nature of this protection is of considerable interest.

We found the same hyper-protection when T2 phage were suspended in thiourea instead of nutrient broth. The advantage of this suspending medium is that it does not froth and can be bubbled with oxygen or nitrogen during irradiation. Bubbling oxygen during x-irradiation supressed the hyper-protection given by cysteamine, even though at least 60% of the cysteamine remained in the reduced -SH form at the end of the irradiation. The inactivation dose of the free phage in 0.01 M thiourea is almost the same in nitrogen, in oxygen and in oxygen + cysteamine, but is 40% greater in anoxic cysteamine. In other words anoxia protects free T2 bacteriophage when the sulfhydryl compound is present.

In their effect upon radiation injury, it appears that oxygen and cysteamine are in competition. We can think of oxygen and cysteamine as competing for reaction at the point of radiation injury in the DNA.

The radiation-induced removal of a hydrogen atom from the phage DNA would constitute an injury which could be modified by chemical environment. Reaction with oxygen would be rapid and lead to an oxidized inactive product, while reaction with a sulfhydryl compound might result in the donation of a hydrogen atom to the phage DNA radical and in a return to the original molecular configuration. Thus anoxic protection may be an active process involving a substance with a labile hydrogen group.

It is tempting to think in terms of intracellular sulfhydryl compounds acting as hydrogen doners and their being responsible for the anoxic protection observed with the phage-bacterium complex. An oxygen effect may be observed as long as both oxygen and the sulfhydryl compound have free access to the phage DNA. The protein coat is put onto the phage DNA late in the period of intracellular phage development. This coat may prevent the access of oxygen and the intracellular sulfhydryl compounds to the phage DNA and so reduce the oxygen effect with the free phage. However, it seems that the smaller sulfhydryl molecules such as cysteamine are able to penetrate at least partially into the phage head and so induce a moderate degree of anoxic protection.

The peroxy radical theory of the oxygen effect is hard to reconcile with the effects of the sulfhydryl compound on the free phage.

Dr. Hutchinson has been developing very similar ideas from his work with transforming principle DNA, which has an advantage over bacteriophage in the present context in that, as free DNA, it is more accessable to chemical substances.

HUTCHINSON: I would like to describe some oxygen experiments of Paul Howard-Flanders' similar to these which Dr. Ebert has just described. Howard-Flanders is aware of the difficulty that Dr. Ebert mentioned that the protein coat is around the T2 virus in solution could certainly modify the way in which the DNA, which is believed to be the vital target, is being affected by any radical reactions in dilute solution.

The system I was working with is the transforming principle in pneumococcus. The pneumococcus cells were irradiated, then the DNA extracted and assayed for transforming ability. The logarithm of the percent survival plotted against dose gives a straight line, showing exponential inactivation. For irradiation wet and in the anoxic state the D_{37} is 1.2 megarads. For dry irradiation in the anoxic state the D_{37} is about 3 megarads. This dry inactivation measures the direct effect. The difference between the wet and the dry curves is a measure of the indirect effect. Knowing how rapidly the transforming principle is inactivated in dilute solution, from work by DeFilippes and Guild, the magnitude of the indirect effect can be used to estimate a radical diffusion distance. This distance turns out to be of the order of magnitude of 10 or 15 Å.

With oxygen the radiation sensitivity in the wet state in the cell increases by a factor of 3. By the results just quoted, most of the damage in the wet cell is caused by the indirect effect. Supposedly, much of the oxygen effect must be an increased indirect effect in the presence of oxygen.

The effect of oxygen on the indirect effect has been studied both by DeFillipes and Guild, and by Latarjet and Ephrussi-Taylor. They agree that there is no effect of oxygen on dilute solutions of the transforming principle. So that the fact that one gets this additional increase with oxygen, you see now becomes inexplicable because we are ascribing most of the damage to indirect effect, yet it has already been demonstrated there is no change in indirect effect with oxygen.

We first found that this increase in radiation sensitivity with oxygen could be obtained with a cell extract. If one merely lyses cells and irradiates, one gets an oxygen effect. On the general radiation-biological principle that any interesting effect must be due to SH groups, we added an SH compound to a TP solution to see what would happen.

In the presence of millimolar glutathione or cysteine, the radiation sensitivity of the TP increases by a factor of about 3 in the presence of oxygen. We find this same increased sensitivity with oxygen for dilute trypsin solutions that have added SH compounds. This is the reverse of the results of experiments with pure solutions of a number of enzymes, including trypsin, which show no change in radiation sensitivity with changing oxygen concentration.

This is the same phenomenon qualitatively that has been obtained by Howard-Flanders with T2 phage in dilute solution. He finds that by adding cysteamine in rather higher concentration, 10 millimolar or something of that kind, there is an increase in radiation sensitivity of the T2 phage in dilute solution. However, the increase instead of being a factor of 3 is only 25 to 40 percent, which may very well be related to the heavy protein coat which exists around the virus in dilute solution.

The mechanism Howard-Flanders has suggested is essentially as follows. If the material we assay is represented by RH, then in pure solution and the absence of O_2, the basic reaction is:

$$RH + OH \rightarrow R + H_2O$$

In the absence of other substances, some slow further reaction, say,

$$R + R \rightarrow R - R$$

takes place, and the attacked material is inactive.

If you have O_2 present, then $R + O_2 \rightarrow RO_2$, which may, of course, be followed by other reactions and again results in the inactivation of the system. Note that although both of these processes lead to inactivation, they will lead to different species so that chemical and physical tests would tend to show different results although the molecules in both cases are biologically inactive. However, with no oxygen but with SH present, the hypothesis is that

$$R + SH \rightarrow RH + S$$

and, therefore, one gets an actual healing effect taking place.

I was hoping that by this time we would have experiments to prove this. We can measure this with enzymes such as trypsin, where assays are quite simple, because enough of the radiation chemistry is understood to correct for the amount of radicals that will be sopped up by the added SH groups.

This, as you see, is an actual healing operation. One is reversing a radiation-biological effect. The data that we have so far is consistent with such a healing effect, but these are only two or three experiments and I would hate to be dogmatic about it.

EBERT: Your interpretation of the action of oxygen on irradiated biological material is certainly consistent with your observations. But it still does not exclude alternative explanations, and so far we have no experimental data to do so. It is very probable that the protein coat of a phage particle suspended in a liquid medium will react with free radicals or peroxides produced in the medium and thus shield the DNA from their attack. This situation is obviously different when the DNA is injected into the host cell.

But I have a question. Did I understand you to say that the transforming principle does not show an oxygen effect?

HUTCHINSON: None in dilute solution.

EBERT: It shows, however, an oxygen effect if you freeze it. That is what Latarjet found. He irradiated the SR factor in the presence of yeast extract, the concentration of which he increased until he had complete protection against the action of radicals. But on adding catalase he reduced the sensitivity of the transforming factor further and this he ascribed both in the liquid and in the frozen state to the presence of some sort of peroxy radical or peroxides which are then taken care of by the catalase. There seems to be some indication for an oxygen effect. It does not occur in pure suspensions of the transforming principle.

HUTCHINSON: You mean just in buffer?

EBERT: The oxygen effect only occurs in rather complex solutions containing organic material, and the results are difficult to interpret under these circumstances. I thought in view of the work of Weiss and his collaborators that one has to admit the formation of organic peroxides or hydroperoxides in these solutions which can interact with DNA and cause further reactions.

KASHA: There has been some evidence for possible organized water structure in biological systems, so it would not be surprising to find a low temperature coefficient in biologically bound water, even to the super-cooled state. But for water in a liquid state, there may be a different temperature dependence as one goes from the super-cooled state to boiling water. I wondered if that could account for radiation chemistry differences?

ALLEN: Radiation-chemical temperature coefficients in water are so low that there is no reason to believe that liquid water at -30° is any different from liquid water at +30°. Solid water at -30° is quite different.

KASHA: Dr. Reid's discussion about the difference between D_2O and H_2O with reference to hydrogen bonding suggests that one might really find a difference at 100° C versus -30°.

ALLEN: I don't think there is any evidence for this effect. It certainly might occur, but there is no evidence for it.

CHARLESBY: The indication is that you go from liquid to solid you get a change in activity. I don't think it follows at all that this must be due to indirect effects. It seems to me quite reasonable that as you go to the solid state, chain mobility alters considerably and we know that this itself is going to have an enormous effect on radiation sensitivity. In fact, I would not be surprised if there is a difference between, shall we say, -70°C and -100°C which is again due to the different configuration of the molecule.

EBERT: The experiments of Wood were planned to exclude this difficulty. He showed that the speed of freezing determines the radiation sensitivity rather than the temperature during irradiation. When cells are frozen very rapidly they show a sensitivity which is quite similar to that in the liquid state. If cells frozen rapidly to -72°C are irradiated at -30°C, their sensitivity is considerably higher than that of cells cooled slowly to -30°C and then irradiated. This precludes a mechanism based on the temperature dependence of the mobility of biological molecules at the time of irradiation.

CHARLESBY: I still don't think this follows at all because when you cool molecules down very rapidly you are going to get them in a different shape. This is taking a very simple case. They will be in a different shape than if you cool them down slowly. In fact, if you take molecules and stretch them they will be different.

DOLE: There are two factors I would like to comment on. One is I believe that in order to measure the effect of radiation on the material, it must be warmed up to room temperature. Therefore, any radicals that might have been produced at low temperature react so that you really don't know exactly what the effect is at -140°C.

The second comment has to do with the mobility of oxygen in the material at -196°C. I believe that the mobility is practically zero so that the oxygen is fixed in the lattice and can't move. I imagine that this might have some effect on the results.

BENINSON: In the group type experiments the difference between the higher temperatures after freezing would be due to indirect effects. I wonder what would happen if you irradiate with high LET radiation.

EBERT: That has not been done.

BENINSON: I think after the curves shown yesterday that the yield of H and OH radicals decreases with high LET, and one might expect a small difference. I think it would be a worthwhile experiment.

AUGENSTINE: I would like to agree with Dr. Charlesby's comment, and I think you have to exercise some caution in interpreting Wood's experiments. There has been a good deal of study done on factors which affect the rupture of red blood cells as you freeze them to lower and lower temperatures.

It has been found, for instance, that you can decrease the rupture and apparently, therefore, change the conditions inside the cell by putting in various types of anti-freeze. It turns out that when Wood tested some of these same compounds in his cells that you could pretty well predict the effect that you would get just from looking at the protective effect of these anti-freezes on the ruptured red blood cells. I think that you might infer from this that you are changing the interior of the cells.

EBERT: As far as I know the action of anti-freezes of the glycerol type is dependent on the replacement of cellular water by the anti-freeze. This has been shown experimentally and is the argument which Wood uses.

MARSHAK: I just wanted to make a comment in connection with the glycerol effect. In some cells if you add glycerol you can see vacuoles appearing. In other words, you may get a dehydration without losing water from the cell, due to a localization of the intracellular water.

HOLLAENDER: I think one should point out it is not sufficient to keep a material liquid at CO_2 temperature but one has to keep it at liquid N_2 temperature. It has been found in our laboratory that the preservation time in CO_2 snow is very limited. The preservation seems to last much longer at liquid N_2 temperature. Also, I should point out that one does not necessarily need glycerine for preservation but one can use certain simple sugars. A review of this material is now in press.

In regard to the low-temperature story, the resistance of cells shows a definite oxygen effect even at temperatures less than -100°C. What has actually happened in these suspensions of microorganisms we do not know. As far as we have tested, this is down only to liquid N_2, there is still some additional resistance in the presence of nitrogen over the presence of oxygen. There is an oxygen effect even at this low temperature.

CHARLESBY: Potentially oxygen acts as an electron trap.

HOLLAENDER: In the solid state, yes.

SZENT-GYORGYI: I find the condition very complicated in frozen aqueous systems. What mostly happens on freezing is that the water crystallizes out leaving solutes, like O_2, behind in high and undefined concentration.

HENDRICKS: I would like to wonder about the question introduced by Dr. Charlesby; namely, as to whether or not the molecular configuration depends upon the rate of cooling. In a slowly cooled system ice separates as crystals and leaves a less hydrous system. As the cooling rate is increased, can the molecular configuration be fixed in a non-equilibrium position or is the rate of change of molecular configuration very much more rapid than is the cooling rate?

CHARLESBY: I cannot comment on any biological molecules but certainly in the case of hydrocarbons you can quench molecules in a random configuration to produce a glass. Slow-cooling results in crystallization. The same thing may not be true of biological systems, but on general grounds I would expect it might occur.

DOLE: Speaking about the effect of temperature on molecular configuration, we have irradiated isotactic crystalline polypropylene and found six peaks in its ESR spectrum at liquid nitrogen temperature. On heating briefly to room temperature, the number changes to nine when at liquid nitrogen temperature. On warming to about -140°, the number of peaks changes from 9 to 17 and if you go back to liquid nitrogen temperature, it goes back again to 9 from 17. We have interpreted this effect in terms of the onset of rotation of the methyl group on warming from -196° to -140°

C. The nine-peak spectrum can be explained if the methyl group is fixed in an appropriate configuration, and the 17-line spectrum on the basis of a mixture of an 8 and a 9 line spectrum.

BURTON: Professor Pollard has suggested that the distinction between direct and indirect effects is fading. I would like to suggest that an attempt at distinction in biological systems may be fundamentally ill-founded. Consider, for example, the case of the radiation-induced decomposition of n-pentane on silica gel sensitizer such as studied by Caffrey and Allen ('58). I understand that in such gels every molecule is within one molecule of the surface. When we note the nature of initial excitation effect in general, we find a real difficulty (much more than a semantic one) in saying that in this case it does or does not involve a pentane molecule. A similar situation exists, so far as I understand, in living cells. One cannot speak of an initial effect in an isolated water molecule or in an isolated protein molecule. The latter is so thoroughly solvated as to preclude the possibility of an excitation not involving water. The water is so thoroughly involved in protein structure, and itself has so extensive a molecular structure, that again it may be improper to speak of an initial excitation which does not involve protein.

11

A. Charlesby
Royal Military College of Science

Recent Advances in the Study of Irradiated Polymers

Long chain polymers owe many of their properties to the structural arrangements of their molecules. These can be profoundly affected by exposure to high energy radiation, and new properties emerge, often difficult to obtain by other means. This technique has provided a valuable tool to polymer scientists, but much of the information developed in these studies may be of value in other scientific fields. In physics for example radiation enables us to modify solids in clearly defined ways, and to a reproducible extent; the effect of these structural modifications on the mechanical and electrical properties can then be traced. The results obtained with long-chain polymers extend those derived from the simple small-molecules studied in radiation chemistry, and help in the discussion of reaction mechanisms. Perhaps of greatest interest here is the ability to irradiate these simple macromolecules under a wide range of conditions - in the solid crystalline phase, in rigid amorphous or liquid systems, in solutions, or even in oriented structures - and to deduce the effect of polymer state on the reaction. The influence of temperature and of oxygen, and of small amounts of additive can also be readily followed. The general conclusions obtained for such simple macromolecules may then serve as a guide to the behavior of the complex macromolecules studied in radiation biology, where it is far more difficult to alter the environment during radiation and to analyze in detail the chemical and structural changes produced.

A number of techniques have been developed to characterize long-chain polymers, and these can be readily adapted to determine many of the changes produced when these polymers are exposed to radiation. Among these techniques one may mention measurements of osmotic pressure, light scattering and viscosity (to determine molecular size), solubility, swelling, elastic modulus (to characterize a network), infrared and ultraviolet spectra (to indicate chemical change in solids), X-ray analysis and density measurements (to determine crystallinity and orientation).

It was early found that the most striking physical changes produced in irradiated polymers could be ascribed either to fracture of the main chain, or of the side chain. In the former case (degradation) the average molecular weight is reduced, and the number of polymer molecules increased, in proportion to the radiation dose. In the second case (crosslinking) reactive side groups are formed which subsequently serve to link polymer molecules together to form branched structures. When the density of crosslinking is sufficiently high, closed loops are formed, and the polymer is partially present in the form of a three-dimensional network or gel, with physical and chemical properties very different from those of the initially linear or branched molecules. For example, this network is insoluble (although it may swell) and it can have highly elastic properties. The point at which the transition to a partial network first occurs is quite sharp, and its determination is a convenient method of assessing the susceptibility of a molecular chain to damage by radiation. *

In addition to crosslinking or degradation, we may also observe changes in main chain unsaturation, gas evolution, decreases in crystallinity, (where initially present) the formation of endlinks (the end of one molecule being attached to the side of another) and possibly the formation of trapped radicals, and of conjugated unsaturation (Charlesby

* The "gel" point corresponds to an average of one crosslinked unit per "weight-average" polymer molecule. As there are two crosslinked units per crosslink, 0.5 crosslinks are required per weight-average molecule.

'52).

Although the mechanisms by which these various radiation-induced reactions occur are still very much in dispute, there is far less doubt as to the magnitude of these changes. It is therefore relatively easy to study the influence of environment, for example, without being directly involved in detailed discussions on mechanism.

In what follows I shall attempt to summarize some of the recent experimental work on irradiated polymers, particularly insofar as this appears relevant to radiobiology. Only the main lines of argument will be indicated; fuller details may be readily found in the published literature. I shall take the opportunity of indicating areas where knowledge is most seriously lacking, and of presenting certain ideas, which have arisen out of past work, but which have not yet been fully tested. No apologies need be made for introducing such unverified theories, if they serve to stimulate interest in neighboring fields of radiation work.

11.1. DEGRADATION VERSUS CROSSLINKING

In the early experiments the vinyl polymers studied were sharply differentiated as between those which degrade under radiation, and those which crosslink (Miller et al. '54). As a general rule, polymers with units of the form

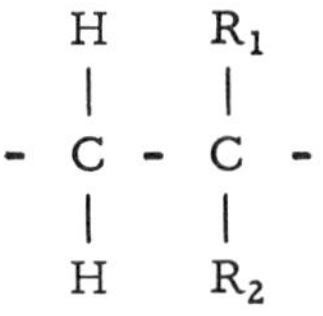

degrade, while those of the form

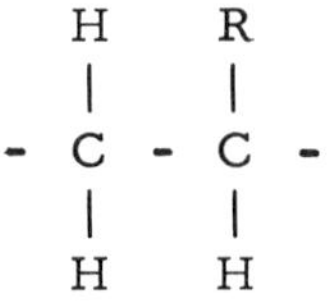

or

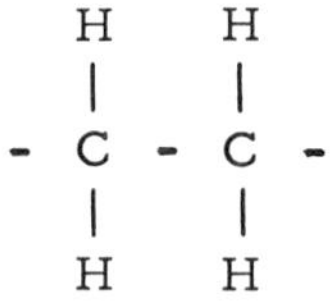

crosslink. (Here R is a side group different from hydrogen). These two very different patterns of behavior have been related to the heat of polymerization (Wall '55). Thus polymers of ethylene, methyl acrylate and styrene crosslink, and have heats of polymerization in the range 17 to 22 kcal $mole^{-1}$; polymers of isobutylene, methyl methacrylate and α-methyl styrene degrade, and have heats of polymerization of 13 to 9 kcal $mole^{-1}$. A comparison may also be made with the monomer yield on pyrolysis; polymers which crosslink under radiation give little monomer on pyrolysis, those which degrade give large yields of monomer.

Subsequent work has extended these results but has shown some departures from these rules. Thus, in the absence of oxygen, Teflon is reported not to degrade (Wall, private communication); polystyrene crosslinks, while polyisobutylene degrades, although the former yields more monomer on pyrolysis. Polypropylene suffers degradation and crosslinking to nearly the same extent, although it has a high heat of polymerization, and yields little monomer on pyrolysis. Polyethylene degrades as well as crosslinks.

An interesting comparison can be made between polyisobutylene and polymethyl siloxane; both have two methyl groups attached to the main chain, but the former degrades, the latter crosslinks. This difference in behavior appears to depend on the steric hindrance between side-groups. If this is high, as in polyisobutylene, excitation or ionization will produce main-chain fracture, which is permanent. In the siloxane on the other hand, the Si-O bond is longer, and the O-Si-O valency angle is greater, so that steric interference between methyl groups is far smaller; in this polymer there is no evidence of main-chain fracture, and it crosslinks readily under radiation.

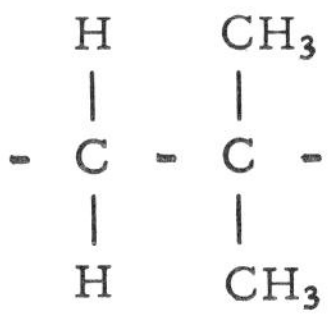

polyisobutylene

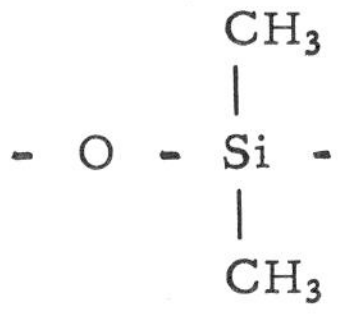

dimethyl siloxane polymer

In isotactic polypropylene the ratio of fracture to crosslinking is approximately unity. In random polypropylene, with less steric hindrance, the ratio is reduced (Dole, private communication).

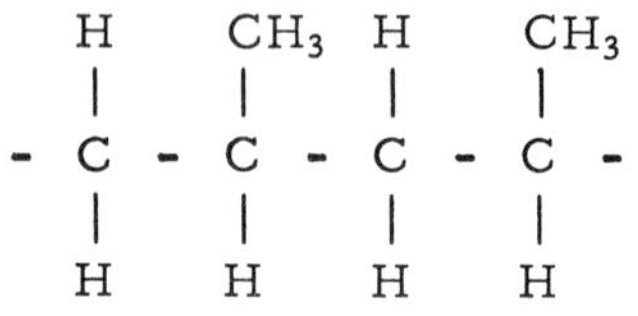

isotactic polypropylene

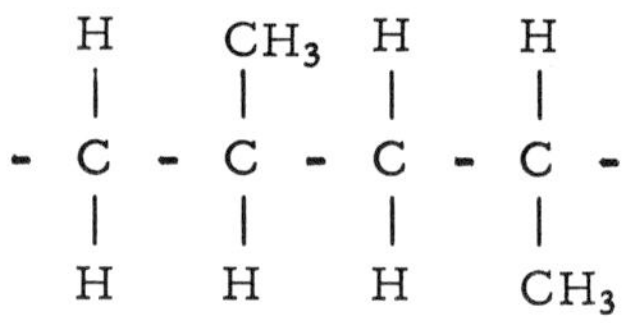

random polypropylene

There are several objections to the view that steric hindrance is the major factor in determining whether a polymer will degrade or crosslink under irradiation. In polyethylene, and even in low molecular weight paraffins, there is evidence for main-chain fracture even in the absence of such steric hindrance. One might also expect to see an increased rate of degradation in oriented or stressed structures, but this has not yet been reported.

By analogy with the behavior of polymers, we may expect the main-chain fracture or scission of irradiated biological molecules to be favored when large side-groups produce steric repulsions, and lead to strain within the molecular chain.

11.2. DEGRADATION

Random scission or fracture of the main chain in a polymer is a frequent consequence of irradiation. The process, termed degradation or dislinking, is to be clearly distinguished from depolymerization. In degradation each break produces two shorter polymer molecules from a single one, and there is a reduction in average molecular weight which can be readily calculated (Charlesby '54). A minor rearrangement around the point of fracture is to be expected, but this constitutes a very small chemical change in comparison with the numerous bonds present in the molecules. The total weight of polymer is essentially unaffected. In depolymerization, on the other hand, a weak point in a molecule breaks (e.g., under the influence of high temperature), and the molecule affected depolymerizes down to its basic constituents, e.g. monomer. The amount of polymer decreases but the average molecular weight of the remaining polymer molecules is unaffected. There is considerable production of monomer with depolymerization, little or none with degradation.

In many polymers exposed to high energy radiation (gamma, electron, pile radiation) degradation, but not depolymerization takes place. This is best shown by the decrease in intrinsic viscosity, (from which the average molecular weight can be deduced). Figure 1 shows that the reciprocal molecular weight, which is a direct measure of the number of main chain breaks, is linear with dose. In general we find that in degrading polymers the number of main chain breaks is directly proportional to the radiation dose, over a wide range of molecular weights, doses and radiation intensity. Typical G values are: 1.6 for polymethyl methacrylate, 3 for polyisobutylene. These G values increase with temperature, and show a marked change in slope at some critical temperature, which may correspond to the glass temperature, below which the molecular chain is rigid. It would be of interest to follow the reaction at higher temperature, to see whether there is a maximum number of breaks, corresponding to the total number of ionizations and excitations, or whether the reaction increases progressively without a limit, passing over to a chain reaction typical of depolymerization.

The nature of the rearrangement of atoms in the neighborhood of the break is not yet completely settled. In flexible polymers, side-groups may be broken off and, after recombination, appear as gases. In rigid polymers, trapped free radicals are detected, and these decay on heating or when oxygen diffuses into the system. It may be argued that the immediate effect of radiation on a rigid polymer (such as polymethylmethacrylate at room temperature) is the formation of trapped radicals, trapped electrons and ions, and it is only when the specimen is heated or dissolved to provide molecular flexibility that these react to

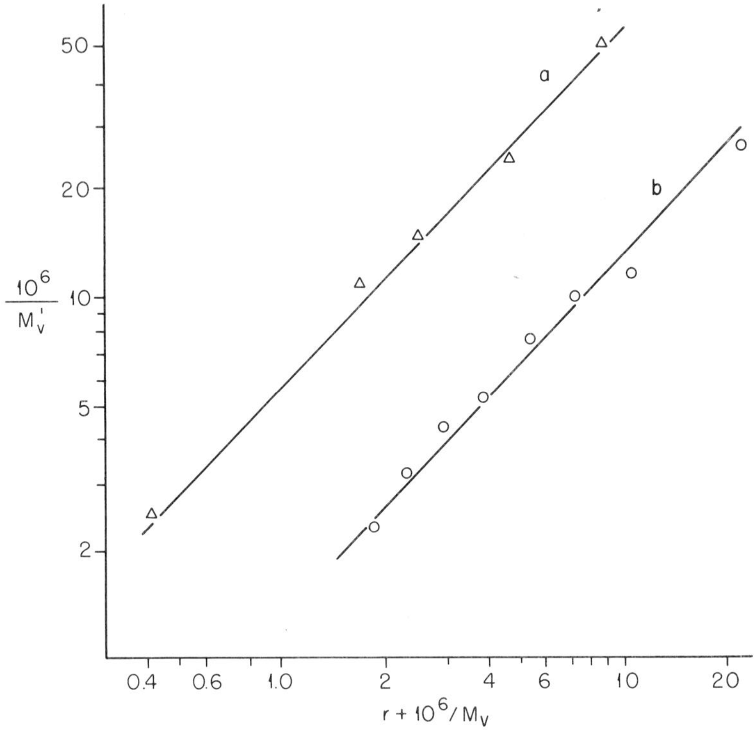

Fig. XI. 1. Degradation of Polyisotubylene by Cobalt 60 Gamma Radiation. The linear plots, of slipe 1, show that the number of breaks is proportional to radiation dose r. The effect is more pronounced in solution.
No oxygen; 800 rads/min

(a) 10% solution in n-hexane

$$\frac{10^6}{M_V'} = \frac{10^6}{M_V} + 5.6r$$

(b) Solid PIB $\frac{10^6}{M_V'} = \frac{10^6}{M_V} + 1.3r$

r = dose in megrads

M_V = viscosity average molecular weight before irradiation

M_V' = viscosity average molecular weight after irradiation give the observed chemical changes including degradation. Thus in polymethylmethacrylate, color changes occur, and these disappear on standing in the presence of oxygen, or when the polymer is dissolved. The precise nature of these color changes is not clear; they may be attributable to changes in chemical structure, to trapped radicals and electrons, or to gaseous fragments.

On heating an irradiated block of polymethylmethacrylate, bubbles are formed; they are caused by the migration of gaseous products arising from the fracture of side-chains. Before heating, these products were presumably held rigidly in the solid polymer matrix. Further work on the precise nature of the rigid irradiated polymer, prior to heating or solution, appears highly desirable (Ross and Charlesby '53).

It has been reported that exposure to ultraviolet light produces crosslinking in polymethylmethacrylate and polyisobutylene (Oster '56). If confirmed, this effect would indicate a fundamental difference between ionizing radiation and exciting radiation since the former results in degradation only. However, these experiments were carried out in the presence of photosensitizers which may themselves take part in the reaction. We have therefore carried out experiments on pure polymethylmethacrylate exposed to ultraviolet light of 2537 Å wave length, which is absorbed very readily in the pure polymer. Further experiments were also carried out with the polymer in dilute or concentrated benzene solution. After varying lengths of exposure the polymer was dissolved in benzene and its intrinsic viscosity determined as a function of exposure time. From this viscosity the average molecular weight could be readily calculated, its reciprocal being a direct measure of the number of main chain breaks.

When this experiment is carried out with electron or gamma radiation there is a linear relationship between the number of breaks and radiation dose. With ultraviolet light a similar relationship is observed (Figure 2), the slope of this curve also being independent of the polymer concentration. Furthermore no monomer is produced at room temperature, showing that little or no depolymeriza-

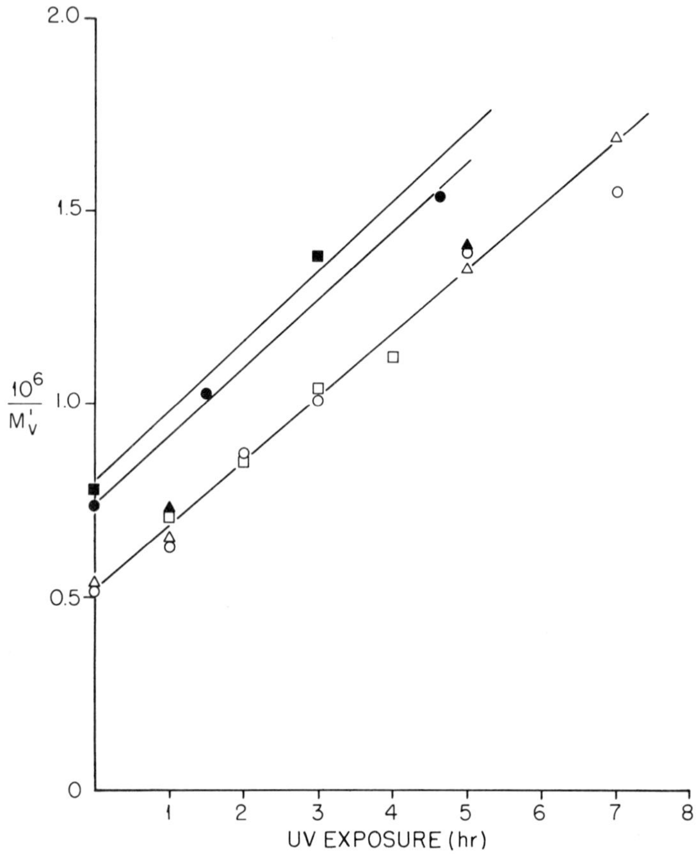

Fig. XI. 2. Degradation of Polymethylmethacrylate by Ultraviolet light (2537 Å).

- ● solid
- ■ 1% solution in benzene
- □ 0.15% solution in benzene
- △ 0.1326% solution in benzene
- ▲ 0.1005% solution in benzene
- ○ 0.0502% solution in benzene

As in Figure 1, the number of breaks is proportional to dose, but is no longer affected by solvent.

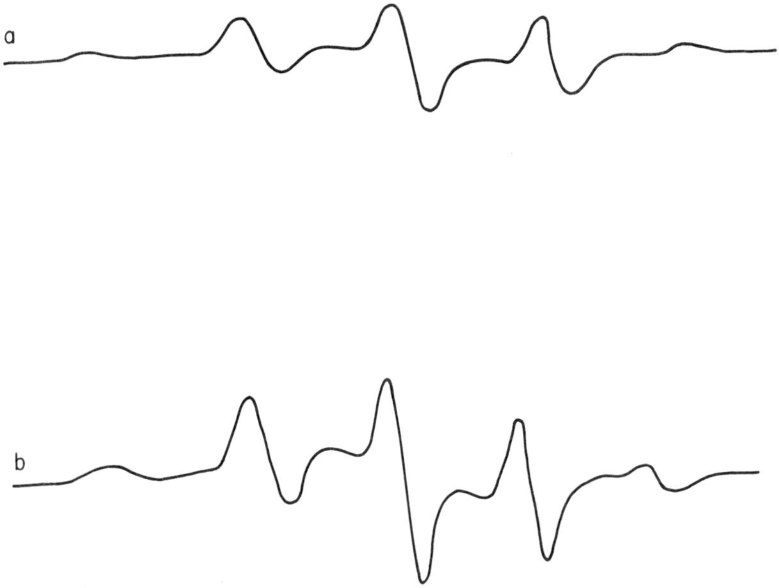

Fig. XI. 3. Electron spin resonance spectra of polymethylmethacrylate (PMM).
(i) Exposed to cobalt 60 gamma radiation.
(ii) Exposed to 2537 Å ultraviolet light.
Ionizing radiation produces same types of radicals as ultraviolet.

tion occurs. In spite of the greater mobility of the polymer molecule in solution, it degrades no more rapidly. The indirect effect arising from solvent molecules is also absent, showing lack of an energy transfer step. Although the quantum efficiency of the process is not known, there is clearly no essential difference between ionizing and exciting radiation as far as the main chain scission of PMM is concerned.

The same conclusion is reached when the ultraviolet spectrum of the irradiated polymer is considered. This mode of study has the advantage that any changes produced when heating or dissolving the polymer are obviated. Following irradiation the absorption in the UV increases; the spectra obtained by gamma irradiation and ultraviolet irradiation are identical. The same is also true of the electron spin resonance patterns obtained with similar specimens of PMM. Although the exact interpretation of the patterns has not yet been agreed, there can be no doubt that the patterns obtained from samples irradiated in these two ways are identical and no different trapped radicals or electrons are formed by ionization (Figure 3). Ionization and excitation to a high energy level appear to have the same effect as excitation to the lower energy level of 5 ev.

11.3. CROSSLINKING

A number of polymers crosslink when exposed to high energy radiation. At first, branched molecules are formed and the viscosity rises; above a certain density of crosslinking a network structure forms, which is insoluble but can swell. The polymer molecules not linked into this structure are still soluble, but the percentage solubility decreases rapidly as the crosslinking density rises. In polyethylene and n paraffins there is also some degradation. In other polymers, e.g., silicones and rubber, no evidence of degradation has been obtained.

Polyethylene has received most attention, although its structure is more complex than appears at first. Polyethylene may have side branches which interfere with crystallinity, and some small degree of unsaturation. One would expect the degree of crystallinity to have a marked effect on the density of crosslinking, but polyethylenes with between 50 and 85% crystallinity show the same G value (Epstein '57, Charlesby et al. '58, Schumacher '58, Waddington '58).

The initial unsaturation, due largely to pendent groups disappears very rapidly (Dole et al. '58) indicating some form of energy transfer within the molecule. At the same time main-chain unsaturation increases to a maximum value. This increase is also observed in n paraffins, and is due to some molecular process which is quite different from the

TABLE XI. 1.

TYPICAL G VALUES FOR POLYMERS

	G(cross-linking)	G(C=C)	$G(H_2)$	G(degrad)
Polyethylene (low density)	2 (1-3)	1-2	3-5	0.6
Polyethylene (high density)	2 (1-3)	1-2	3-5	0.6
Paraffins C_7 (liquid)			4.2	
C_2 (gas)			5.0	
Polystyrene	0.04		0.2-0.4	
Polybutadiene	2-10	1000		
Rubber	1	65		
Polyacrylates	0.5			
t-Polybutylacrylate	0.15			
Polydimethylsiloxane	2.5			
Polymethylphenylsiloxane	0.8			
Polyisobutylene	0	5		3
Polymethylmethacrylate		3		1.6
Cellulose				10

reaction leading to crosslinking; for example, unlike crosslinking, it shows little dependence on temperature.

Other polymers with a high degree of initial unsaturation (e.g. rubber), show a very rapid decrease. The G value for this reaction may attain several hundred (Table 1) and must therefore be a chain reaction, but the precise mechanism is unknown.

Many theories have been advanced to account for crosslinking, but none have been fully accepted. These theories assume either an ionic mechanism, the formation of two adjacent radicals per ionization or excitation (which crosslink immediately), the formation of mobile radicals (which move around until they find a partner to give a crosslink) or a reaction via unsaturation. It is therefore of considerable interest to compare crosslinking by high energy radiation with crosslinking by ultraviolet light.

To produce crosslinking in polyethylene with ultraviolet light it is necessary to incorporate a sensitizer such as benzophenone, which has a strong absorption peak at 2540 Å. The effect of irradiation can be assessed both from the disappearance of benzophenone and from the change in solubility of polyethylene. In our experiments a thin film of polyethylene, incorporating some benzophenone, was placed between the ultraviolet lamp and a photocell. The transformation from benzophenone to benzpinacol could be readily measured from the increased transmission of 2537 Å light through the film. It was found that this increased transmission reached a maximum value depending on the amount of benzophenone present initially. The maximum degree of crosslinking was also directly proportional to the initial concentration of benzophenone. The changes in solubility follow those observed with ionizing radiation, and lead to the same limiting solubility. This is a measure of the degradation: crosslinking ratio. Thus, there is no difference in either crosslinking or degradation induced by exposure to ultraviolet light (plus sensitizer) or to electron or gamma radiation.

In polyethylene a new absorption peak at about 2100 Å is found when the polymer is exposed to ionizing radiation. This new peak is best ascribed to the formation of conjugated unsaturation. A similar peak appears in crosslinking, induced by exposure to ultraviolet light, and we must therefore conclude that in this case also there is no essential difference between ultraviolet and high-energy radiation. However, we have been unable, as yet, to obtain evidence of trans-unsaturation with UV. If confirmed, this result would indicate one real difference between ultraviolet and high-energy radiation.

We have carried out further experiments on paraffins of varying molecular weight, sensitized by benzophenone and subjected to ultraviolet light either in absence or presence of oxygen. In the absence of oxygen the quantum efficiency of crosslinking increases rapidly as the molecular chain length and viscosity decrease. Oxygen suppresses the disappearance of benzophenone, particularly in the liquid paraffin, in which oxygen can readily diffuse (Figure 4).

An interesting extension of this work would be to compare crosslinking for similar materials in the solid and liquid phase, thereby providing information on the importance of molecular mobility in these radiation-induced reactions.

11.4. POLYMER SOLUTIONS

In spite of its considerable interest, comparatively little work has been carried out on the behavior of solutions of polymers subjected to ionizing radiation. Here we may expect

(a) a direct effect of radiation on the polymer (comparable to its behavior in the solid) and

(b) an indirect effect due to the formation in the solvent of active species, which subsequently react with the polymer molecule.

Experiments on polymers dissolved in a number of organic solvents show some dependence on concentration and character of polymer. For example, in the irradiation of

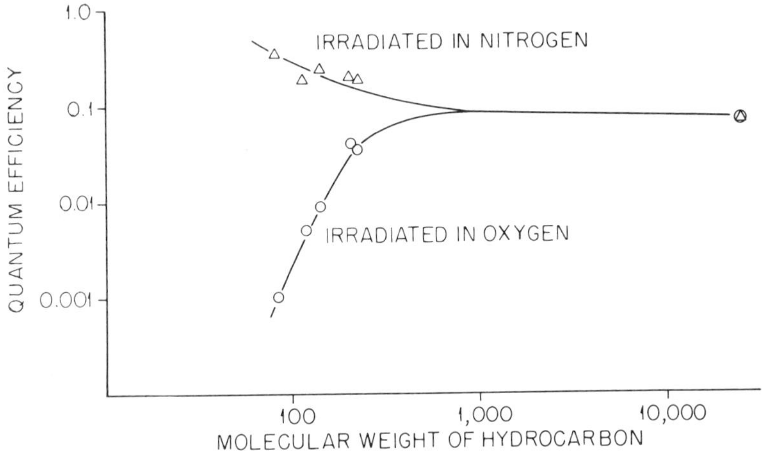

Fig. XI. 4. Quantum Efficiency for Decomposition of Benzophenone in Hydrocarbons by UV at 2537 Å.

Oxygen is more effective in reducing the effect in the lower hydrocarbons. The decomposition of benzophenone results in the crosslinking of the hydrocarbon.

polystyrene solutions, the presence of the solvent may increase radiation damage, as in heptane, or decrease it in unsaturated solutions such as di-isobutene. The indirect effect is, however, relatively small and does not usually alter the direct effect by more than a factor about three (Henglein and Schneider '59). Such interaction may possibly be explained by energy transfer from a solvent, as distinct from a direct attack of solvent radicals on the polymer chain.

Degradation can also be increased by the presence of a solvent. Figure 1 shows the reciprocal of the average molecular weight (i. e., the density of breaks) of polyisobutylene plotted against dose. The plot is linear, showing that the number of breaks is proportional to dose. In the presence of a solvent, the number of breaks is increased fourfold.

The most interesting effect occurs with water-soluble polymers such as polyvinyl alcohol and pyrollidone. Exposure to either electron or gamma radiation produces crosslinking and leads to the formation of a network structure which retains the shape of the vessel in which the solution was first irradiated. The point at which this network is first formed corresponds to the introduction of one crosslinked unit per weight-average molecule (i. e., 0.5 crosslinks). The dose r_g required to achieve this density of crosslinking is very readily measured since the transition from the soluble linear or branched polymer chain to an incipient network structure occurs very sharply, and can be observed merely by inverting the container.

If the dose to form this network is plotted against polymer concentration, it shows a rapid fall as the concentration is decreased. This means that crosslinking occurs more readily the farther the molecules are apart. Such behavior is to be expected from the indirect effect. Radicals formed in the water promote radicals on the polymer. The lower the concentration of polymer molecules in a gram of solution the greater the number of water radicals available per polymer molecule for a given dose. If there is no other cause of loss of radicals, the dose r_g for gel formation should be proportional to polymer concentration.

A remarkable change in the gelation dose vs. concentration curve occurs at about 0.5% polymer concentration, and below this concentration the gelatin dose rises rapidly. For concentrations below about 0.3% it is no longer possible to crosslink the polymer chains however large a dose is used. This critical dependence on polymer concentration has been observed repeatedly and may perhaps serve as a warning to radiation biologists of the importance of even slight changes in concentration (Alexander and Charlesby '57).

Studies have also been carried out on the change in viscosity of irradiated solutions. The intrinsic viscosity shows a rapid rise at high concentration, but a more complex variation at intermediate concentrations. It is difficult to interpret the viscosity of branched polymers; thus, a decreased intrinsic viscosity is not necessarily an indication of degradation (Berkowitch et al. '57, Danno '58).

Various theories have been propounded to account for this remarkable change in behavior. One theory assumes that active sites are formed on the polymer molecule and may eventually lead to crosslinking. However, these active sites have only a limited life-time and, if no other polymer molecule becomes available to permit formation of a crosslink within this time limit, the reaction is changed and the active sites give rise to degradation. It can be shown that this theory leads to a sudden change in the crosslinking: degradation ratio in accordance with the observed data. One possible objection to this theory is that if each crosslink arises from the interaction of two sites (e.g., radicals) it may be expected to show some dependence on radiation intensity and this is certainly not found.

A second theory assumes competition between internal and external crosslinks. In solution, polymer molecules take up a random configuration and each unit will be surrounded by other units both on the same polymer chain and on other polymer chains. Whether a crosslink is formed between units belonging to the same molecule or to different molecules will depend only on the relative concentrations. Internal crosslinks do not help in forming a network structure and energy absorbed in this way is therefore effectively lost.

In itself this theory would not explain any sudden transition. However, molecules are not in a static condition, but are constantly coiling and uncoiling. Internal crosslinks will be formed only when one particular part of the chain is strongly coiled on itself. The formation of a crosslink at this particular moment will help in retaining the coiled structure so that the overall size of the molecule is permanently reduced. Formation of such internal crosslinks will therefore effectively increase the local density of units belonging to the same molecule and also increase the distances between adjacent polymer molecules. Thus, when internal crosslinking occurs, it may have a drastic effect on the ability of molecules to link one to the other to form a network.

On this theory the critical concentration will show a marked dependence on the configuration of molecules within a solution and we may expect the behavior to change very rapidly with minor changes in pH for example.

It has been found that relatively small amounts of additive can produce a very marked increase in gelation dose. This result is to be expected because radicals formed on the polymer have a long life-time before they meet other radicals. Experiments on the effect of additive concentration on the gelation dose may be expected to provide interesting data on their reactivity (Figure 5).

It may be of interest to add that polyvinyl alcohol does not crosslink when irradiated in the dry state. If partially swollen in water it crosslinks in the amorphous regions only. This model may be useful also in studying the behavior of more complex biological molecules.

Results have been reported on the irradiation of very dilute solutions of polymers (e.g., polymethacrylic acid) which degrade. In this case very small doses suffice to produce marked degradation. The dose needed to produce a given degree of degradation is inversely as the polymer concentration. This result accords with the predominance of the indirect effect.

11.5. RADIATION PROTECTION AND THE EFFECTS OF ADDITIVES

Although ionization and excitation are generally assumed to occur at random throughout an organic system, the final chemical reactions show a considerable degree of specificity. Changes in chemical structure of one part of a molecule can lead to greatly enhanced radiation resistance of the whole polymer molecule, as for example in polystyrene. Small amounts of additive molecules can also change the nature and amount of the final chemical product. Where the physical or chemical change measured is greatly reduced one refers to radiation protection, although damage may have ensued at some other site (as in the additive) which has not been examined.

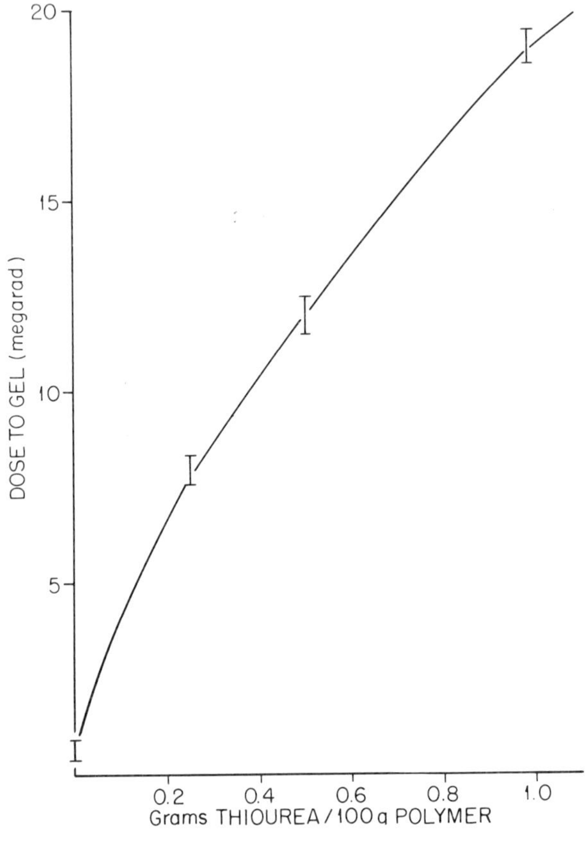

Fig. XI. 5. Effect of thiourea additive on dose required to gel a 2% aqueous solution of polyvinyl alcohol. Even very small amount of additive can greatly increase the amount of radiation needed to form a network.

We sometimes use the term "energy transfer" rather indiscriminately to cover any change in location of damage from the position of initial ionization or excitation to some other site within the molecule or even in a neighboring molecule. In some cases little permanent change may result. The use of the blanket term "energy transfer" covers a number of possible modes of reaction, from dissipation of the excitation or ionization energy without chemical change, to a simple transfer of radicals. It is only our lack of information on the precise mechanism involved which justifies the widespread use of this term. As, however, experimental information becomes available it will be possible to distinguish more clearly between the various reactions involved, and confine the term to its strict meaning.

Radiation protection is readily offered by the phenyl group with its great multiplicity of energy levels. Polymers containing this group show remarkable resistance to radiation; examples are polystyrene and Terylene (Dacron). Rough estimates of the extent of protection offered show this extends for about four or five carbon atoms along a main chain. On the other hand, double bonds are particularly sensitive; this is shown, for example, in the rapid decrease in unsaturation in polyisoprene (rubber), and in the pendent unsaturated groups in certain forms of polyethylene. Associated with these reactions there may be changes in isomerism. One way of looking at the latter effect is to assume that radiation removes a bonding electron, thereby allowing parts of the molecules to rotate about the common bond.

At the other extreme, one may envisage a modification of the final chemical structure by a different reaction of the radicals formed. This can be readily demonstrated by the addition of small amounts of additive capable of reacting with the radicals formed by radiation. This technique has been used to measure the number of radicals produced in simple organic molecules such as heptane (Forsyth et al. '54). Iodine and diphenylpicrylhydrazyl (DPPH) have been used as efficient radical scavengers (Prevot-Bernas et al. '52). Radicals formed by irradiation react with the additive instead of with each other, and the number of these radicals can be deduced both from the amount of iodine etc. removed from the solution and from the decreased amount of cross-linking. Studies have shown a linear decrease in iodine concentration with dose, even at very low concentrations. The slope of this curve gives the G value for the formation of radicals capable of reacting with iodine.

At very low concentrations of additive one would expect a departure from this simple dependence, since the radicals then have a significant probability of reacting with each other before they meet an additive molecule. In fact, such a departure has not been observed with iodine as an additive, presumably because the required concentration of additive is too low. We have, therefore, carried out experiments on a less efficient radical trap, namely anthracene (Charlesby et al. '59).

Dilute solutions of anthracene in hexane or cyclohexane were irradiated for varying doses and at various intensities. The loss of anthracene was found to depend on total dose, and also on radiation intensity at low anthracene concentrations. The results follow closely those expected from the two competing reactions:

$$R\cdot + R\cdot \rightarrow RR \qquad (1)$$

$$R\cdot + A \rightarrow RA + H \quad \text{etc.} \qquad (2)$$

At high concentrations of additive (A) and at low radiation intensities, I, reaction 2 dominates, while at sufficiently high radiation intensities and low concentration of additive radicals disappear largely through reaction 1. Protection against crosslinking is effective only above a certain value of $(A)/\sqrt{I}$ (Figure 6).

Our experimental results, therefore, agree with the assumption of a random distribution of the radicals produced by irradiation. In the absence of any additive, these radicals disappear by a bimolecular reaction to give crosslinks. From the rate of loss of anthracene at various values of (A) and I, we can deduce some of the reaction constants.

A very different result is obtained in dilute solutions of anthracene in silicone fluids. The disappearance of anthracene follows the same pattern of behavior with dose as was observed for hexane and cyclohexane, but there is no significant dependence on radiation intensity. In this system, therefore, one cannot consider the loss of radicals as being due to bimolecular reactions $R\cdot + R\cdot \rightarrow RR$, competing with

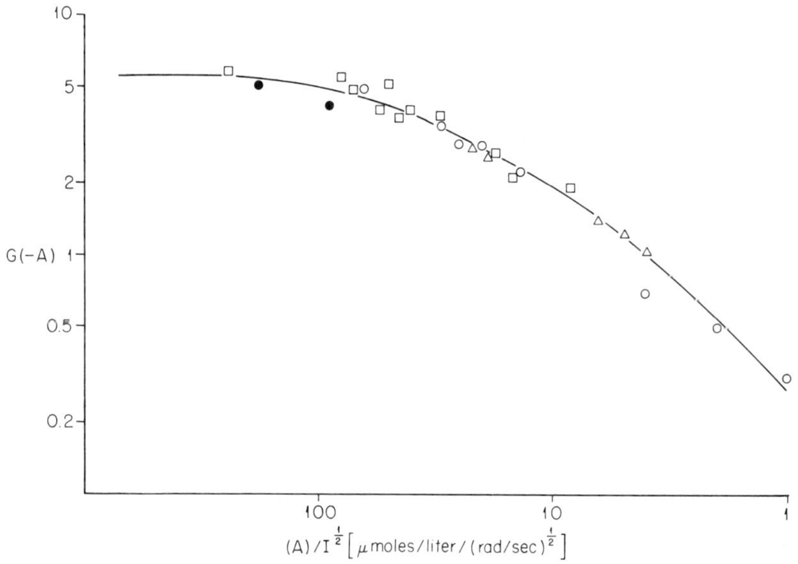

Fig. XI. 6. Destruction of anthracene in cyclohexane. Effect of additive concentration (A) and radiation intensity I. At low concentration (A), the radicals R· formed in cyclohexane react with each other, and not with the protecting additive. G(-A) is low. At high values of (A), and low values of I, almost all R· react with additive, and G(-A) reaches a maximum value.

the first order reaction R· + A → RA + H. We can, however, envisage an entirely different type of reaction, in which a radical can react directly or indirectly with a neutral environment,

$$R\cdot + P \rightarrow RR + H \quad (3)$$

$$R\cdot + A \rightarrow RA + H \quad (4).$$

With these two competing reactions we need not expect any intensity dependence. Equation 3 could be replaced, e.g., by certain types of ionic reactions which have been proposed to account for crosslinking.

The studies just described were carried out on liquids in which radicals could be expected to move around. A further study on the effect of additives was made on a partially crystalline solid in which extensive movements of individual molecules were impossible. In these experiments polyethylene (approximately 60% crystalline) was mixed with a number of additives and then irradiated. The degree of crosslinking of the polymer was determined by elastic modulus measurements, while the change in the additive could be followed by infrared measurements without melting or dissolving the specimen. Although a wide range of additives was used there was little protective effect, except in the case of acids. With these there was both a reduction in crosslinking and a removal of labile hydrogen. The extent of destruction of the acid was considerably higher than could be accounted for by the direct absorption of energy within the acid molecule. This result is clear evidence of either energy transfer from polyethylene or, more probably, a reaction of polymer radicals with the acid (Table 2).

In spite of the very high concentration of acid (compared with iodine or anthracene in the previous experiments) only a limited degree of protection was offered. This fact seems to offer some support for a view that several types of radicals are formed which may lead to crosslinking and that, of these, only one can be suppressed by reaction with the acid. It is known from electron spin resonance experiments that irradiated polyethylene gives rise to a complex structure which changes in character with time, as one type of radical decays more rapidly than the others (Libby, private communication).

Protection against degradation in irradiated polymers has also been observed (Alexander et al. '54, '58). Fairly high concentrations of additive in polymethylmethacrylate are needed to reduce the amount of degradation. The dependence on additive concentration follows the dependence expected for either of the theories suggested above for hexane or cyclohexane and for silicones, but the maximum degree of protection is always less than 100%. Here again there is evidence for a fast and a slow reaction, only pro-

TABLE XI. 2.

PROTECTION AGAINST CROSSLINKING OF POLYETHYLENE BY MEANS OF ADDITIVES

Additive, %		G(crosslinking)*	Protection, %
Nil		1.44	-
Paraffin Wax	10	1.50	0
Cesium Chloride	2	1.46	0
Ethyl Stearate	10	1.56	0
Sodium Stearate	10	1.69	0
Stearic Acid	10	1.09	25
	2	1.06	27
	1	1.14	21
Unsaturated Fatty Acids	10	1.14	21
Erucic Acid	10	1.02	30
Behenic Acid	10	1.06	27

* G(Crosslinking) deduced from elastic modulus of crosslinked polyethylene, and corrected for wasted crosslinks and direct energy absorption by additive.

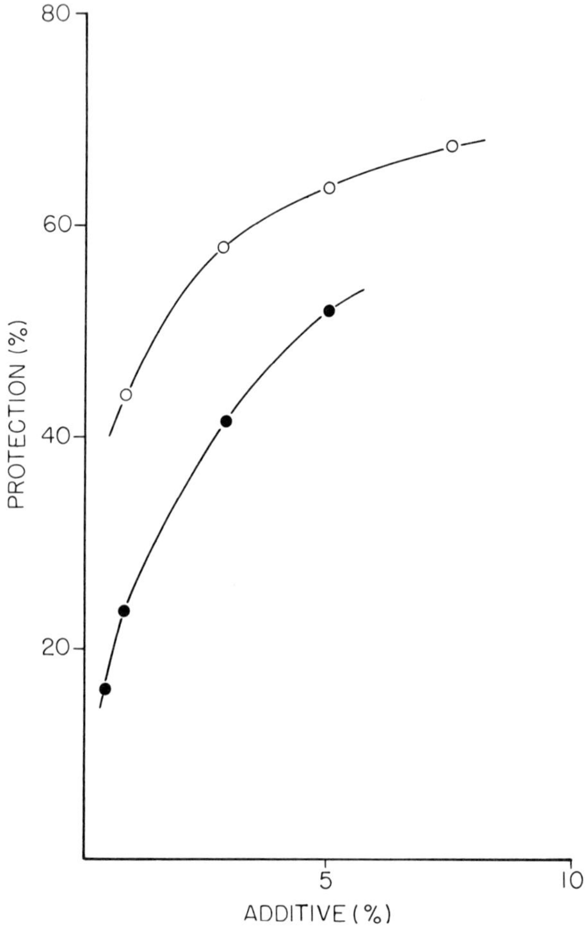

Fig. XI. 7. Protection of polymethylmethacrylate against degradation by pile radiation.

O 8-hydroxyquinoline additive

● naphthalene additive

— calculated from theoretical relation: protection =

$$80\% \frac{(A)}{(A) + c}$$ where (A) is percent additive and c = 1% for hydroxyquinoline, 2.5% for naphthalene. The curves indicate that even at the highest additive concentration, only 80% of the degradation can be avoided.

tection against the slow reaction being possible (Figure 7).

11.6. RADIATION-INDUCED IONIC REACTIONS

Although the primary effect of exposure to high-energy radiation is the production of ionization and excitation, there has been remarkably little evidence of ionic reactions. Where detailed investigations of radiation induced reactions have been possible, they have led to the conclusion that the primary chemical change is the production of radicals, the ultimate chemical reaction being due to the reaction of these radicals with each other and with the environment. The best evidence for these conclusions arises in polymerization work. The kinetics of polymerization of many vinyl monomers have been studied in detail, and comparisons made between initiation by radiation and by conventional catalysts known to initiate radical polymerization. The similarity in kinetics indicates clearly that the initiating species in this case is radical in character. A further experiment consists of polymerizing an equimolar mixture of methylmethacrylate and styrene, and studying the proportions of these two monomers in the polymer form. It is known from conventional chemical experience that anionic or cationic polymerization leads to a predominance of one or other type of monomer in the polymer, while radical reactions give an equimolar copolymer. When the reaction is initiated by radiation, it is the equimolar copolymer which is found.

To determine whether an ionic reaction can be initiated by radiation we attempted to polymerize at low temperature a monomer which could only react by an ionic reaction. The monomer chosen, isobutene, was irradiated in vacuum at various low temperatures and the amount of polymer formed was studied in terms of temperature, radiation dose and intensity, etc. The observation that this reaction increases in rate at the lower temperatures and that it is independent of radiation intensity are both indications that the reaction is ionic in character (Davison et al. '57).

Having established the existence of ionic reaction in isobutene and other polymers, it becomes of interest to determine the G value for initiation. The only data available are the average molecular weight of the polymer formed

TABLE XI. 3.

EFFECT OF ADDITIVES IN IONIC POLYMERIZATION OF ISOBUTENE AT - 78.5°C.

(Dose - one megarad gamma radiation)

Additive	%	Conversion %	G(initiation)	Molecular weight* $M_v \times 10^{-3}$
Nil		5.3	0.19	513
Calcium oxide	61.8	72.7	1.6	303
Magnesium oxide	47.9	40.1	0.8	469
Zinc oxide	13.5	28.9	1.5	297
	24.8	55.4	1.8	404
	41.0	95.9	2.6	388
Aerosil	1.9	12.9	0.32	607
	5.7	21.1	0.48	923
Glass 0.025 cm diam.	62.2			
500 cm²/g monomer		20.5	0.7	508
0.0025 cm diam.	12			
300 cm²/g monomer		21.1	0.45	798
0.025 cm diam.	63.4			
1000 cm²/g monomer		44.7	1.5	508

* Viscosity - average molecular weight.

and the conversion, or overall yield. Division of G(overall) (i. e., number of monomers converted per 100 ev) by the average number of monomers per polymer molecule should give the number of polymer molecules initiated per 100 ev, i. e., the G value for ionic initiation. Unfortunately, the situation is confused by the possibility of chain transfer (whereby the single initiation eventually leads to several polymer chains). Furthermore, the polymer chain once formed will be degraded by subsequent irradiation. Finally the yield is not linear with dose, and this points either to a reduction in the G value for initiation or, alternatively, to the formation of an inhibitor which reduces polymer chain length. The detailed analysis of all results available has led to the conclusion that the decreased yield is due to shorter chain length and that the G value for initiation is approximately constant at about 0. 2. This number is very small when compared with the usual G values for radical reactions, which are in the neighborhood of five.

A very interesting feature of the ionic polymerization of isobutene is that the yield can be greatly increased by the addition of relatively small amounts of inorganic powders (Worrall and Charlesby '58). A wide range of these has been used, but the most detailed results are those for zinc oxide, Aerosil and powdered glass. The results for powdered glass shown in Table 3 indicate that the surface area exposed is of considerable importance, although the G value is not proportional to surface area. With relatively small amounts of Aerosil for example, the G value can be increased approximately ten-fold, when it approaches the G value for radical production. There is little effect on the average molecular weight of a polymer showing that it is increased initiation rather than an increased chain length which accounts for the higher yield. The cause of this increased G value is by no means clear. It could be due to the capture of the electron ejected during ionization, giving a charged solid particle which stabilizes the growing polymer chain. The data for the glass powder shows that the distance which this electron must travel to reach a solid particle is quite considerable, and one would expect it to become completely free of the molecule from which it initially arose. However this conflicts with the requirement that the charge in ionic polymerization remains bound in the neighborhood of the growing chain. (This view is in any case necessary to account for the observed lack of any dependence on radiation intensity; each individual ionization leads to polymerization of a chain with no intervention from products of other ionizations.) The whole process may perhaps be compared with the stereospecific polymerizations studied by Ziegler and Natta. One can reasonably expect this kind of investigation to lead to better knowledge of the fate of electrons ejected during ionization.

11.7. CONCLUSIONS

After the initial spate of experiments on the crosslinking and degradation of various polymers, fuller attention is being paid to subsidiary reactions, such as unsaturation changes and the formation of trapped radicals. Electron spin resonance patterns from not very dissimilar polyethylenes for example show widely differing spectra and these are being investigated at low temperatures to reduce subsequent chemical reactions. Polymer orientation has also been found to influence the e.s.r. pattern. Sufficient data on the effect of temperature on G values may soon be available to enable us to relate these to polymer state and flexibility.

We may expect to see further work on additives in polymers, and especially the influence of temperature and radiation intensity on their interference with the primary reactions.

All these are attempts to study changes in irradiated polymers prior to the final chemical reaction. One further approach is the study of radiation induced conductivity at low radiation intensities. This conductivity has been ascribed to the effect of impurities, but this idea is difficult to sustain in view of the relatively small influence of very high doses, which produce very high concentrations of crosslinks, unsaturations etc., which can be considered as impurities but nevertheless have little effect on conductivity. One may perhaps consider this radiation-induced conductivity as one aspect of energy transfer. The electron vacancy left after excitation or ionization can be taken as mobile, and the molecule may then be considered to have hole conduction. The delocalization of the site of excitation or ionization will occur far more rapidly than any changes in nuclear arrangement. This may be one method of accounting for the specific nature of radiation-induced chemical changes, in spite of the random location of the initial interaction.

DISCUSSION

BENINSON: In the last case of polyvinyl alcohol, is the polymerization independent of dose rate?

CHARLESBY: It is not polymerization (which is a chain reaction, starting from reactive monomer) but a crosslinking reaction. The crosslinking is essentially independent of dose rate.

BENINSON: Over what range of dose?

CHARLESBY: Pretty wide; over a range of 10^4 to 1 there is no obvious effect of dose rate, certainly not to within a factor of two, probably far less.

BENINSON: Would you expect a different probability of external crosslinking with different dose rates?

CHARLESBY: Yes, the amount of detailed work which has been carried out on these systems is not really very adequate, but the general position is reasonably clear. I am hoping to continue this work, particularly on the effect of additive on the gelation dose.

CONGER: Does gel formation increase linearly with dose?

CHARLESBY: In the absence of a radiation-protective additive, the effect of small doses is merely to increase the average molecular weight, but the polymer remains completely soluble. Above a critical dose, corresponding to incipient gelation, a fraction of the polymer is converted to an insoluble network or gel. As the dose increases, so does the insoluble fraction, but the exact relation between dose and insoluble fraction depends on the molecular weight distribution of the original material as well as on other factors. In the presence of additives this dependence will be altered, but no detailed results are yet available.

ALLEN: I don't understand how, on your theory of competition between internal linking and crosslinking, the dose can decrease more than proportionally to concentration.

CHARLESBY: This is only true at concentrations below about 0.5%; i.e., below those corresponding to the minimum gelation dose. What we may assume is this: If two radicals on the same molecule come together, they form a permanent link. The molecule instead of being able to rearrange itself by thermal vibration, is now permanently constrained to a smaller volume. Once it starts to form internal links, it is going to be easier to form further internal links because it occupies a smaller total volume. The bias toward a smaller size results from the fact that internal links are formed in regions where the density of packing of molecular chains is above average; once a link is formed this higher local density is retained permanently. At the same time, the average distance between separate molecules is increased.

FANO: Your suggestion looks extremely plausible and attractive. Since it is so critical I wonder whether any attempt has yet been made to make an analytical calculation to test whether the phenomenon could really go like this. I suspect that it would be somewhat slower.

CHARLESBY: No, I should imagine it is almost the other way around. By measuring the dose vs. concentration curve one gets information about the molecular arrangement of polymers in solution.

FANO: No, it is just a matter of trying to get statistics.

CHARLESBY: One has to know a lot about the flexibility of the molecular chains in solution to do this. It may well be that by carrying out this work we can get a better idea of the flexibility of molecular chains.

BENINSON: Have you used high LET radiation? This should favor internal crosslinking.

CHARLESBY: The difficulty would be to get radiation of high LET with adequate penetration.

DOLE: I wonder whether you have tried this experiment: after the solution has been irradiated for some time at a low concentration, with a high probability of internal linking, it is concentrated by evaporating the solution.

CHARLESBY: This experiment has not been tried in detail. There has been a question as to whether polyvinyl alcohol forms aggregates which are unstable. However, this suggestion has been disproved as an explanation. Once the network is formed, it swells in the water. As irradiation continues, the network becomes tighter and water is extruded. If the water is evaporated, a small solid blob of of the crosslinked network is left. If water is readmitted, this swells again to the same extent as before. The process is completely reversible, so that the structure of the crosslinked network is quite specific.

KALLMANN: How do you determine the number of double bonds?

CHARLESBY: Infrared and chemical titration with iodine or bromine.

KALLMANN: How do you determine the degree of crystallinity?

CHARLESBY:. This is an old problem. There are lots of ways of doing it; X-ray diffraction, density measurement, specific heat. I suppose that people working on polymers have probably spent more time measuring the degree of crystallinity in polyethylene than anything else. The information is pretty good.

REID: In gel formation I imagine that the rigidity of the gel depends on water being held in the individual polymer cells, does it not?

CHARLESBY: Well, if you drive the water out, then, of

course, it becomes a much firmer material.

REID: On top of the effect that you have mentioned then, is it not true that as the concentration goes down you are trying to stretch the polymer into bigger and bigger cells and you might reach some critical point at which water can escape from the cell even though you have a network. It seems to me that this might be a point that you might reach more sharply than this transition from external to internal linkage.

CHARLESBY: No, why shouldn't the molecule be stretched?

REID: Well, you have less and less polymer material and it is trying to hold the same amount of water in the crosslinked network. Consequently, you are trying to hold the same amount of water with less crosslinked material.

CHARLESBY: It does not have to hold the same amount of water. In fact, it does extrude extra water. So eventually you find a swollen gel floating inside the excess water.

DOLE: A remarkable difference between radiobiological effects and the effects of ionizing radiation on solid high polymers is seen in the doses required to produce observable effects in the two systems. In the case of polymers a dose of about 0.1 Mrad or more is required whereas the effect of only one roentgen has been detected in certain biological systems. Therefore, any process occurring in high polymers with a very high G-value is of interest. Working with my associates, T. F. Williams and H. Matsuo, we have found that 100 ev of gamma radiation from a Co^{60} source causes the production of about 1.5 crosslinks and 3.1 trans-vinylene double bonds, yields comparable to those of the radiolysis of monomers. However, the absorption of the 100 ev of energy prevent 1300 methylene groups from crystallizing. Thus G(inhibition of CH_2-groups from crystallizing) is 1300!

It would be interesting to study the effect of radiations on the helix coil transitions in such substances as poly-Lglutamic acid (cf. P. Doty '59).

BURTON (communicated): In Section 11.3 of his paper, Professor Charlesby implies the existence of only four possible modes of crosslinking. I should like to suggest alternative possibilities which are wholly speculative but which, at this stage in our considerations, seem to have some good theoretical foundations.

(a) The initial excitation or ionization travels freely and with high velocity along the units of a chain until it is trapped and essentially localized. It is suggested that such traps may, in specific cases, be in sites well adapted geometrically for crosslinking.

(b) Excited sites of such nature react in Stern-Volmer reactions (zero or low activation energy) to yield H_2 and cross or end links.

$$M_1^* + M_2 \rightarrow R_1R_2 + H_2$$

(c) Ionized sites may capture an electron and react similarly (Burton '58)

$$M_1^+ + M_2 + e \rightarrow R_1R_2 + H_2$$

The occurrence of reactions of these types depends both on rapid transfer of excitation (cf. Dreeskamp and Burton '59, Burton et al. '59, Magee, Chapt. VII) or ionization and the proper geometrical location of the trapping sites.

HAMILL (communicated): Professor Charlesby's results in Figure XI. 10 require an abrupt change in kinetic mechanism with changing composition. The following conventional kinetic considerations can account for the unusual features which were reported if we consider that the polymer free radical P· diffuses much more slowly than H· and HO·. Radical-radical combination reactions, which are presumed to occur upon every properly exiented encounter, can be predicted to have very much greater rate constants when (P·)(H·) or (P·)(HO·) is involved than when

(P·)(P·) is involved.

For convenience let us consider a one-radical model for water. The simplest mechanism is

$$2R\cdot \rightarrow H_2O_2,\ H_2,\ H_2O \qquad k_1$$

$$PH + R\cdot \rightarrow P\cdot + RH \qquad k_2$$

$$2P\cdot \rightarrow P{:}P \qquad k_3$$

$$P\cdot + R\cdot \rightarrow P{:}R \qquad k_4$$

where R· is H· or HO·, PH is polymer and P:P is cross-linked polymer. The probability of crosslinking, Π_X, is simply the ratio at step 3 to the sum of all radical combination steps, viz.

$$\Pi_X = \frac{k_3\,(P\cdot)^2}{k_3\,(P\cdot)^2 + k_4\,(P\cdot)(R\cdot) + k_1\,(R\cdot)^2}$$

At sufficiently high concentration of polymer practically all R· radicals are removed by step 2:

As (PH) steadily diminishes, the term $k_3(P\cdot)^2$ can be dropped from the denominator of equation 1 since $k_1 >> k_4 >> k_3$ because the rate constants are diffusion-controlled. Depending upon (PH) we may expect equation 1 to assume either of the limiting forms 2 or 3.

$$\Pi_X = k_3\,(P\cdot)^2/k_4\,(P\cdot)(R\cdot) \qquad 2$$

or

$$\Pi_X = k_3\,(P\cdot)^2/k_1\,(R\cdot)^2 \qquad 3$$

The stationary state concentration (P·), obtained in the usual way, gives

$$d(P\cdot)/dt = O = k_2(PH)(R\cdot) - k_3(P\cdot)^2 - k_4(P\cdot)(R\cdot)$$

and neglecting $k_3(P\cdot)^2$ we find

$$(P\cdot) = k_2 k_4^{-1} (PH) \qquad 4$$

Similarly, for $(R\cdot)$ one obtains

$$d(R\cdot)/dt = O = \phi - k_1(R\cdot) - k_2(PH)(R\cdot) - k_4(P\cdot)(R\cdot) \qquad 5$$

where ϕ is proportional to dose rate. When equation 2 applies (i. e., $k_4(P\cdot)(R\cdot) >> k_1(R\cdot)^2$), nearly all $P\cdot$ radicals from step 2 are removed by step 4 and $k_2(PH)(R\cdot) = k_4(P\cdot)(R\cdot)$. Then $\phi = 2k_4(P\cdot)(R\cdot)$ and from equations 2 and 4

$$\Pi_x = \frac{k_3(P\cdot)^2}{1/2\ \phi} = \frac{2k_3k_2\ (PH)^2}{k_4\ \phi} \qquad 6$$

When equation 3 applies and $k_1(R\cdot)^2 >> k_4(P\cdot)(R\cdot)$, by equation 5 we have $k_1(R\cdot)^2 \simeq \phi$, since by postulate nearly all $R\cdot$ radicals are removed by step 1. By equations 3 and 4 we have

$$\Pi_x = \frac{k_3 k_2^2 (PH)^2}{k_4^2\ \phi} \qquad 7$$

Therefore, cases 3 and 4 cannot be distinguished.

The dose D required for a given degree of crosslinking varies as Π_x^{-1} and for small (PH)

$$D = \frac{k_4^2\ \phi}{a k_3 k_2^2 (PH)^2} \qquad 8$$

where a = 1 or 2. Clearly $k_4/k_2\ k_3 >> 1$ and the large negative slope of the curve for D vs. (PH) also varies as $(PH)^{-3}$ at sufficiently small (PH), and is likewise expected to be dose-rate dependent.

12

Franklin Hutchinson

Yale University

Radiation Effects on Monolayers Considered from the Biological Point of View

From the biological standpoint, the effect of radiations on monolayers is important in three ways: (1) the radiation sensitivity of molecules which are adsorbed to some surface in a cell may be completely different from the radiation sensitivity found in the more commonly studied forms such as dilute solution or dried purified preparations; (2) monolayers of biological material can at times be a convenient form to use in studying their response to radiation; (3) the destruction of layered structures in the cell may be of great importance in radiobiology.

Each of these three facets will be considered in turn.

12.1. SENSITIVITY OF MOLECULES IN THE ADSORBED STATE

It can probably still be said, as Cheesman and Davies said in 1954, that "it is difficult, despite a wealth of hypotheses, to name a single case in which the physical state of an intracellular protein has been established." Nevertheless, because of the abundance of intracellular interfaces, many molecules of biological significance must be found in the adsorbed state within the cell.

12.1.1. Sensitivity of dry monolayers. -

Currently the most interesting results to discuss here are those of Okada and Fletcher* ('60) who find that the enzyme DNase adsorbed

* The author is indebted to Dr. J. H. Stuy and to Drs. S. Okada and G. Fletcher for their kindness in permitting quotation from manuscripts which have not yet appeared in print.

on cellulose or ion exchange resins and then dried is about ten times more sensitive to ionizing radiation than it is in the dry pure state. If true, this result could well have important biological consequences, and provide a clear indication of energy transfer over quite significant distances under certain conditions. The possible experimental complication is that the adsorbing material might hold large quantities of water which could give rise to a considerable indirect effect by diffusion.

In this connection, it should be noted that Caffrey and Allen ('58) have found a large increase in the radiochemical dissociation of pentane when adsorbed on silica gel. In this system there is no possibility of radical diffusion in absorbed water. The working out of a simpler system such as this would be of immense help in understanding more complicated adsorbed systems involving biological molecules.

We have found that bovine serum albumin when assayed for biological activity by serological methods is about ten times less sensitive to ionizing radiation in the dry state when spread on a monolayer on a solid surface than when irradiated in the dry solid state. This result is equally true whether the bovine serum albumin is first spread as a monolayer on a water interface and picked up on the solid surface (Hutchinson '52) or adsorbed directly from dilute solution on a uranyl acetate-treated slide (Hutchinson and Mosburg '54). In both the monolayer cases the assay was conducted with the molecule still firmly adhering to the surface, by determining the thickness of antibody which could be specifically adsorbed to the underlying antigenic albumin.

The reason for these results turned out to be simple. For irradiation in the solid state, the observed loss of serological activity was caused mainly by loss of solubility in the buffer solution in which the assay was conducted (McNulty and Hutchinson '54). From the way in which assay of the monolayers was carried out, the solubility of the molecule was not a factor. This conclusion is confirmed by the data of Fricke et al. ('57), who show that precipitates of egg albumin still have serological activity.

These results emphasize the rather obvious conclusion that a loss of solubility caused by irradiation will probably not affect the biological activity of adsorbed molecules. It seems likely that, on irradiation, most biologically important molecules such as enzymes lose the ability to perform a specific biological function more rapidly than their ability to dissolve.

12.1.2. Sensitivity of monolayers in contact with water. Molecules in monolayers in contact with water might be very sensitive to ionizing radiation because of diffusion of radicals through the water layer to the surface. Mazia and Blumenthal ('48, '50) reported that 100 r can strongly decrease the activity of pepsin in mixed pepsin-albumin films spread at an air-water interface. However, this result apparently could not be repeated by Rossi and Blumenthal (see comment on this point in the paper by Augenstine and Ray, '57). Smith ('54) was unable to get such extreme radiosensitivities with adsorbed layers of catalase or bovine serum albumin, and required hundreds of kilorads to inactivate to 37% of the initial activity. Augenstine and Ray ('57) observed no inactivation of a trypsin monolayer at a water-air interface with 1000 r.

If it is assumed that radicals will react with a molecule on its first encounter, then Smith's results can be interpreted in terms of a radical having a half-life the order of microseconds, and diffusing a distance the order of 1000 Å in solution. This is the same order of radical diffusion distance and half-life which can be observed in water purified by ordinary multiple distillation. Mazia and Blumenthal's results are consistent with a mean diffusion distance the order of 100,000 Å, and a life time the order of 10^{-3} sec.

12.1.3. Radiation and "enzyme alteration". - In a discussion of adsorbed enzymes in the cell, the topic of "enzyme alteration" deserves a brief mention. Many years ago Euler and Blix ('19) found that treatment of cells with physico-chemical agents which would cause lysis would increase the apparent activity of certain enzymes over that which could be measured in the intact cells. This result was first ascribed to permeability effects, until Kaplan ('55) proposed that the phenomenon was caused by a change in the state of absorption in the enzyme molecule to an interface. He proposed this because (1) the treatments increase the activity of catalase in yeast cells, but not that of invertase, and (2) the thermal inactivation rate of catalase in the treated cells was different from that in the intact cells.

It now seems clear that the original explanation based on permeability is correct. Clayton ('59) has found that Phodopseudomonas cells poor in catalase show no increase in enzymatic activity on lysing, whereas those rich in catalase show a large increase. Furthermore, the observed increment in activity on lysing with increasing concentration of intracellular catalase is in quantitative agreement with that which would be expected if permeability were the cause of the phenomenon, with no change in the specific activity of the enzyme. Kaplan's first argument is explained by the work of Preiss ('58 and '59) with slow electrons. He finds that the invertase in yeast cells is rapidly inactivated by even quite low voltage electrons, indicating that the enzyme is located near the surface of the cell. For catalase, on the other hand, higher energy electrons are required to produce equivalent inactivation, and the enzyme is therefore more evenly distributed throughout the cell. Kaplan's second argument assumes that thermal inactivation is independent of the immediate surroundings of a molecule, a view which is not in good agreement with many experiments.

Aronson, Fraser, and Smith ('56) show that catalase activity of intact yeast cells may be increased with a few hundred thousand rads. This increase is presumably caused by changes in the permeability of the cell; their ultraviolet photographs of irradiated cells are consistent with this interpretation, and the experiment therefore becomes another one on the effects of ionizing radiation on membrane permeability.

12.1.4. Ultraviolet light on monolayers. - The effect of ultraviolet light on protein films spread at an air-water interface causes initially an increase in film pressure, and then a drop in film pressure at higher doses (see Kaplan and Fraser '52 for summary of earlier data). The explanation seems fairly clear that at low doses the major effect is the breaking of intramolecular bonds, which, from the work of Setlow ('55), in some cases probably represents S-S bond breakage. This breaking allows an increasing degree of molecular spreading at the surface. At high doses photolysis of other bonds takes place with the escape of fragments into the water phase (as deomonstrated directly by Mitchell and Rideal '38 with films of insulin) and the consequent decrease of material in the film.

Augenstine and Ray ('57) measured the quantum yield for the inactivation of trypsin activity at an air-water interface, and found a value of .005. This quantum yield almost surely refers to molecules in the film in globular form, since it seems fairly certain that enzymatic activity is lost by molecules which are fully spread at an air-water interface (Cheesman and Davies '54). The authors felt that this number was significantly lower than the values for the quantum yield quoted for trypsin in dilute solution, 0.010 (Goldenberg and McLaren '51) and 0.017 (Uber and McLaren '41). They believe this lower yield is caused by the stabilizing effect of surface pressure on S-S bonds which would otherwise be broken. In view of the experimental errors and the variation in quantum yields with such factors as pH, I would take the result as showing that the quantum yield in the adsorbed case was approximately the same as that in dilute solution.

12.2. MONOLAYERS AS A RADIOBIOLOGICAL TOOL

It is occasionally very useful to have the material which is to be irradiated spread out in a thin two-dimensional array. I have irradiated monolayers of bovine serum albumin with electrons of energies between 1 ev and a few hundred ev (Hutchinson '54). A rapid increase in radiosensitivity above 10 ev is taken as an indication that ionization is the main process responsible for inactivation.

Similarly, Setlow et al. ('59) have irradiated very thin layers of the enzymes trypsin, chymotrypsin, and ribonuclease with ultraviolet radiation of wave lengths from 2537 Å down to 1100 Å. Very thin layers of enzymes were needed to allow the radiations to penetrate. They found that the quantum yield for enzymatic inactivation was essentially constant from 2537 Å down to about 1400 Å, with a value the order of 0.01. However, at 1215 Å the quantum yield is the order of 0.1, and for ribonuclease the yield at 1100 Å is the order of unity. This again indicates that ionization is the mechanism by which enzymes are inactivated by ionizing radiation, and that the yield for inactivation by a single ionization is the order of unity.

Recently, a very simple technique for adsorbing the enzyne ribonuclease onto a glass surface has been developed independently by Shapira ('59) and by R. A. Deering (manuscript in preparation). The procedure is simple, and it has the additional advantage that ribonuclease may be assayed in extremely small quantities, as little as 10^{-9} g. Because the adsorbed layer contains the order of 10^{-7} g cm^{-2}, only a fraction of a square centimeter of surface is required. This matter is brought up to emphasize an important point: working with monolayers is not impossibly hard, and experimentalists could well make more use of them in specific problems.

Spreading treated samples of material at interfaces is a practical method of determining physico-chemical changes in molecules as a result of irradiation. The average moleular weight, forces between adjacent molecules, the area occupied by the average molecule, and the ability to be displaced from an interface are all properties which can be studied in this way. This technique has recently been applied by Korgaonkar and Desai ('59) in a study of irradiated histones.

12.3. MONOMOLECULAR LAYERS IN THE CELL

Many membranes of molecular thicknesses have been recognized in the cell - to name only a few, the outer cytoplasmic membrane, the nuclear membrane, the membrane that surrounds the mitochondria, and the membranes within the cell cytoplasm variously called the "ergastoplasm" or the "endoplasmic reticulum." It seems generally agreed (Danielli '58, Sjöstrand '59) that these membranes comprise a lipid double layer, as shown in Figure 1, with the hydrocarbon groups facing towards the interior of the membrane

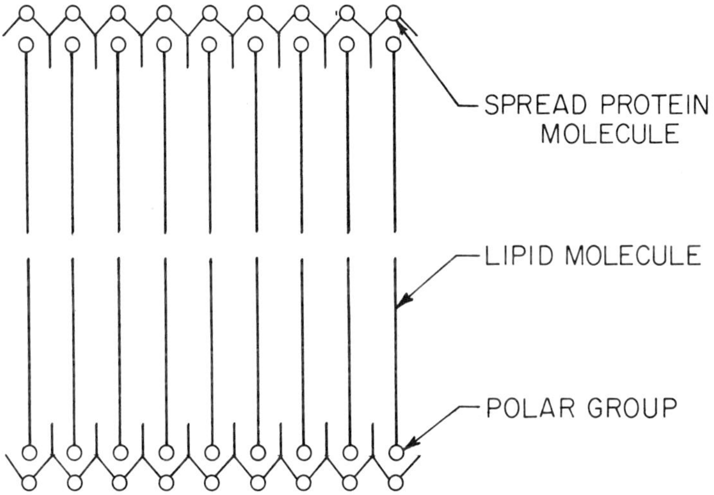

Fig. XII. 1. Diagrammatic structure of a membrane in a cell, after Danielli ('58).

and the polar group pointing outwards, with a total thickness the order of 50 Å. The lipid layer is lined on both sides with protein molecules.

12.3.1. Radiation-induced leakage from cells. - The effects of ionizing radiation on such membranes have been studied mainly by measuring the leakage of cell components from irradiated cells, or in other words, radiation damage to the outer cytoplasmic membrane. It seems quite clear that 30,000 to 100,000 rads will cause leakage of potassium ion from red blood cells (Sheppard and Stewart '52, Ting and Zirkle '40), yeast cells (Rothstein '59), or nucleic acid fragments (Billen '57). Indeed, an effect on rubidium ion transport into chlorella by as little as 100 r has been reported (Barber, Neary and Russell '57).

It is doubtful that such results indicate breakdown of the cytoplasmic membrane to the extent of permitting unrestricted leakage for the following reasons.

1. Billen finds that although nucleic acid fragments and adenosine triphosphate are released from E. coli B

by irradiation with 60,000 r, no release of peptides is found. Also, the escape of nucleic acid fragments could be prevented by (a) withholding an energy source such as glucose, (b) withholding phosphate, or (c) incubating at 5°C. It seems clear that metabolic processes are involved in the leakage of these fragments.

2. The flow of relatively inert molecules such as water, ethylene glycol, or diethylene glycol through the cell membrane of arbacia eggs is not affected by 100,000 r (Lucke, Ricca and Parpart '51).

3. Ting and Zirkle ('40) report that 33,000 r given to red blood cells modify potassium and sodium ion transport but not the membrane permeability to magnesium ions.

There is some evidence that radiation effects are localized, since Buchsbaum and Zirkle ('49) have shown that with localized alpha particle irradiation, membrane damage is also localized. However, they used enormous doses (seven million rads) to get large effects within minutes.

Three kinds of processes are recognized by which a molecule can pass through a cell membrane (Danielli '58).

1. Diffusion which takes place by solution of the diffusing molecules into the lipid layer and escape into the water phase on the other side of the membrane.

2. Facilitated diffusion, a process as yet poorly understood, by which diffusion takes place more rapidly than is possible by ordinary diffusion as defined in the first paragraph. The units involved in this facilitated diffusion occupy only a small area of the membrane, and in some cases, probably involve pores.

3. Active transport, in which ions are transported through the membrane against a concentration gradient. The transport of sodium and potassium ions in red blood cells with a high internal potassium concentration

and low internal sodium ion concentration is a good example. It is well established that metabolic processes are involved in such active transport.

A possible speculation on the basis of the available data is that radiation may well affect the process of active transport at much lower doses than it affects the other two processes.

12.3.2. Radiation injury and membrane rupture. - Although the direct evidence of irradiation effects on membrancs is by no means easy to interpret, many have suggested that an important primary effect of ionizing radiation is to break down membranes within a cell. A most forceful and direct statement of this point of view is by Bacq and Alexander ('55). They cite the following radiobiological observations in support of this view.

"1. A small dose of radiation affects many more molecules than could be explained on the number of free radicals formed by allowing enzymes and substrates normally kept separate to interact, i. e., the ionizing radiation opens the "flood gates" of a dammed up system.

"2. Enzymatic activity in the blood and tissue fluids is in many cases increased and not decreased immediately after irradiation.

"3. The relevant biological effect generally increases with an increase in specific ionization of the radiation used because the intracellular barriers cannot be destroyed by a single ionization and it is necessary for much energy to be deposited in a small area, i. e., a number of ionizations widely dispersed have no effect and are wasted."

Spectacular evidence that unexpectedly large numbers of molecules can be broken down by small doses of radiation is given by the work of Stuy ('60). Figure 2 shows the amount of extractible deoxyribonucleic acid (DNA) in irradiated hemophilus cells. The DNA is measured either by the total DNA extracted, or by the transforming ability of this DNA. The initial drop, directly after irradiation, in amount of DNA is quite small. However, incubating cells at 37° C produces a rapid drop in extractable DNA as

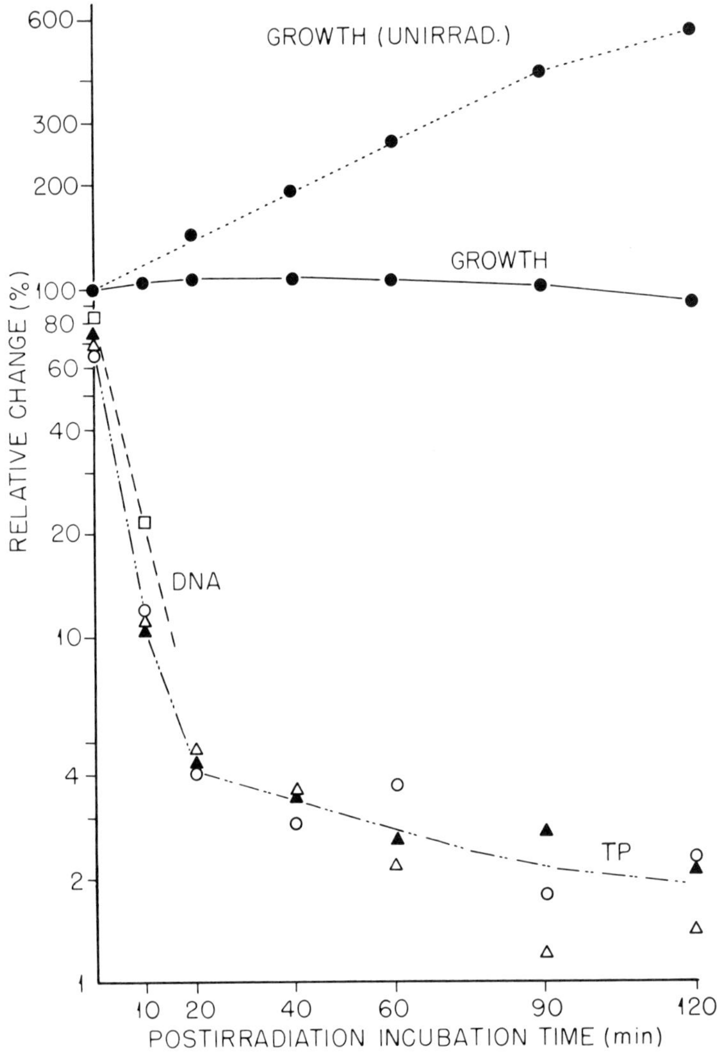

Fig. XII. 2. The disappearance of DNA on incubation of Hemophilus cells given 50 kilorads. The curve marked "DNA" indicates the quantity of DNA demonstrable by a diphenylamine test. The curve marked "TP" shows the amount of transforming activity found for three different markers. Reproduced from J. H. Stuy ('60).

measured by either method, showing that some additional process is at work. This additional process is completely stopped by the addition of citrate, which fact is a strong indication that the enzyme deoxyribonuclease is the agent which is depolymerizing the DNA.

The observed DNA breakdown by doses in the range 10-50 kilorads shows that energy release within a large volume will produce the effect. If it is assumed that only a single ionization will trigger the event, it can be calculated by target theory (Lea '46) that the sensitive volume is the order of 5×10^7 in molecular weight. This calculation assumes that diffusion of water radicals will not increase the sensitive volume by a significant amount, say an order of magnitude; this assumption appears to be true (Hutchinson '57). The dissociation of DNase from an inhibitor would require energy transfer over a large distance in the cell, a process for which there is no solid evidence. The breakdown of a membrane which separated enzyme and substrate could readily explain the observed results.

The small drop in DNA at short times is in good agreement with our data (Hutchinson and Arena '60) which show that, in irradiated pneumococcus cells, 400,000 rads is required to reduce the transformation ability of the DNA which can be extracted to 37% of the original activity. A large initial drop in DNA with as small a dose as 50 kilorads would be quite out of line with our present understanding of the action of ionizing radiation on molecules in cells. The drop in total DNA after incubation measured by Stuy is in good agreement with the results of Chantrenne ('58), who finds that 100,000 r given to yeast cells at room temperature over a period of hours completely depolymerizes the DNA in these cells.

With reference to the third point made by Bacq and Alexander, the relative biological effect for the inhibition of cell division certainly increases with linear energy transfer, but the efficiency of membrane breakage as a function of linear energy transfer does not seem to have been measured. We are beginning experiments measuring the leakage from cells which have been irradiated with heavy ions such as 160 Mev oxygen ions, which have a rate of energy loss of about 30 ev $Å^{-1}$ or 2,000 ev in a 60 Å thick lipid membrane. We do not have any experimental results as yet.

It might be noted that Sheppard and Stewart ('52) looked for an oxygen effect on the radiation-induced leakage of potassium from erythrocytes, but did not find it.

Thus the speculation that significant radiation injuries are caused by membrane breakage is within the realm of possibility, but without much convincing experimental backing at present.

12.4. DISCUSSION

The hypothesis that the rupture of intracellular membranes is an important form of radiation injury is little more than informed speculation at the moment, owing to the vagueness of the data which are presently available. Needed most of all are better and more specific criteria of membrane damage.

Careful studies on the changes in membrane injury with the rate of energy loss of the radiation would be important. It is presumed, without much evidence to be sure, that a single ionization is not effective in rupturing a membrane, and that only a cooperative effort involving a number of simultaneous closely spaced ionizations will have such an effect. From membrane studies with radiations of various LET it should be possible to gain some information on this score. From such information it would then be possible to calculate the number of spots in which membrane injury is possible, as the result of a given dose of X-rays, from the number of densely ionizing electron tails that are produced.

Considerable study might also be devoted to the mechanisms by which the membrane can be so injured. Nothing really seems to be known about the immediate effects of radiation on lipids in cells. The lipid constituents of many membranes are known to contain unsaturated fatty acids, such as arachidonic acid in the red blood cell membrane. It is known from the work of Mead ('52) that aqueous solutions of such acids can undergo a chain reaction of the following kind in the presence of oxygen.

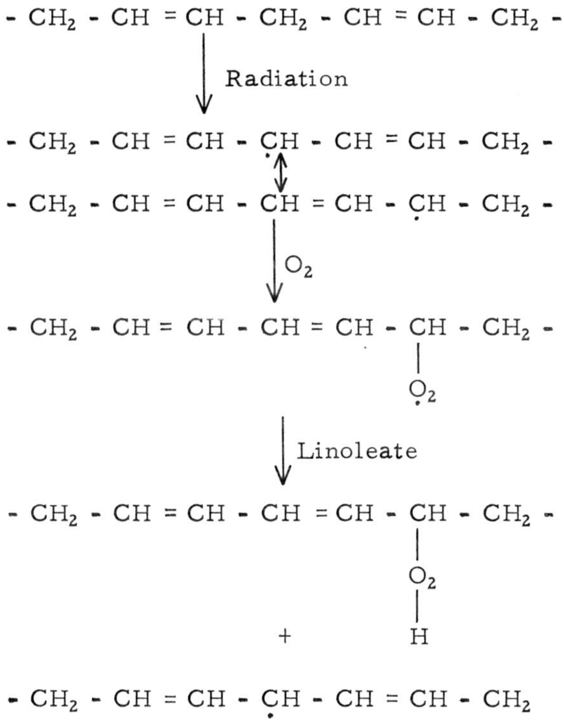

etc.

Although astonishingly high yields of fatty peroxides have been reported in irradiated mice (Horgan and Philpot '53), there now seems to be general agreement between Mead and Philpot (personal communication from J. S. Mead) that these high yields were mistaken, and that there is, as of now, no evidence as yet for any kind of lipid chain reaction in living systems.

The possible connection between lipid damage and oxygen concentration should not be overlooked, in view of the well established effect of oxygen in increasing the sensitivity of the living cell to ionizing radiation. Thus, the change in membrane damage with oxygen concentration could also be looked at more extensively.

Lastly, irradiation studies on biological molecules adsorbed at interfaces other than air-water surfaces would be informative. The conditions at an air-water interface are completely different from those found within a cell. A number of studies have been made at water-oil interfaces (E. Hutchinson '54), but no irradiation work seems to have been reported. Irradiation of artificial membranes resembling those found in cells would also be of interest.

DISCUSSION

BURTON: In the quantum yield determination for trypsin monolayers by Ray and Augenstine, may I ask if the possibility was taken into account that the absorption spectrum of the unimolecular layer was different from that of the massive material.

AUGENSTINE: We took the absorption to be the same in the monolayer as in the solution.

HUTCHINSON: I would like to say one point relative to this. I think that although at an air-water interface one can get spread layers in which the polypeptides have obviously unfolded to an extent, I would say that the weight of evidence is strongly in favor of the fact that the biological activity is shown only by molecules that are unspread in the layer. Probably one is measuring only the adsorbed globular trypsin molecules in the surface in this particular case where a biological assay was used.

KASHA: Was the angle of incidence studied carefully for a monolayer under tension, in Dr. Augenstine's light absorption work? The electric vector of the light wave for normal incidence might be perpendicular to the polarization axis for molecules in an oriented monolayer.

AUGENSTINE: The answer to that question is "No", although we have some studies planned in which we hope to do this. It turns out that by carefully compressing the monolayers you can note a shift in the film pressure which indicates that you initially have fairly long, cigar-shaped molecules lying flat. At higher pressures you can detect evidence that they have flipped up. What we hope to do next is to irradiate these films under different film pressures and study exactly the thing that you are mentioning.

PLATT: In the phenomenon of "enzyme alteration", is this a case where every cell shows increased activity or is it a case where they are selected for increased activity after radiation? Is the increase caused by damage to repressor or inhibitor enzymes?

HUTCHINSON: There is just no evidence of this whatsoever. The evidence is completely consistent with the hypothesis that the effect is a permeability change.

PLATT: No, I did not ask if it was consistent. I asked if damage to repressor or inhibitor enzymes could be excluded.

HUTCHINSON: I think you would have to say there is not enough evidence to exclude them. I find it is always a very dangerous thing to say a thing cannot happen.

PLATT: It is more important to know what is excluded than what is consistent.

HUTCHINSON: It is not excluded I would say, but none of the original arguments for enzyme alteration are valid. Someone will have to bring up new evidence in order to demonstrate its existence.

PLATT: But the original argument concerned enzyme alteration, not the damage to inhibitors or repressors?

HUTCHINSON: That is right.

PLATT: I feel hesitant as a physicist to say anything biological, but I think it is important to mention some ideas of Professor Leo Szilard which are going around Chicago at the present time. Two of his recent papers are in the February 1960 Proceedings of the National Academy of Science, one of them on antibody-antigen effects and another on enzymes.

The central idea in these papers, if I understand correctly, is perhaps not a new one, but he has stated it very cogently and very provocatively. The central idea is that the cell, at least in a multicellular organism, contains a tremendous number of potentialities, genetic potentialities which are normally unrealized. They are kept in check, until perhaps some of them become realized for some cells in the course of differentiation. Some of them become realized in one organ and some of them become realized in another organ, etc.

The mechanism for keeping these genetic potentialities in check until needed is an elaborate system of inhibitor and repressor enzymes. One set of developmental enzymes is controlled and inhibited perhaps by another set of enzymes, and this perhaps even by another, making a very complex system, at least for the physicist.

Nevertheless, the Szilard theory deals with a system kept in check, "dammed up", I believe is the phrase from the book, Dr. Hutchinson mentioned. Yet its potentialities may be released by a very small perturbation such as a single antigen or hapten molecule entering the cell; one molecule of the right kind can cancel out one inhibitor in this set of coupled inhibitors, thus releasing another enzyme which releases a third, etc.

And one of the important conclusions from these ideas of Szilard's is that almost every biological cell can be a "flip-flop", to use the electronic term. It can be a system in which either (a) there is a low level of enzyme A in stable equilibrium for the lifetime of the cell, or in which (b) a single trigger event in the cell — which may be the appearance of a single molecule of a certain type — can raise the level of enzyme A up to a very high value which the cell then maintains for the rest of its lifetime.

In short, the cell "remembers". You have a biological flip-flop which can be either "on" or "off". And the theory suggests these may be a total of the order of, say, 30,000 enzymes per cell, each of which can be either on or off (although this figure might be wrong by a factor of 100 in either direction for a particular cell). The relevance of this to Dr. Hutchinson's experiments is that the effect of radiation on a single inhibitor enzyme in such a coupled chain might be just as important as the appearance of a single antigen or hapten molecule in producing a flip-flop over to a new state of the cell.

Now, presumably radiation is somewhat random in its effect, and some irradiated cells in a solution might have one enzyme, say, enzyme A, increased greatly. Other cells might have enzyme B increased greatly, or decreased and suppressed greatly. In others, enzyme C could be affected. And then there is some unnoticed selection taking place in the experiment — some selection for one kind of cells before you count them or measure their activity, say a selection for those which have been activated with respect to enzyme A — then one would have a kind of an explanation of some of these results, perhaps without intruding such questions as permeability and perhaps talking entirely in terms of normal cell potentialities which have been released by the radiation.

MARSHAK: I would like to say a bit about Stuy's experiments in which the degradation of DNA in irradiated cells was followed. I think there is very good evidence that at least some of the DNase is enclosed within particles called lysosomes and these particles can be ruptured, releasing DNase. I think in any studies such as these which are related to the biological condition this factor should be taken into account.

CHARLESBY: A point was raised regarding an electrostatic effect in membranes. I wonder if you have a membrane and you put on it a few electric charges whether this would not greatly affect the permeability of the membrane. In fact, is the damage essentially due to chemical reaction? May it not be due to a purely electrostatic effect which would allow certain ions to diffuse more freely?

If you charge up the membrane, does it affect mobility? After all, it does not require many electrons to alter the electrostatic potential of the membrane or anything else significantly.

POLLARD: I don't think you can simplify the membrane down to the variation of charge on it. It would be nice if you could.

SZENT-GYORGYI: I would just like to call attention to a very nice material for this permeability problem. If you buy an apple and happen to drop it, the next day you will find a brown patch. This is an expression of a chain reaction. There is a polyphenoloxidase which oxidises a phenol to a quinone. If you mince apple tissue and add ascorbic acid, the ascorbic acid will become oxidised by the phenol, while the phenol stays reduced till ascorbic acid is present. By titrating the vitamin concentration you can easily follow the activity of the phenoloxidase. Damage increases the activity of this enzyme many hundred fold.

It struck me that most of your protective agents had a very low ionization potential or a high electron affinity. This may be connected with their effects.

PLATT: May I ask Dr. Szent-Gyorgyi if this reaction which he has described in the case of apples is one which has survival value, as the apples drop from the tree? That is, is this a normal expression of a physiological potentiality which apples need to have, or is this something associated with an actual type of membrane damage which is nonbiological?

SZENT-GYORGYI: It may have a natural role in oxidation but I am positive that it has value in protecting the apple against damage because as soon as you make a little cut then the quinones are formed and quinones combine with the protein on the surface and make a sort of barrier to the damage. It does not allow the damage to penetrate deeper.

I think it is a very important protective agent against physical damage. Quinones, to which the phenols are oxidized, are also strongly bacteriacidal.

AUGENSTINE: I would like to describe two or three very short experiments which I think may bear on this problem of the effect of radiation on membranes. I am very much happier with Szilard's explanation which Professor Platt was describing, than the idea that radiation punches holes in membranes so that very large molecules can wander around rather freely in the cell. I think that the three experiments which I propose to describe argue against the notion that actual "holes" in membranes are important in radiation damage. Changes in permeability would, of course, not be ruled out by these arguments.

The experiments are briefly as follows: (1) When surface films of trypsin are formed on 15% ammonium sulfate, no enzymatic activity can be recovered if less than 10 μg is spread over 100 cm^2. To a first approximation, the amount of trypsin in excess of this (which is spread over the same area) can be recovered. Our interpretation was that the first molecules deposited completely unfolded as a result of exposure to the surface forces. After unfolding they presumably lie flat on the surface with a vertical thickness of one polypeptide chain. Once 10 μg has spread over 100 cm^2 a coherent film is established and a sufficient film pressure built up so that molecules deposited subsequently would not unfold and thus would retain their activity and presumably their normal configuration. (2) The film pressure changes observed during and after irradiation depend upon the surface concentration. For films containing less than 10 μg/100 cm^2 the film pressure only decreases. Undoubtedly this reflects a decrease in total film material as a result of solubilization of small molecular fragments produced by the action of the radiation on the unfolded molecules. At large surface concentrations there is a considerable rise in film pressure followed by an eventual decrease with sufficient dosage of ultraviolet. We interpreted this as an indication that some of the molecules which had not previously unfolded do so as a result of the irradiation. Thus, the rate of the rise and its extent are determined by the number of molecules unfolding and the associated increase in film pressure balanced against the amount of fragmentation and solubilization with the associated decrease in film pressure. (3) The third set of experiments was carried out by Dr. Trurnit at RIAS. He formed barium stearate films at very high levels of compression (18-20 dynes/cm) and then injected protein underneath. When he expanded the stearate films at a constant rate, the film pressure dropped continuously until a value of about 3 to 4 dynes/cm was reached. Continued expansion (by a factor of at least 2) did not cause a decrease in film pressure and in some cases a rise was observed. This seems to indicate that at the large film pressures the underlying protein is excluded from the film. However, once the pressure drops sufficiently the protein can penetrate into the film so as to maintain the film pressure approximately constant.

If we assume that these reactions at the liquid-air interface are at all representative of those in cellular membranes then it might be expected that radiation would produce some of the following effects. Fragmentation and solubilization of some of the molecules in the membrane would likely occur. However, such "holes" would undoubtedly be filled immediately as the result of penetration of other cellular material or the unfolding of neighboring molecules. Very likely this would change the permeability in that region, but I think the probability that large holes would be maintained for considerable lengths of time is quite small.

HOLLAENDER: I would like to describe a phenomenon which might be pertinent to this discussion. I have reference to the high sensitivity of the grasshopper neuroblast in regard to the rate of mitosis. The work was done by Dr. M. E. Gaulden of our laboratory.

If you irradiate the neuroblast with moderate amounts of radiation in tissue culture, you invariably notice the slowing down of the rate of mitosis and the recovery overshooting and finally then going to normal. It is possible to prevent this slowing down of the rate of mitosis if radiation is not excessive by using a hypertonic salt solution immediately after exposure, about 1.2 molar. If this is done inside sixty seconds, one can prevent the entire slowing down of the rate of mitosis, and the cell will behave like a normal cell. If the hypertonic salt solution is not given, it will take the cell considerable time to recover from radiation damage, or the process of damage may continue and finally lead to an interference with the synthetic processes in the cell.

That a 1.2 molar salt solution is sufficient to counteract serious radiation damage is interesting to me because it shows that the initial steps in radiation effects induced by X-rays possibly are due to an imbalance of ions in different locations in the cell which we can counteract by this hypertonic solution. I had hoped that this discussion today would go into some of these points because such simple processes must be involved which should be possible to understand or possibly also to control if one could get into the cell immediately after irradiation. What we usually recognize in interference with nucleic acid or protein synthesis is a much later step which comes in after probably several intermediate steps have taken place. It should also be possible to recognize other processes which might be interfered with by radiation which go rather rapidly and which could be studied with the newer techniques which are available.

REID: In this connection I would like to draw to your attention some experimental work which is not yet published which I think would be extremely valuable for the kind of experiment which Dr. Hollaender is talking about. This is concerned with the immediate effects of irradiation on ion balance. There are some people in England now who have succeeded in simply recording with sodium and potassium electrodes the sodium potassium ion balance in single squid axons. So far as I know they are not concerned with radiation work, and are not planning to do this under irradiation at present.

I think that anyone who wants to pick up this technique has a wonderful way of measuring immediate effects of ions in the system which may very well affect chromosome orientations.

LINSCHITZ: I would like to introduce the idea that many of these membranes can have the action of nerve fibers.

The structure of the rod apparatus of the eye, as shown for example by electron microscopy of Sjöstrand, involves a stack of layers each of which seems to have the sandwich structure that has been indicated by Dr. Hutchinson. There seems also to be structures connecting each layer with processes leading down to the base of the rod. Now, in vision one has the problem of how any given rhodopsin molecule can trigger a wave impulse at the base of the cell. It is very tempting to conjecture that each of the layers or discs in the stack has some nerve-fiber properties, and that any action in the center of the disc, for example, will change the state of polarization of that layer, and act in some way as a kind of flip-flop mechanism that Dr. Platt discussed.

HUTCHINSON: With reference to the suggestion made by Dr. Linschitz, it is a most attractive speculation for transmitting radiation effects through a cell. The other case, of course, is that of nerve membrane breakdown. The fact that you can get impulses propagating along a nerve, which is believed to have a membrane rather similar to this, again makes you wonder if you have a purely localized impulse sent along the membrane, and perhaps here you have an energy transmitting mechanism. Who knows? It is very complicated.

I would like to say this relative to Dr. Marshak's suggestion implicating lysosomes in the radiation degradation of DNA in cells. I think you have here a situation where it is difficult to see how ionizing radiation affects the membranes which might separate the DNA and DNase, but it is equally difficult to see how ionizing radiation would cause lysosomes to break down.

So I think you can choose between them. I think they are both feasible mechanisms to talk about.

POLLARD: In conclusion I would just like to make a little remark. Dr. Platt told us of some very interesting ideas of Szilard's and I would like to put in the record just a little word or two about Dr. Szilard. His influence is much greater than his published work would suggest. As far back as 1927 he began a very perceptive investigation of the "Maxwell demon" and this resulted in essentially the beginning of the correlation between entropy and information theory. He is well known for his work on bacterial mutation and for the chemostat which he and Novick developed, but far more than these few papers and publications that he has put out he has been a very powerful influence in stimulating thought. His critical mind and his keen discussion have been definitely a strong influence on the post-war development of biophysics. I feel very sorry indeed that he has not been here with us to stimulate us further with his brilliant ideas.

13

Th. Förster
Laboratorium für physikalische Chemie

Excitation Transfer

In the last few years a large number of biologists have become interested in mechanisms which might be able to transfer quanta of energy between different molecules or between different sub-units of a macromolecular framework without the necessity of close contact between them. With such mechanisms energy might be transferred from an aromatic amino acid of a protein to a prosthetic group attached to the same protein molecule or in a chloroplast from some auxiliary pigments to a chlorophyll molecule and from one chlorophyll molecule to another. In the respiratory system energy transfer between different cytochromes might occur. Different mechanisms of such transfer have been discussed and investigated. This paper will be restricted to one of these mechanisms, namely to the transfer of electronic excitation energy. As we shall see, it is a very efficient mechanism leading to transfer over appreciable distances. By its nature, it requires the energy to be present primarily as electronic excitation energy, so that it must be considered, mainly, in connection with photo- or radiobiological reactions. In ordinary biology, electronic excitations rarely occur so that other transfer mechanisms are more likely. However, the importance of photobiology and, perhaps even more, that of radiobiology seems to justify the consideration of a mechanism which might be restricted to these fields.

The experimental background of excitation transfer has been discussed in detail within the last two years at several meetings which have been attended by many participants in this symposium. I give, therefore, only a short survey of these experiments and concentrate more on theoretical aspects. This emphasis on theory seems to be necessary because there are different theories of excitation transfer and their mutual relations are not always well understood. Furthermore, our theoretical considerations may be used to decide whether excitation transfer is to be expected in any system whether it is biological or not.

13.1. EXPERIMENTS

Most of the existing information on excitation transfer comes from experiments on sensitized fluorescence. In typical cases two kinds of molecules are present at comparatively low concentrations either in the gas phase or in liquid or solid solution. One we call the sensitizer and the other the acceptor. It may be assumed that the first excited singlet level of the acceptor is somewhat lower than that of the sensitizer. If the acceptor concentration is not too low, light absorbed by the sensitizer will then be transferred to the acceptor and emitted from that molecule. This process occurs independently of the trivial mechanism of reabsorption of sensitizer fluorescence by the acceptor; the latter process can be eliminated in such experiments. From the acceptor concentration necessary for transfer an average transfer radius may be calculated on the assumption of random molecular distribution, which is justified in most cases. More detailed discussions of this subject have been given by Livingston ('57) and by Förster ('59).

Excitation transfer between like molecules may be investigated by means of the phenomenon of concentration depolarization of fluorescence in viscous or solid media. This depolarization results from transfer between differently oriented molecules at moderate concentrations. Transfer radii up to 50 Å or more have been found in suitable cases in both kinds of experiments. Usually, the transfer radii are less between like molecules so that excitation transfer between sensitizer molecules does not essentially interfere with transfer to the acceptor.

A different situation prevails in experiments on sensitized

fluorescence with high concentration of the sensitizer, e.g., when the sensitizer is the solvent. In such cases, sensitized fluorescence often occurs by a number of transfer processes between neighboring sensitizer molecules followed by one final transfer process to an acceptor molecule. Host sensitization of impurity fluorescence in organic crystals, e.g., of naphthacene in anthracene, is a typical example. Similar systems include organic scintillators and biological systems, where high energy radiation primarily excites the bulk material and transfer occurs to other components present at lower concentration.

Instead of fluorescence emission, a chemical reaction of the acceptor may be the final result of excitation transfer. Experiments of this kind have been made with biological systems under UV irradiation (Bücher and Kaspers '47, Lautsch et al. '58).

Further evidence on excitation transfer has been derived in a less direct way from electronic absorption spectra. In concentrated solutions may dyes form dimers or polymers the absorption spectra of which are considerably different from those of the monomers. This difference can be interpreted in terms of a splitting of the original electronic levels because of resonance interaction between associated molecules. The splitting indicates the distribution of excitation among these molecules or a rapid exchange of excitation if one of them can be conceived as primarily excited. Some more recent publications on that subject are those by Lavorel ('57), Mataga ('57) and by Förster and König ('57). Resonance interaction also occurs in molecular crystals of aromatic compounds but in such case it is generally different insofar as it leads to a splitting of the vibrational levels only. Examples are given in two recent review articles by McClure ('59) and by Wolf ('59). In crystals with more than one molecule in the lattice cell the transfer rate is approximately equal to the frequency splitting.

13.2. THEORY

13.2.1. General considerations. - As has been mentioned before, different theories have been developed for these effects. Excitation transfer in molecular aggregates or in molecular crystals and the band splitting effects connected with it are generally described by Davydov's theory ('48) or by versions of it as in the papers by Simpson and Peterson ('57), Fox and Schnepp ('55), McClure ('59), Wolf ('59) and others. On the other hand, fluorescence effects connected with excitation transfer have been treated by Vavilov ('43, '50), Förster ('46, '48) and Dexter ('53) in an apparently different way. It is intended to consider here these different developments from a more general aspect and to demonstrate their mutual relations. For simplicity, our treatment will be confined mainly to the special case of two molecules of the same kind, because the extension to more general cases is more or less obvious.

In order to occur at all, excitation transfer requires some interaction between excited and unexcited molecules. More specifically, if we call these molecules M and M' and denote electronic excitation by an asterisk (*), this interaction is one between the two configurations M* M' and M M'*. Neglecting exchange interactions, this interaction energy is of the type

$$u = \int\int \psi_M^*(x)\, \psi_{M'}(x')\, H(x,x')\, \psi_M(x)\, \psi_{M'}^*(x')\, dx\, dx' \qquad \text{XIII. 1.}$$

where x and x' represent the electronic coordinates, ψ_M etc. the molecular electronic eigenfunctions and H(x, x') the interaction operator. By expansion into a series with decreasing powers of the intermolecular distance R, neglecting less important terms, the interaction energy can be represented by a sum of multipole interactions between the transition charge distributions of both molecules which may be calculated from the products $\psi_M(x)\psi_M^*(x)$ and $\psi_{M'}(x)\psi_{M'}^*(x)$ of excited and unexcited wave functions of the same molecules.

In Figure 1, such charge distributions are represented for the transitions from the ground states to different excited states of some aromatic hydrocarbons. These charge distributions have dipole moments of largely varying magnitude or even higher moments only. Depending on the character of these transition moments, the interaction energy decreases with a certain power of the reciprocal intermolecular distance R and depends also on the mutual orientation of both molecules. At considerable distances, only dipole-dipole interaction will be strong.

In aromatic hydrocarbons, only the transitions to higher excited states have large dipole moments. The weakly absorbing ^1L_b states, which are the lowest ones in benzene, naphthalene and many other hydrocarbons have only small, or even no, transition dipoles and will interact, therefore, only weakly with each other. Mixing with other states, due to vibrational perturbations, may however enlarge this interaction to some extent. Large dipole moments in the lowest transition are characteristic of typical dye molecules, which are, therefore, especially suited for the demonstration of strong interaction or of wide range excitation transfer.

Triplet states are not considered here, because for molecules with singlet ground states their transition moments are extremely small, resulting only from some mixing with closely lying singlet states. Moderate interaction is possible only at very small distances, where the electronic clouds of both molecules begin to overlap, so that electron exchange is possible. Nevertheless, excitation transfer can occur at moderate distances during the long lifetime of such triplet states in rigid media, as has been observed by Terenin and Ermolaev ('52, '56).

The interaction between two excited configurations as defined before, allows a non-radiative transition from one to another, if both have approximately the same energy or are in resonance with each other. For this reason, that kind of excitation transfer is often called resonance transfer or transfer by inductive resonance. The sharpness of this resonance, or of the energy equality of the two configurations participating in the transfer, is determined by the magnitude of the interaction energy. Therefore, the detailed mechanism of transfer depends on this magnitude in relation to some inherent energy parameters of the molecules themselves. These parameters are represented schematically in Figure 2. The largest one is the Franck-Condon energy $\Delta\varepsilon$ which results from the difference between nuclear equilibrium configurations of ground and of excited states

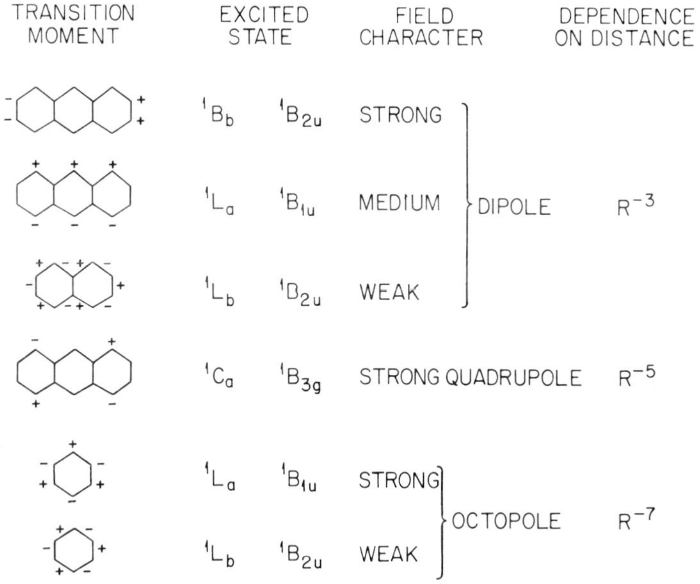

Fig. XIII. 1. Transition charge distributions and transition moments of different excited states of some aromatic hydrocarbons.

and which determines the width of the entire band system of the isolated molecule in absorption or in emission as well as the Stokes shift between the two kind of spectra. A wave number of 3000 cm^{-1} may be regarded as typical for aromatic molecules. Another quantity, which is somewhat smaller, is the energy of vibrational quanta (e. g., about 1000 cm^{-1} for highly excited vibrations in aromatic compounds) but which may for our present rough considerations be regarded as of the same order of magnitude. Another essential quantity which is distinctly smaller is the width $\Delta\varepsilon'$ of a single vibronic level. In liquid or solid systems, where molecular rotations are more or less suppressed, $\Delta\varepsilon'$ should be determined mainly by the redistribution of vibrational energy, which restricts the lifetimes of the individual vibronic levels. In crystals, this quantity can be

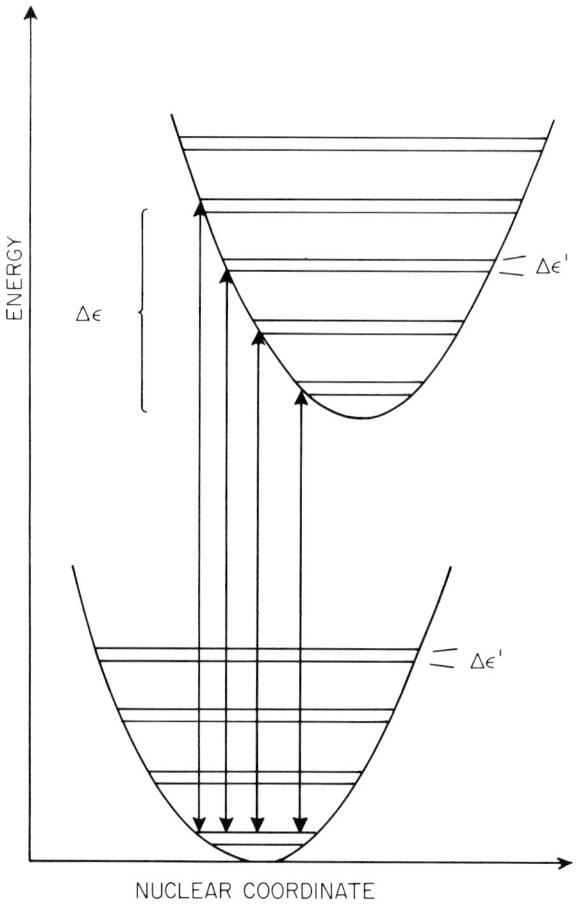

Fig. XIII. 2. Illustration of the quantities $\Delta\varepsilon$ and $\Delta\varepsilon'$.

interpreted as an effect of lattice vibrations. A value of about 10 cm^{-1} (corresponding to a vibrational lifetime of 3×10^{-12} sec) may be considered as typical for any condensed system near room temperature.

Depending on intermolecular distances and on the multipole character of interaction, the interaction energy u may vary considerably. We expect, therefore, three different cases which are defined by the following inequalities and will be designated as the cases of strong, medium and weak interaction

a	$\lvert u \rvert >> \Delta\varepsilon >> \Delta\varepsilon'$	strong
b	$\Delta\varepsilon >> \lvert u \rvert >> \Delta\varepsilon'$	medium
c	$\Delta\varepsilon >> \Delta\varepsilon' >> \lvert u \rvert$	weak

Essentially the same classification has already been used in an early paper by Franck and Teller ('38), in which systems belonging to cases a, b and c were named exciton scatterers, phonon scatterers and absorbers, respectively. All later work on excitation transfer may be traced back to the qualitative statements made in that paper. The original Davidov ('48) theory is concerned with case a; on the other hand Fox and Schnepp ('55) and Simpson and Peterson ('57) have discussed cases a and b.* Case c has been treated in detail by Förster ('46, '48) and by Dexter ('53).

13.2.2. Strong interaction. - In case a, the interaction between both excited configurations M*M' and MM'* is so strong that all vibronic sublevels of the first one are in resonance with those of the other. We may, therefore, neglect the vibrational quantization entirely and describe the system by the following time dependent wave function.

$$\Psi(x, x', t) = c(t)\psi_M{}^*(x)\psi_{M'}(x') + c'(t)\psi_M(x)\psi_{M'}{}^*(x') \qquad \text{XIII. 2.}$$

With suitable normalization, the squares of the coefficients $c^2(t)$ and $c'^2(t)$ describe the probabilities of finding M or M' in their excited states at a certain time t. For primary excitation of M, a simple calculation (cf. Hellmann '37 or Bohm '51) leads to the following results:

*Simpson and Peterson ('57) who did not consider case c have called our case b the weak coupling case.

$$c^2(t) = \cos \frac{2\pi|u|t}{h} \; ; \; c'^2(t) = \sin \frac{2\pi|u|t}{h} \qquad \text{XIII. 3.}$$

This corresponds to an oscillation of excitation forth and back between both molecules, in which M' is excited exclusively the first time at $t = \frac{h}{4|u|}$. We may consider the reciprocal value of that time as the transfer rate of this process:

$$n^{(a)} = \frac{4|u|}{h} \qquad \text{(case } \underline{a}\text{)} \qquad \text{XIII. 4.}$$

Some restrictions are necessary, however, because the transfer is oscillatory and, therefore, different from a kinetic first order process. Furthermore, because of the speed of such transfer, it seems impossible to excite one molecule alone. Instead, the system attains either one of its two stationary states, in which the excitation is distributed equally between both molecules. The nuclear equilibrium positions are, then, intermediate between those of ground and excited states in the single molecule. For systems consisting of a greater number of molecules a description in terms of excitons of the Heller-Marcus type ('51) is adequate. Electronic transitions in such extended systems are not accompanied by vibrational excitation, because the nuclear equilibrium positions of the individual molecules remain nearly unchanged. The width of the electronic band systems is then no longer determined by vibrational excitation and may be extremely small, if the symmetry of the arrangement allows transitions only to one of the borders of the exciton band. This circumstance has been recognized for crystals by Frenkel ('31) and Peierls ('32). In the related case of dye polymers it has been discussed further by Franck and Teller ('38), Förster ('46) and McRae and Kasha ('58).

Independent of the size of the system the complete alteration of absorption and fluorescence spectra is a characteristic feature of case a interaction. Examples for this case are the visible absorption regions of dye dimers and polymers and the strong absorption regions of molecular crystals like that of the ^1B_b transition of anthracene (Bree and Lyons '56, Craig and Hobbins '55).

13.2.3. Medium interaction. - In systems with weaker interaction, belonging to case b, the transfer must be slower. However, the transfer rate decreases even more than the interaction energy because the transfer mechanism becomes different. Following excitation, the nuclei perform several vibrations around their new equilibrium positions before transfer is possible. As the interaction energy is less than the vibrational separations, resonance exists only between individual vibronic levels of both molecules, i.e., between such configurations as $M_v^* M'_{v'}$ and $M_{v'} M'^*_v$ where v and v' denote vibrational quantum numbers. The transfer rate can be calculated as in case a, but one must keep in mind that the interaction energy is no longer u, but only the matrix element $u_{vv',v'v}$ of the interaction operator belonging to these vibronic eigenfunctions. With the usual separation of the molecular eigenfunctions ψ_M etc. into their electronic and vibrational parts, the matrix element can be written as follows (cf. Simpson and Peterson '57).

$$u_{vv',v'v} = u \cdot S_{vv'}^2 \qquad \text{XIII. 5.}$$

The $S_{vv'}$ are product integrals of the vibrational eigenfunctions v of the excited and v' of the unexcited molecule and satisfy the relationships

$$S_{vv'} = S_{v'v} \quad \text{and} \quad \sum_v S_{vv'}^2 = \sum_{v'} S_{vv'}^2 = 1 \qquad \text{XIII. 6.}$$

The separation of the eigenfunctions into electronic and vibrational parts restricts the validity of our following considerations to some extent, because it implies the neglect of vibrational induced transition moments. Our treatment seems, however, safe not only for states with strong dipole transitions but also for others which are not too close to other states with larger transition moments or with such of lower order.

By applying the same considerations as in case a, the

individual transfer rate $n_{vv'}$ is now calculated as

$$n_{vv'}^{(b)} = \frac{4|u|S_{vv'}^2}{h} \qquad \text{(case } \underline{b}\text{)} \qquad \text{XIII. 7.}$$

Although in this case the transfer rate depends on the vibrational quantum numbers of both molecules, we may calculate an average transfer rate by summing up the individual transfer rates $n_{vv'}$ each of them multiplied by the factors g_v^* and $g_{v'}$, representing the prevailing population of the vibrational level v in the excited molecule and of the level v' in the unexcited one. Thus we obtain

$$n^{(b)} = \sum_{v,v'} g_v^* \, g_{v'} \, n_{vv'} = \frac{4|u|}{h} \sum_{v,v'} g_v^* \, g_{v'} \, S_{vv'}^2$$

$$\text{(case } \underline{b}\text{)} \qquad \text{XIII. 8.}$$

According to their definition, the g_v^* and $g_{v'}$ satisfy the relations

$$\sum_v g_v^* = \sum_{v'} g_{v'} = 1 \qquad \text{XIII. 9.}$$

From these relations together with the corresponding relations in equation 6 it is easily derived that the double sum in equation 8 is always less than unity, so that the transfer rate is less than that calculated for case a in equation 4.

One arrives at a simple interpretation of the double sum in equation 8 if one reconsiders the fact that the $S_{vv'}^2$ together with the g_v^* and $g_{v'}$ also determine the relative intensities of the individual bands in an allowed electronic transition. With this fact in mind, the double sum is equal to the probability (normalized to unity) that, under the prevailing distribution of vibrational levels, an allowed emission transition to the ground state occurs just to a vibrational level which is occupied in the ground state of an unexcited molecule. This equality allows an estimation of the double sum from spectral data in individual cases. The $g_{v'}$ represent the vibrational population of an unexcited molecule and correspond, therefore, to a Boltzmann distribution of the prevailing temperature. This situation is not necessarily true for the g_v^*. If the molecule M* has been excited directly by an absorption process, the distribution g_v^* will depend on the spectrum of the exciting light. If it has been excited indirectly by a sequence of transfer processes, thermal equilibrium may have been established so that the g_v^*, also, represent a Boltzmann distribution. At very low temperature, only the zero vibrational levels of both molecules will then be occupied, so that the transfer rate for these latter stages is given by

$$n^{(b)} = \frac{4|u|}{h} S_{00}^2$$

This rate may be very small because the 0-0 bands have only low intensity in molecules of moderate size. A decrease in transfer rate is, therefore, expected in case b at low temperature. This effect might be the reason for the inhibition of fluorescence sensitization in anthracene crystals containing naphthacene at low temperature such as has been reported by Lyons and White ('58). Multiple transfer between anthracene molecules can be suspected to occur in this system before the final transfer to a naphthacene molecule occurs.

Under the conditions of case b, absorption and fluorescence spectra of the system show vibrational structure. This structure is similar to that of the single molecule, but may display additional splitting. The weaker transitions in molecular crystals, like ^1L_b of benzene or ^1L_a of anthracene, are typical examples (cf. McClure '59 or Wolf '59). From the appearance of vibronic bands with higher quantum numbers one may conclude, that in such systems the excitation can not be distributed over a large number of molecules. Instead of such distributions, the excitation should be approximately localized, because an excited molecule must perform several vibrations before it can transfer.

13.2.4. Weak interaction. - With weak interaction (our case c), where the interaction energy is less than the vibrational band width, only small regions of corresponding bands are in resonance with each other. The transfer must be treated now as a non-radiative transition between two configurations with continuous energies. The quantum mechanical treatment of this case (cf. Pauli '33 or Hellmann '37) leads to an individual transfer rate* of

$$n^{(c)}_{vv'} = \frac{4\pi^2 u^2_{vv',v'v}}{h\Delta\varepsilon'} = \frac{4\pi^2 u^2 S^4_{vv'}}{h\Delta\varepsilon'} \qquad \text{(case c)} \quad \text{XIII. 10.}$$

Contrary to the foregoing cases a and b, the transfer rate is now proportional to the square of the interaction energy matrix element and decreases, therefore, more rapidly with the mutual distance R between both molecules. It also decreases with increasing vibrational band width $\Delta\varepsilon'$, i. e., with increasing deviation from exact resonance. The transfer rates calculated from equation 7 for case b and from equation 10 for case c would become equal for $\pi^2|u|S^2_{vv'} = \Delta\varepsilon'$; i. e., near the border between these two cases. In the case of weak interaction the redistribution of vibrational energy occurs more rapidly than excitation transfer, so that it is justified (more than in case b) to calculate the total transfer rate by summing up the individual transfer rates of equation 10. The result is

$$n^{(c)} = \frac{4\pi^2 u^2}{h\,\Delta\varepsilon'} \sum_{v,v'} g^*_v\, g_{v'}\, S^4_{vv'} \qquad \text{(case c)} \quad \text{XIII. 11.}$$

---

*Strictly speaking, the transfer rate calculated here has a meaning different from that in cases a and b, because in case c the transfer is no longer oscillatory. This difference, however, is not essential for our present considerations.

Here both g_V^* and $g_{V'}$ are Boltzmann distributions so that $n^{(c)}$ does not depend on the kind of primary excitation. From the relations 6 and 9 it is easily concluded, that the double sum in equation 11 is less than unity and even less than the double sum in equation 8.

Furthermore, the double sum in equation 11 can be interpreted as follows. It represents the probability (again normalized to unity) that under thermal equilibrium conditions an allowed emission transition in an excited molecule happens to occur between the same vibrational levels as an allowed absorption transition in an unexcited molecule. Thus, it represents the mutual overlap between the fluorescence and absorption spectra of the same molecule. For continuous spectra, the double sum in equation 11, with inclusion of the factor $1/\Delta\varepsilon'$, degenerates into an overlap integral over the suitably normalized fluorescence and absorption spectra. From more general considerations the equivalent of equation 11 has been derived for dipole-dipole interaction by Förster ('48) and for more general cases by Dexter ('53). In both cases, the treatment has been extended to the transfer between different molecules.

The equivalence of both treatments can be demonstrated as follows. For dipole-dipole interaction, the interaction energy must be specified as:

$$u = \frac{\kappa \, |\mathfrak{M}|^2}{n^2 R^3} \qquad \text{XIII. 12.}$$

Here $\mathfrak{M}$ is the transition moment between ground and excited state, κ is a numerical factor which depends on the mutual orientation of both molecules and n^2 is the square of the refractive index of the surrounding medium (which replaces the dielectric constant for this kind of interaction). Because the transfer rate is proportional to u^2, it decreases with the 6th power of R; we may, therefore, define a critical transfer distance R_0 by

$$n^{(c)} = \frac{1}{\tau} \left(\frac{R_0}{R}\right)^6 \qquad \text{XIII. 13.}$$

τ is the lifetime of the excited state, so that at $R = R_0$ there are equal probabilities for transfer and for internal deactivation of the primarily excited molecule by a radiative or non-radiative process. From equations 11, 12 and 13 the following expression for R_0 results.

$$R_0^6 = \frac{4\pi^2\kappa^2\tau\,|\mathfrak{M}|^4}{n^4\,h\,\Delta\varepsilon'} \sum_{v,v'} g_v^* \, g_{v'} \, S_{vv'}^4 \qquad \text{XIII. 14.}$$

If we suppose, for our rough consideration, that the absorption and emission bands are concentrated at an average wavelength λ, the electronic transition moment $\mathfrak{M}$ can be expressed by the radiative lifetime τ_e of the molecule by the well known relation

$$\frac{1}{\tau_e} \sim \frac{64\pi^4\, n\, |\mathfrak{M}|^2}{3\,h\,\lambda^3} \qquad \text{XIII. 15.}$$

If we consider the following relation between τ, τ_e and the fluorescence quantum yield η

$$\tau = \eta\,\tau_e \qquad \text{XIII. 16.}$$

and further replace the double sum in equation 14 by

$$\frac{h}{\Delta\varepsilon'} \sum_{v,v'} g_v^* \, g_{v'} \, S_{vv'}^4 = \frac{1}{\Delta\nu} \qquad \text{XIII. 17.}$$

we arrive at

$$R_0 \sim \frac{\lambda}{2\pi n} \sqrt[6]{\frac{9}{16} \; \frac{\kappa^2\,\eta^2}{\tau\,\Delta\nu}} \qquad \text{XIII. 18.}$$

For not too viscous solvents where both molecules are in rapid Brownian rotation, the average value of κ^2, which according to an earlier calculation (Förster '51) is 2/3, must be used so that our final result is

$$R_0 \sim \frac{\lambda}{2\pi n} \sqrt[6]{\frac{3}{8} \frac{\eta^2}{\tau \Delta\nu}} \qquad \text{XIII. 19.}$$

With $1/\Delta\nu$ representing the overlap integral of the absorption and fluorescence spectra, both normalized to unity on a frequency scale, this approximate result has already been derived for molecules with continuous spectra (Förster '48). From equation 17, together with the normalization relations 6 and 9 it is easy to see that $1/\Delta\nu$, as defined here, is essentially the same overlap integral. In a similar way it can be demonstrated that equation 11 is exactly equivalent to the more rigorous expressions which have been derived previously for excitation transfer under dipole-dipole interaction between molecules with continuous spectra.

In this case of weak interaction, the interaction leads to no recognizable alterations at all in the absorption or fluorescence spectra because all splitting effects are less than the vibrational band width. Furthermore, the transfer rate is independent of the exciting wavelength even in the primary step, because it occurs after the establishment of thermal equilibrium in the excited molecule. The decrease in transfer rate between like molecules at low temperatures should be even stronger than in case b, because the $S_{vv'}$ appears in equation 11 to the fourth power instead of the second. There is no doubt, that the excitation is temporarily entirely localized in one single molecule.

13.3. NUMERICAL COMPARISON AND CONCLUSIONS

In order to compare the transfer rates in our three cases of strong, medium and weak interaction, these have been calculated with the numerical values

$$\sum_{v,v'} g_v^* g_{v'} S_{vv'}^2 = 0.1; \quad \sum_{v,v'} g_v^* g_{v'} S_{vv'}^4 = 0.01$$

$$\Delta\varepsilon = 3000 \text{ cm}^{-1}; \qquad \Delta\varepsilon' = 10 \text{ cm}^{-1}$$

which may be fairly realistic. The results are depicted in

Figure 3 on logarithmic scales. The straight lines corresponding to equations 4, 8 and 11 have been connected more or less deliberately in the intermediate regions, but the general picture should not be too far from reality.

For a given pair of molecules, the interaction energy is a function of the distance R as well as of their mutual orientation. We consider, therefore, interaction energies of the following kind:

$$u = \frac{a^2e^2}{R^3} \qquad \text{and} \qquad u = \frac{a^4e^2}{R^5}$$

where a is a length of molecular dimension and e the electronic charge. Neglecting angular dependence, these functions represent dipole-dipole and quadrupole-quadrupole interactions. With a = 1 Å both may be regarded as typical for strong interactions of that kind (for a wave length of about 3000 Å, this value corresponds to an oscillator strength of unity in the dipole case). With these functions and with the same values as used in Figure 3, the rates are depicted in Figure 4. In addition to the distance, a molar concentration scale is used according to the relation

$$c = \frac{3000}{4\pi NR^3} = \left(\frac{7.35}{R(\text{Å})}\right)^3 \text{ mole liter}^{-1}$$

N is Avogadro's number; c represents that concentration for which, on the average, a sphere of the radius R contains one single molecule. For random distribution, there is a probability of $1 - 1/e$ for every molecule to find its nearest neighbor within that distance.

The lifetimes of short lived excited states, in the absence of essential internal quenching, are of the order of 10^{-8} sec so that the minimum necessary transfer rate is about 10^8 sec^{-1}. Necessarily, this transfer must be of the slow interaction type <u>c</u>. As may be seen from Figure 4, such a transfer is possible over distances of the order of 100 Å, or at concentrations of about 10^{-3} molar, with strong dipole interaction. With increasing concentration the transfer be-

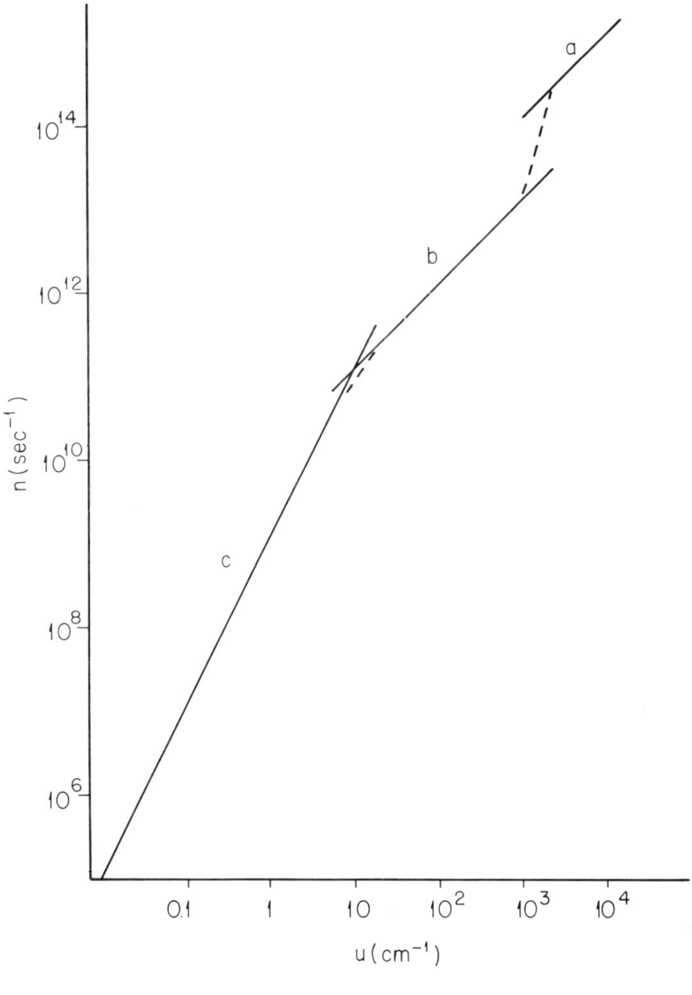

Fig. XIII. 3. Rate n of excitation transfer as a function of the interaction energy u in cases a, b and c. Broken lines represent intermediate cases.

comes faster, but between 10^{-2} and 10^{-1} molar it goes over into type b transfer.

With quadrupole-quadrupole interaction the radiative transfer is necessarily forbidden, so that the lifetime may

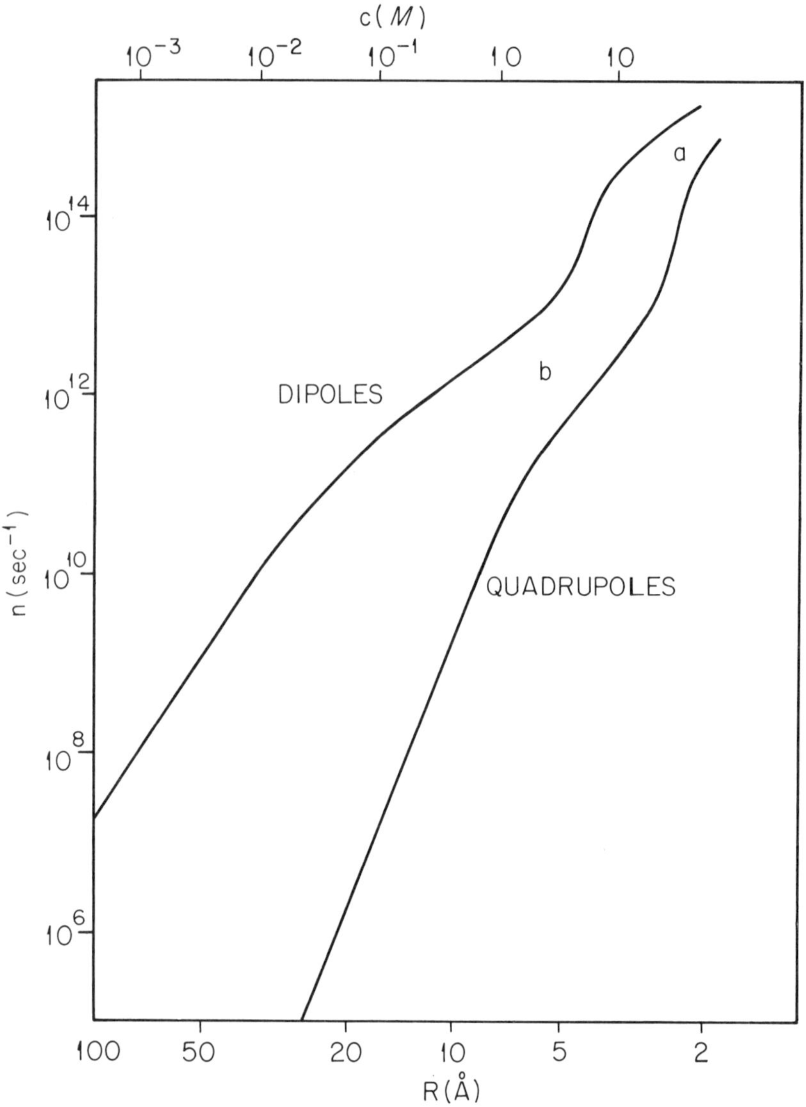

Fig. XIII. 4. Rate n of excitation transfer as a function of the intermolecular distance R and of the molar concentration c for random molecular distribution. Calculated for dipole-dipole and quadrupole-quadrupole interaction.

be essentially longer, say 10^{-5} sec (Kasha and Nauman '49, have measured a lifetime of that magnitude for naphthalene, where the lowest excited singlet state is ^1L_b). However, even so, quadrupole interaction allows transfer only for short distances, corresponding to concentrations of about 10^{-2} molar, and the conditions are still worse for higher multipole interaction. Nevertheless, even then rapid transfer is possible in pure liquids or solids, where the concentrations are high and the transfer rates according to Figure 4 depend less on the kind of interaction.

The preceding theoretical developments are by no means complete and might need much improvement for quantitative applications, especially near the borderlines of our cases of strong, medium and weak interaction and for vibrationally induced transitions. Furthermore, an exact treatment of multistep processes in these different cases would seem necessary. My main purpose in this presentation, however, is not to present a complete new theory, but to demonstrate the existing relations between different current theories of excitation transfer and to propose some classification of typical cases. Such classification should be useful - even without complete theoretical treatment - for an understanding of the processes which undoubtedly occur in biological systems as well as in other ones.

DISCUSSION

PLATT: At this conference there are a number of people who have worked on problems of energy transfer in simple systems. I think, that, in connection with Dr. Förster's paper, they may want to discuss some issues which have arisen in this field about which those outside the field may be uninformed.

One of these issues is the question of a molecule to molecule picture of energy transfer versus a collective picture. Dr. Forster's picture is a picture of energy transfer through space over distances which are of the order of 50 Å from molecule M to molecule M'. An alternative description — which Dr. Förster has mentioned, and which is in many ways equivalent to this scheme — is a description in which one speaks of M and M' as forming a "collective system", a system in which you can never say that "the radiation has been absorbed by M and then has been transferred to M' " because such a statement is operationally meaningless.

There is no way in which you can establish which molecule M absorbed radiation and which molecule M' emitted it. The various molecules of a given species absorb and emit as a collective group over distances comparable to the electromagnetic coherence length which is of the order of $\lambda/2\pi$. In the experimental evidence for energy transfer, all we really observe is that some fluorescence, either emitted or quenched, was first absorbed by some molecular species in the solution. Dr. Förster interprets this as transfer from one particular molecule to another, over distances of the order of 50 Å, with the effectiveness of the transfer depending on the amount of overlap of the emission and absorption spectra, of the two molecules.

In a collective model, on the other hand, one has a sharing of the energy between molecules of the same species, but then one may have to have a separate kind of "sink" for the energy to go into before fluorescence takes place. Such a sink could be a foreign molecule complexed to one of the M molecules. Because it contains an M molecule, the complex is coupled to the rest of the M's to some degree but, being different, it can pick up and localize the energy in the course of non-radiative degradation. The energy can pour into the complex as into a "sink", and can then be emitted as radiation, or can possibly be degraded and quenched.

In some respects these two schemes are mathematically identical because, as Dr. Förster said, you can write down Ψ_M at M and then $\Psi_{M'}$ at M' and show that there is transfer between one Ψ_M and another $\Psi_{M'}$ at a certain rate which depends on the separation of the two energy states produced by the coupling. Yet, if one writes down the energy states of the whole system, they are represented by states of the form, $\Psi_M \pm \Psi_{M'}$, which is mathematically equivalent to the sum of the two Ψ's.

One of these schemes is like the "atomic orbital" scheme of describing a molecule; and the other is somewhat like the "molecular orbital" scheme of describing a molecule. You never say that a single atom does the absorbing or emitting in the molecular orbital scheme; and likewise one would never say that a single molecule does the absorbing or emitting in a crystal or a collective or aggregate scheme.

It is interesting to look for experiments which would distinguish between these two cases or schemes. One possible experiment would be one where all the molecule M and M' are just about 50 Å apart, or somewhat less, but in which there are no dimers or complexes present to serve as "sinks" in the collective model. The Förster model might then predict easy transfer from M to M' or to a second species N, while the collective model would predict sharing of the energy among all the M's and a poor transfer to N. We could make this situation experimentally perhaps by putting the chromophores M and M' or M and N on the ends of an inert "molecular stick" such as a saturated steroid unit. In dilute solutions, the M and M' or M and N groups would never come in contact with each other, and one could see experimentally whether energy is or is not transferred from M to M' or from M to N.

The converse experiment would be to have a very dilute solution in which every M is, say, 100 Å from every M', but in which there are M•N complexes in any region of dimensions of the order of $\lambda/2\pi$. In this case, the Förster theory would predict poor transfer from most M's to the N's, but the collective model would predict good transfer.

So it seems that it will be possible to make experiments which will tell which of these two possibilities is correct: (1) the case of collective absorption with a "sink", where Förster's conditions are violated, or (2) the case where Förster's conditions are satisfied, but where there is no dimer or complex present to serve as a sink.

AUGENSTINE: How important is the dielectric constant of the material separating the two molecules in distinguishing these two cases?

FÖRSTER: My opinion is that only the high frequency dielectric constant is relevant. It is represented by the square of the refractive index.

AUGENSTINE: The reason I ask is it might be possible to do such a crucial experiment by building up multilayers. The best material for separating the molecules M and N would be something like barium stearate, which has a reasonably high dielectric constant. One might be able to orient these molecules on barium stearate layers, or even change the overlapping if you desire.

FÖRSTER: Can you make such a multilayer without holes in it?

AUGENSTINE: Not too often, but we can try.

PLATT: A multilayer would be an excellent kind of "molecular stick". I believe there is a group at the Naval Research Laboratory that has been making multilayers of the order of 50 Å thick without any holes in them for nuclear beta ray absorption experiments. So I think it may be technically possible even though very difficult to do reliably.

The reason I made the elaborate remark on the difference between the Förster model and a possible collective model is that there are in this room several people who have acquired evidence over the last few years bearing on one or the other of these schemes. Drs. Lipsky and Burton had a scheme which was a quasi-collective model for benzene.

Dr. Reid also did some experiments a few years ago, which indicated that you have to have actual complex formation in order to have energy transfer. His experiments seemed to show that it is not sufficient to have the molecules M and N 50 Å or so apart in the solution. Dr. Kasha has a student, Dr. El Bayoumi, who has performed somewhat similar experiments; and I think Dr. Kallmann, Dr. Charlesby and Dr. Fano have other evidence on these questions. And Dr. Livingston did one of the crucial papers with Bowen and Brocklehurst on these questions.

I would like to start by asking Dr. Lipsky if he is willing to make a few remarks on this subject from his point of view.

LIPSKY: If I may, I would prefer to make my comments in the form of a question to Dr. Förster. Perhaps Dr. Förster is aware of some of the experiments in which dilute solutions of terphenyl in benzene are excited either with light or gamma rays and fluorescence from terphenyl is measured. For both types of excitation, evidence exists for transfer of electronic energy from the benzene to terphenyl. Furthermore, the addition of low concentrations of oxygen or bromobenzene has been shown to quench the terphenyl emission not by reaction with terphenyl but by interfering with the transfer process itself, presumably by reaction with excited benzene. The strange thing about all this is that the specific rates both for energy transfer to terphenyl and for quenching by oxygen are too large to be explained either by material diffusion or by the diffusion of excitation from benzene to benzene — at least within the weak coupling approximation. To explain our data we have suggested that in liquid benzene a strong coupling model is applicable — not necessarily over the entire liquid but rather in small regions of perhaps 10 to 15 molecules. The diffusion of either terphenyl or oxygen to such a region is postulated to lead either to transfer or quenching with a cross section characteristic of the size of the excited region rather than that of a single molecule. We wonder though, is it even feasible to consider strong coupling in a liquid like benzene with its small oscillator strength to the first excited state and relatively large Stokes shift?

FÖRSTER: My opinion is that excitation transfer in liquid benzene cannot occur by the way of strong coupling because the absorption spectra do not show any indication for this. But it may belong to the medium-coupling case even though the excited state (^1L_b) of benzene involved belongs to an electronically forbidden transition with a very low f-value. This is because at the small distances present in the pure liquid considerable coupling results even from the interaction between the non-vanishing octopole moments of that transition. This might allow a fairly rapid excitation transfer from one benzene molecule to another and finally to a solute molecule.

In any case, I should not recommend calculation of the transfer efficiency in liquid benzene by use of the formulae published earlier because these correspond to weak coupling with dipole-dipole interaction only.

LIPSKY: Would exchange interaction contribute significantly to transfer efficiencies in these liquid hydrocarbons?

FÖRSTER: Exchange interaction certainly needs some overlap between the electronic clouds of both molecules. But for any kind of resonance transfer much less overlap is required than that usually considered in bond formation or even in contact separation of molecules.

In the present case this possibility should not be overlooked. As I have mentioned already in my foregoing talk, Terenin and Ermolaev ('52, '56) have observed that though excitation transfer radii are less than in more favorable cases, they still exceed the usual molecular separations. In order to account for this, the authors have proposed a transfer by exchange interaction which in this case seems to be very reasonable.

BURTON: But that is a slow transfer, isn't it?

FÖRSTER: It is a slow transfer, but over distances of 13 or 14 Å (Ermolaev and Terenin '58).

FANO: What does Dr. Lipsky mean that the direct transfer was not sufficient? Does he mean that the original benzene would have to shoot the excitation directly into the sink? That the excitation can't travel from benzene to benzene, a very short distance, very fast?

LIPSKY: It might be possible to explain the high specific rate for energy transfer from benzene to terphenyl by invoking a long range transfer. But then one is still left with the puzzle of why benzene loses energy about as efficiently to oxygen as it transfers it to terphenyl. A long range effect for oxygen quenching is hard to imagine. As far as benzene to benzene transfer is concerned, the difficulty here is that within the weak coupling approximation this transfer is just not fast enough.

FANO: Why?

LIPSKY: Because in benzene the oscillator strength is too low and the Stokes shift too large. These are the factors that contribute strongly in calculating the excitation jump frequency.

FANO: But you go in weak coupling over very short distances but very fast.

FÖRSTER: May I say only that the weak-coupling case cannot apply to a fast multi-stage transfer. And I agree completely with Dr. Lipsky insofar as the weak-coupling case seems to be quite improbable in the case of liquid benzene.

BURTON: There is one point that seems to have been overlooked. In general, quenching constants for solvents are 10 to 100 times as great as quenching constants for scintillator solutes. So far as I know, no one has concerned himself with this apparently anomolous situation. It seems to me that there is a very simple explanation. Scintillator solute molecules are always present in only very small concentration. They are effectively isolated; they do not exist in domains. When the solvent is otherwise pure, the solvent molecules do exist in domains. The assumption offered is that quencher molecules are effective only when they collide with an initially excited molecule or some other molecule in the domain to which it belongs. From this point of view, the ratio of quenching constants of solvent and scintillator respectively should be about the same size as the number of molecules in a domain. This view does not eliminate the possibility of other (apparently less important) types of quenching. Under certain conditions the latter can become very important (Nosworthy, Magee and Burton '60).

PLATT: I am worried about the number of mechanisms we have here. We are not going to be able to explain anything; or, more exactly, we shall be able to explain absolutely anything.

LIVINGSTON: If anyone deserves credit for our experimental proof of the occurrence of energy transfer by inductive resonance, it is E. J. Bowen. He picked out a pair of compounds which made the proof possible. The fluorescence efficiency of the donor compound, while appreciable, was much less than that of the acceptor. Using an equimolar mixture of the two substances, the total fluorescence emission of the solution increased as the total concentration was increased. This eliminates the "trivial" mechanism. The independence of the efficiency of energy transfer from the viscosity of the solvent eliminates any diffusion-controlled mechanism. We also attempted to show that mixed dimers did not form to any appreciable extent.

I would like to ask Dr. Förster about an observation on self-quenching. In the case of another of my favorite molecules, chlorophyll, self-quenching goes just as well in a highly viscous medium as it does in a non-viscous medium, as would be expected in terms of Dr. Förster's equation. In the case of self-quenching of anthracene and of substituted anthracenes, it occurs only in non-viscous solvents and stops entirely in rigid media.

FÖRSTER: Unfortunately, as you know, there are different mechanisms of self-quenching. One case is that of so-called static quenching which results from the formation of a certain amount of dimers or of higher aggregates in more concentrated solutions. If, as usually, these dimers are non-fluorescent, this leads to concentration quenching.

There is another case where no aggregation occurs, but where an excited and an unexcited molecule come together by a diffusion process. But there is also a third case which seems to be identical to that in chlorophyll solutions mentioned by Dr. Livingston. Many dyes show concentration quenching even in highly viscous solvents and without any indication of considerable aggregation. Migration of excitation over a large number of molecules, terminating at one of a few number of dimers, of impurity molecules or of other molecules with some preference for non-radiative deactivation would be the most reasonable explanation for this kind of concentration quenching.

May I make another remark. Dr. Livingston has been very modest in describing his results. He and Dr. Bowen also have demonstrated in their experiments that excitation transfer can be interpreted neither by a diffusional mechanism nor by formation of dimers. Their very careful confirmation of Beer's Law in these solutions is very essential in this connection. Actually I am used to referring to these experiments as the most reliable spectroscopical demonstrations of excitation transfer.

PLATT: There are a very few examples of such crucial experiments, where pain has been taken to exclude all the trivial cases. It is surprising that in a field as interesting as this one there have been so few crucial experiments in the last five years. Dr. Reid, would you care to make some remark about the need for molecular association?

REID: Well, I don't think that I should say very much because the work that we did is now considered old hat. It was done before the Bowen-Livingston experiment, which I think is reasonably convincing, but perhaps I should outline what the problems were.

We tried to observe excitation transfer in systems of non-polar pairs of aromatic hydrocarbons under such conditions that there would not be association under any circumstances. In fact, both molecules were associated more easily with solvent alcohol molecules than with each other. Under these conditions we were unable to find any transfer at all down to distances around I think 30 or 40 Å.

However, these are just intermediate coupling cases of aromatic hydrocarbons with each other and so this is perhaps not contradictory. The most significant thing is that as soon as we went over to systems which became slightly heterogeneous on cooling, the collective effect immediately asserted its reality and was much more easy to verify. I still feel that long range transfer in solutions — or solid solutions — is something which we know must occur because we all agree with Professor Förster's mathematics, but we have really surprisingly few experiments to back up these assertions.

DOLE: I don't know whether these results which I would like to mention can be interpreted in terms of energy transfer or not. So far there has been much talk about energy transfer, but very little discussion of its connection with radiation chemistry.

The experiments I refer to are some work we did on mixtures of polyethylene and polybutadiene. If 95 percent of a linear polyethylene is mixed with 5 percent of polybutadiene in the liquid state, and then the system allowed to crystallize, polybutadiene exerts no protective action over the polyethylene. But if the material is melted so that a presumably homogeneous system exists then the polybutadiene exerts a very marked effect on the radiation chemistry of polyethylene and we have interpreted these results as indicating that in the solid state the polybutadiene is segregated into the amorphous regions of the polyethylene, and therefore, there is no significant energy transfer from the crystal region of the polyethylene to the polybutadiene.

AUGENSTINE: The other day Kasha pointed out that in a DNA molecule with nicely hydrogen-bonded base pairs you will have good energy transfer; whereas, if you disrupt the hydrogen bonding, points of discontinuity will occur which inhibit efficient energy transfer. Dr. Förster has emphasized that for his particular scheme to work that you have to have a good "overlap" between the two molecules. I wonder if someone would explain what quantities are involved in this "overlap factor" and how this may be involved in the phenomenon Kasha described?

FÖRSTER: I would like to ask what do you mean by overlap?

AUGENSTINE: Does the overlap you speak of involve only the two spectra or is orientation of the molecules involved?

FÖRSTER: What I meant here was overlap of the absorption and fluorescence spectra of both molecules.

PLATT: You have to have overlap between the absorption of M_2 and the emission of M_1. It is this overlap which is referred to in the Förster theory. (It is the same overlap integral that you would get in radiative transfer but it does not mean that the mechanism is radiative transfer; it is radiationless.)

FÖRSTER: Thank you for this interpretation. But concerning molecular orientation, I should like to mention that in specific coupling cases — mainly for dipole-dipole coupling — the mutual orientation of molecules is also essential. Parallel orientation is favorable in any case. I have tried to describe this in my paper by the dimensionless factor κ which sometimes for very bad orientation might even be zero so that no transfer would result.

KASHA: At the beginning of this discussion Dr. Platt made an analogy which I think may be worthy analytically, but may be misleading in terms of physical theories, and I would like to clarify this point. Valence-bond and molecular-orbital mathematical approximations are used by theoretical chemists to treat a single molecular problem: molecular structure. But the various aspects of exciton theory treat <u>different physical</u> phenomena. There are several theoretical coupling cases, in the exciton or energy transfer theory, and I don't think that the problem is to define a crucial experiment to show which one is valid. All three are valid and may occur in nature. In fact, there are spectroscopic criteria for distinguishing between these coupling cases which Dr. Förster has elaborated on, and which other people (Simpson, McClure, Davydov) have elaborated on in the literature. However, there may be special situations, such as in the N-heterocyclic base pairs of DNA, in which a combination of energy transfer mechanisms may be involved. In DNA there may be intra-base pair and inter-base pair energy transfers. The work which we have been doing at Florida State University has been on spectroscopy and intra-base pair energy transfer of model N-heterocyclic dimers. These experiments have been done by Mr. A. El Bayoumi, and for energy transfer studies he designed an elegant experiment. In this, he showed that it was in fact essential for efficient energy transfer to get the formation of a hydrogen-bonded dimer in the case studied (acridine-carbazole), in the dilution at which they were used.

CHARLESBY: I want to ask a simple question. You say that it is possible to transfer excitation over a distance of 50 Å from M to M_1. Is it possible to transfer the excitation another 50 Å to M_2 etc. ?

FÖRSTER: No, because the limiting value which I have given here of 50 Å means transfer during the lifetime of the excited molecule and in this case one would expect only incomplete transfer because we are just at the limit. The first molecule would then transfer to the second one with a probability of one-half, etc., so that the probability of migration via a considerable number of such large single transfer steps would be very low. Furthermore, transfer works only from higher to lower excited states, so that the next partner must have a still lower excited state, etc. The only exception is that of like molecules. In this case multistage transfer is certainly possible but also limited by the lifetime of the excited state.

RUPERT: With regard to your last remark, if there were successive transfers between like molecules, the excitation transfer would proceed more or less like diffusion or a random walk of energy between the like molecules.

FÖRSTER: In the weak-coupling case such a random walk diffusion-like process is to be expected but in the strong-coupling case something of the kind we have learned from Dr. Magee's talk. In the medium interaction case the behavior should be intermediate.

FANO: The collective excitations I was talking about the other day differ from those being discussed now in the sense that the former involve a much higher energy. So there is a big energy gap and to it corresponds a still bigger gap in our knowledge. I mean we don't know what happens between an initial large excitation and the smaller excitation that remains after substantial degradation.

There is one remark in particular I would like to make on the questions of localized versus collective excitation, or time-dependence versus time-independence. I don't think that either of them is a question to be decided by experiment or a question of point of view, such as, "Do you like to think in one way or do you like to think in another?"

In any specific physical situation, one of the two alternatives should obtain.

This is a question that came up in private conversation. Typically when you have an ionic crystal excited directly by electromagnetic radiation, a free exciton results, which clearly constitutes a collective mode of excitation. On the other hand, in the weak-coupling case, where each molecule interacts more closely with vibrations than with the electrons of other molecules, the collective picture is inappropriate and one may visualize a time dependent migration of energy. This is a problem of coherence. Coherence has the feature that in principle there should not be any difficulty but any single practical question offers a trap. So it is a question to be discussed not as a general principle but on specific application.

FÖRSTER: This problem is actually a very difficult one even in the weak-coupling case where it can be treated approximately by analogy to a diffusion process. I have tried this in an early paper ('48) concerning the problem of concentration depolarization. A somewhat better approximation has been used very recently by Ore ('59) and another paper to appear in the same journal. It is very complicated to calculate multiple transfer processes correctly for a random distribution of molecules in solutions.

MAGEE: It would be appropriate to say a word about how the linear model of my paper ties in with the present discussion.

Let me say first that it came as something of a surprise to me when I worked out this model to find that the excitation is propagated with constant velocity. I had thought that excitation migration was diffusion-like, and that the region involved would increase as the square root of times. As it turns out, the constant velocity of excitation propagation is proportional to the absolute value of the coupling constant.

I would also like to say that I agree with Dr. Fano very strongly that it is frustrating to think about problems in general, because there is very little you can do in general, and unfortunately specific problems are very hard to shape up. It might be worthwhile to say a few things about my thoughts on the relation of energy transfer to radiation chemistry although there is very little rigor involved and as yet I do not know how to set up definite problems to solve. In the first place, I think that the excitations that are most important in radiation chemistry are initially very high electronic states. They have various types of coupling with their surroundings and due to the nature of excitation they are general distributed, i.e., non-localized, and I always think of this using a collective model, as a matter of fact. These highly excited electronic states are certainly very, very rapidly converted into states which have much lower energy and the latter are generally like the states that are excited by photochemical means. There are two important considerations now.

First, the correlation functions for the positions at which the lower states are formed are important. The several excited electronic states formed from one highly excited state must be located quite close together. This situation is the result of the nature of the interactions, which must be between neighboring molecules. Although the initial highly excited state was formed as a distributed state because of its mechanism of formation, the lower excited states must form essentially in neighboring molecules.

The second important matter is the rate at which the several excited states run away from the site of their formation. The time-dependent treatment of my paper was intended to apply to this problem. Of course, the spreading competes with conversion into other forms, dissociations, etc.

I would also point out that in addition to the problem that Dr. Fano was talking about, I really think an equally important problem is the one of the lifetime of states. It is not only the radiative lifetime that Dr. Förster was talking about several times, but the various conversion processes.

KASHA: Since Dr. Fano asked a question about the status of theoretical literature, I thought I should point out that probably between Förster's and Magee's papers, most of the literature has been cited, except for the very extensive work done by A. S. Davydov ('48) of which only one paper has appeared translated into English. There are 24 or 25 papers over the last 10 years, plus a book which is on the exciton model applied to many different molecular situations. This book is being published in English this summer. Its exact title is "The Theory of Molecular Excitons", McGraw-Hill 1960, translated by Oppenheimer and Kasha.

I think probably some survey could be made of what important ideas for radiation chemistry have come out of his work. A large part of his work has been on crystal spectroscopy but definitely a very substantial part has been on general exciton theory, especially involving this kind of problem of the coupling cases and the time of transfer.

LINSCHITZ: I just wanted to make one comment that might perhaps clarify a difference between these two ways of looking at the problem. One can indeed speak of a clear-cut localization of excitation energy at the instant of excitation and a definite time lag in the energy transfer between an excited and unexcited molecule which comes about because of the time needed to establish resonance. The original molecule may have to undergo some change in shape in its excited state, to establish the resonance condition for transfer. Under these conditions, there will be no change in its absorption spectrum due to proximity of the "receiver" molecule. This is the type of resonance that comes into the Förster theory. Although these two ways of putting down the mathematics may look alike, nonetheless the physical meaning of the resonance is quite different. In the second case, one cannot say at the beginning where the energy goes because there is no need for any re-distribution of the nuclei to bring about resonance. The second case does require that the absorption spectrum change with concentration. There is a very clearcut difference in these two mechanisms.

FANO: Why does the second involve a smaller shift?

LINSCHITZ: Well, there is some ambiguity to be sure as the shift gets smaller, but it is quite possible to say that one can localize these things into two generally different classes, depending on how much the absorption spectrum of M will change with concentration. If there is a strong Stokes shift in fluorescence, you can still maintain a high probability of energy transfer without any change in the spectrum of the system as a function of concentration.

PLATT: Do you need to use retarded potentials for this discussion? Is the electrical field transmitted with a finite velocity?

FANO: One way of looking at this question, the way I think Dr. Förster looks, is this: You can always — you can, you don't have to, but you can — always split any electromagnetic interaction into an interaction which is electrostatic and seemingly instantaneous and an interaction transported by electromagnetic radiation. In the phenomenon which Dr. Förster discusses the former intensity is almost everything.

There will always be some transfer by electromagnetic radiation. The superposition of the two interactions gives the effect of retardation.

RUPERT: Dr. Kasha mentioned earlier that some actual experiments with DNA showed a transfer of energy up and down the helical array. Can any of this general discussion be applied to that particular case?

KASHA: Well, this model of a collective exciton or in Davydov's language, the free exciton, is a mechanism that can be demonstrated by several different methods. Using the exciton calculations, it is possible to account for spectral changes (cf. McRae and Kasha '58), rotary-dispersion changes (Moffitt) as well as the hypochromism phenomenon (Tinoco '60). Any of these by themselves indicates that the exciton interaction is present. Now, as regards radiation inactivation, I quoted an experiment which was communicated to me by J. Marmur and K. Berns (Harvard University, unpublished, '60) which indicated that denatured DNA was less sensitive to ultraviolet (2537 Å) radiation inactivation than was native DNA. This work suggests that not only is the exciton interaction manifested by optical experiments, but with radiation it seems to provide a very easy path by which radiation migrates to a certain point. There are experiments (Marmur and Berns '60) where the DNA is cut into segments by sonic action, and then it is found that the ability to bring the absorbed ultraviolet energy to the "lesioning" center, the weak point, is reduced or is absent.

MARSHAK: I would like to know whether we can have some kind of a visual picture of the nature of this exciton energy transfer. Is it transfer of electromagnetic radiation? Is it a transfer of vibrational energy, or both types?

KASHA: It involves an interaction of electronically excited states of molecules, in all the cases which are pertinent to radiation chemistry and biology. A simple account of some aspects of exciton theory has been published (Kasha '59) in the Biophysical Science Study Program Symposium.

SWANSON: May I, as a biologist, ask how energy can be held either to be removed and dissipated later without biological damage or changed or translated into damage or changed at some future time? I am speaking in terms of biological time where minutes or hours instead of microseconds are involved.

The time gap of which I speak is a very large one, I realize, but I cannot make the transition from what you were talking about physically and chemically and what we see happening at a biological level at a much later time.

PLATT: It is a tremendous gap. I am glad you mentioned it. Sometimes biologists talk about things happening "instantaneously" in cells; and then it is found that what they mean is "within 30 seconds" or "within 5 seconds". These are times a million to a million million times longer than the lifetimes of excited states and the energy transfer times we are talking about here. In most cases at room temperature, energy cannot be held for such long times in what the physicist calls an excited state or a triplet state. It can only be held in the form of induced alteration in chemical structure, i. e., in a chemically altered molecule.

FANO: It is held in some sort of metastable state.

PLATT: But it is a chemical metastable state, not an optical metastable state. Optical metastable states in solutions at room temperature all decay in time of the order of 10^{-5} seconds. Only chemically metastable states can last longer than that.

MARSHAK: You are referring then to the kind of energy transfer that Gordon is talking about? He has some kind of state that is produced by ionizing radiation, that will remain in some detectable condition for some time.

PLATT: Such a chemical state may be a free radical as Dr. Burton says. It is not an optical excitation. It is a new chemical species.

KALLMANN: But you have states which live for 100 days or so. Why don't you call these optical states?

PLATT: Because they are definable chemical species, such as peroxides for example.

KALLMANN: No.

PLATT: You mean in inter-stellar space?

KALLMANN: Yes.

PLATT: Well, if anybody here works in inter-stellar space this may be important to them.

KALLMANN: Many states you can get live for a half hour.

PLATT: They are not optical states in my book.

KALLMANN: Don't you call a singlet an optical state?

FÖRSTER: How long?

KALLMANN: Ten minutes.

PLATT: Perhaps you are talking about cases like oxygen molecule which is in a triplet ground state all the time?

KALLMANN: No, in the laboratory you can get triplet states which live very long.

BURTON: What kind of substance?

KALLMANN: Naphthalene.

PLATT: Excited naphthalene has a lifetime of a few seconds at the temperature of liquid helium.

KALLMANN: No, no. In liquid nitrogen, if isolated in a good environment, you can get still longer.

PLATT: Naphthalene at liquid nitrogen temperature has a lifetime of its forbidden triplet which is of the order of 5 - 10 seconds; this is perfectly true.

HENDRICKS: What about the case discussed by Kaplan, of so-called active nitrogen emission?

PLATT: Do you mean in inter-stellar space?

HENDRICKS: No, in the laboratory.

PLATT: Active nitrogen is now supposed to consist of nitrogen atoms; the atoms carry around the energy of possible recombination until they hit other nitrogen atoms or other atoms. But I call this a separate chemical species and not an excited state of N_2.

LIPSKY: I believe that the phosphorescence of certain dyes adsorbed on silica gel have lifetimes of the order of a few tenths of a second even at room temperature. Trypaflavine is a good example. The room temperature emission has most of the characteristics of what G. N. Lewis called an alpha emission and presumably has its origin in a triplet state.

PLATT: I had always supposed that was in identifiable species such as a free radical.

KASHA: Temperature is not the important parameter determining the lifetime of an excited state. Long intrinsic lifetimes are observed in many molecules if a high viscosity medium is used. One of the longest observed mean lifetimes observed for a molecular triplet → singlet emission is 15 seconds (triphenylene, EPA, 77°K, McClure).

I would say that probably 100 seconds might be possible for some spin-forbidden dipole transition, but there are electronic transitions that are magnetic dipole or electric quadrupole in mechanism which may have longer mean lifetimes.

PLATT: The trouble is not the optical radiation lifetime but the speed of various competing quenching processes at room temperature in solution.

KASHA: I am saying that if the other processes which shorten the mean lifetime are absent, then one could observe the very long intrinsic mean lifetime of say, triplets, but I would not say 30 minutes could be possible.

KALLMANN: I know somebody has observed lifetimes of 30 minutes.

KASHA: I don't believe molecular triplet states would have a mean lifetime as long as that. I would say about 100 seconds might be the longest.

RUPERT: Perhaps this is only saying the obvious, but I think it's evident that Dr. Swanson's gap isn't being closed here. At this stage of the game, at least, there is a kind of uncertainty principle operating. We want to understand a living thing, but it's too complicated. So we try to simplify matters by concentrating on some isolated little piece of it, and if we're fortunate this can be analyzed in beautiful detail. However, we've left out so much in the simplification process that it isn't clear how our findings apply to the original living thing. Sometimes the better we can analyze, the less relevant the analysis may be to what we were originally studying.

PLATT: So we proceed step by step.

KASHA: Well, the direct answer to Swanson is "No". No, there is no connection between the exciton processes, the slowest of which occurs in 10^{-6} seconds (molecule to molecule transfer) and long-time storage of damaged sites in any system. Those longer lived species, whether they are free radicals, or chemically altered species, are not directly involved in the fast mechanism for transferring energy. The exciton or energy transfer processes are just a way to get energy from one place to another fast, and the slowest one has 10^{-6} seconds for its mean life.

HENDRICKS: I would like to ask a question about energy transfer bearing on the formation of anthocyanins. In the discussion of control of plant growth by phytochrome, evidence was advanced that anthocyanin formation depends upon transfer of energy from excited phytochrome to some other molecule. Do I understand correctly that energy transfer to this other molecule depends upon its having an absorption spectrum overlapping that of phytochrome?

FÖRSTER: Yes.

HENDRICKS: I have been inclined to explain the energy transfer by a collision mechanism involving molecular contact. In the resonance transfer, is contact necessary or can the transfer take place over an appreciable distance?

What sort of quantum efficiencies could be anticipated for energy transfer if the molar absorption coefficients of the two molecules are of the order of 10^3 or 10^4 in the overlapping region?

FÖRSTER: Undoubtedly, the transfer efficiency may amount to unity if the actual distance is sufficiently below the critical transfer distance given in my paper. By analogy to this quantity one might be able to define another transfer distance for (practically) complete transfer which then would depend in a similar way on absorption intensity overlap, etc.

I should like to make a final comment on the question raised by Dr. Swanson: Certainly excitation transfer by itself cannot store energy but it may lead to the same result by bringing the energy to a place where a chemical process, perhaps the formation of radicals, can take place. In this way, excitation transfer might facilitate the storing of energy.

KASHA: (communicated): It is worthwhile to emphasize the importance of the novel contribution of Förster's paper, since it removes an obscurity from the field of inter-molecular energy transfer.

Up until recently an apparent conflict of criteria has existed in the energy transfer process as treated by free excitons (Frenkel, Davydov, etc.) for like molecules, and the energy transfer process treated by the "resonance transfer" phenomenological theory (Förster, Perrin, etc.). The multiplication of names for the second process (sensitized fluorescence, inductive resonance, dipole transfer) added to the confusion.

In his first theoretical paper in 1948 on the theory of resonance transfer, Förster closed with the comment that it was to be looked upon as a "localized exciton" theory.

Now, by applying the Simpson and Peterson criteria for strong, intermediate, and weak coupling, Förster has been able to demonstrate an independent derivation of his resonance transfer equation by the use of localized exciton theory (excitons with vibrational relaxation faster than rate of energy transfer).

14

H. Kallmann
New York University

Energy-Transfer Processes

The discussion following Förster's presentation referred several times to a molecular or atomic picture of energy transfer on the one hand, and on the other hand to a collective picture in which the excitation energy is spread over a whole group of molecules. There seems to be no indication that for systems without order, such as plastics and liquids, these two viewpoints lead to different formulas for the dependence of energy transfer on various parameters. The non-collective model will be used in our approach to the problem. This is reasonable because the interaction is very weak or the dilution of the molecules in question rather high in all systems treated. Many questions raised by the discussion can be answered by suitable experiments. Some of these will be treated in this contribution.

The particular problem to be dealt with is the way in which energy is transferred to an acceptor when the bulk material is excited. Three ways have been discussed (Förster '59; Brown, Furst and Kallmann '58, '59a): the excitation energy reaches the accepting molecules by material diffusion, by migration of excitons or by direct jumps over larger distances according to the theory of Förster ('51). In rigid systems such as plastics the first process is excluded and discrimination between only the last two is

*Supported by the U. S. Army Signal Supply Agency, Fort Monmouth, N. J., under Contract DA 36-039 SC-75043 and by the Wright Air Development Center, Wright-Patterson Air Force Base, Ohio, under Contract AF 33(616)-6119.

required.

The diffusion and migration processes in liquid media lead to the following formula for the fluorescence, I, of the acceptor at concentration, C_A, when it is energized from the bulk material.

$$I = F(C_A) \cdot \frac{1}{(R + C_A)} \qquad \text{XIV. 1.}$$

where

$$F(C_A) = \frac{PC_A}{Q + C_A}$$

$1/(R + C_A)$ describes the concentration quenching of the light emitting acceptor while the first factor describes the energy transfer. The process in which energy is transferred by jumps leads to a different formula (Dexter '53). The actual formula is probably more complicated and may be a combination of the diffusion-migration and the jump mechanisms. It would be of value in order to clarify this situation to determine more accurately the first factor in equation 1, describing the transfer, particularly at high concentration. This determination is difficult with liquids, using the common methods, because of the second factor, but is not so serious a difficulty in a rigid material since practically no concentration quenching occurs in such materials.

Another way to get information about the transfer function $F(C_A)$ has been begun in this laboratory. The well known method of adding quenchers to the solution and measuring the degree of quenching is used. This method is modified to give information about $F(C_A)$ in certain cases. It has been pointed out that quenching is connected to energy transfer insofar as the two processes, material diffusion and migration, can lead to quenching as well as to transfer of energy to the acceptor (Lipsky and Burton '59). Only the third process (jumps) is limited to energy transfer since in most cases there are no resonances in the quencher molecules to induce transfer to the quencher.

14.1. THEORETICAL CONSIDERATIONS

If N_0 excited states are produced instantaneously in the bulk material and, after time t, N remain in the bulk material, N changes according to the well known equation:

$$\frac{dN}{dt} = -N\left(\frac{1}{\tau_B} + \frac{1}{\tau_A} + \frac{1}{\tau_D}\right) \qquad \text{XIV. 2.}$$

$\frac{1}{\tau_B}$ is the probability of decay of N in the bulk material.

$\frac{1}{\tau_A}$ is the average probability of transfer of excitation energy from bulk to acceptor. (The diffusion and migration theory yields $\frac{1}{\tau_A} = \alpha C_A$. The jump theory gives a more complicated expression for $\frac{1}{\tau_A}$ as a function of the acceptor concentration.) $\frac{1}{\tau_D}$ is the average probability that the excitation of a bulk molecule is dissipated by an added quencher of concentration C_D. (Diffusion or migration of excitation energy leads to a linear dependence on C_D so that $\frac{1}{\tau_D} = \delta C_D$ has the same form as the probability of energy transfer to the acceptor by these processes.)

There is still another way in which quenching can occur, which is particularly important when diffusion and migration of energy are insignificant, as for instance in certain rigid media. Only when the excitation energy happens to be within the sphere of influence of the quencher is it dissipated; but then almost immediately, i. e., before anything else can happen to it. The last assumption is, of course, approximate. Under this assumption the equation for $\frac{dN}{dt}$ would be the same as that without quencher, namely

$$\frac{dN}{dt} = -N\left(\frac{1}{\tau_B} + \frac{1}{\tau_A}\right)$$

but with the important difference that from the N_0 excited molecules created by the exciting radiation only the fraction $N_0 [1 - \exp(-V_D C_D)]$ remains with V_D the volume of the quencher sphere of influence.

For the diffusion and migration case the quenching is given by the following equation:

$$\frac{I_0}{I} - 1 = \tau_B C_D \delta \frac{N_A(\infty) - N_A(C_A)}{N_A(\infty)} \qquad \text{XIV. 3.}$$

I_0 is the light intensity without added quencher, I the intensity obtained with quencher concentration C_D. I and I_0 are still functions of the acceptor concentration C_A. What is actually measured is the ratio I_0/I. $N_A(C_A)$ is the total number of excited acceptor molecules produced at acceptor concentration C_A without quencher, and $N_A(\infty)$ is the respective number for infinite acceptor concentration. It is the expression

$$\frac{N_A(\infty) - N_A(C_A)}{N_A(\infty)}$$

which one wishes to determine. This factor with $\frac{1}{\tau_A} = \alpha C_A$ is just $\frac{1}{Q + C}$ according to our theory; then equation 3 is equivalent to the expression of Lipsky and Burton ('59). Formula 3 makes it possible to determine whether our simple theory of energy diffusion, or migration, holds or whether refinements are to be made. Another advantage of this formula is that it allows the determination of $N_A(\infty) - N_A(C_A)$ directly, undisturbed by concentration quenching. When the fluorescent light emission is measured, $N_A(C_A)$ multiplied by a function of C_A describing the solute concentration quenching is obtained. Furthermore, formula 3 allows a determination of the difference $N_A(\infty) - N_A(C_A)$

rather accurately at high acceptor concentrations. This determination is not possible with the fluorescence method which measures $N_A(C_A)$ and $N_A(\infty)$ independently; their difference is difficult to determine at large C_A.

Under the assumption of no diffusion or migration but rather instantaneous quenching, the following formula is derived:

$$\frac{I_0}{I} - 1 = 1 - \exp(-V_D C_D) \qquad \text{XIV.4.}$$

which is independent of acceptor concentration. Equation 4 yields the same proportional dependence upon C_D as equation 3 at low quencher concentration. However, when the left side becomes larger than 1 a strong deviation from the proportionality law should occur. Measurements at high quencher concentrations, C , and $(\frac{I_0}{I} - 1)$ approaching 50 have shown proportionality for various solvents and solutes. This result shows that the diffusion or migration mechanism holds for quenching. Lipsky and Burton ('59) obtained the same results for their lower C_D values. It may be added that in all these cases acceptor quenching is negligibly small compared to bulk quenching.

14.2. DISCUSSION OF EXPERIMENTS IN LIQUID SOLUTIONS

The bulk systems investigated were xylene, 1-methylnaphthalene, and both solvents diluted seven to three with tri-n-butylphosphate. A number of well known fluorescent solutes were used. The formulas will be applied to some special cases. Lipsky and Burton ('59) have already stated that the quenching parameter δ is of the same order of magnitude as the energy transfer constant α. This approximate equality indicates that transfer to the acceptor and to the quencher occur via the same process. It seems to be difficult to account for the magnitude of α and δ by simple diffusion theory. Their values appear to indicate that transfer to the quencher or acceptor occurs via migration of excitation energy rather than by diffusion.

This view would also explain the observation that excited xylene and naphthalene molecules as transferring solvents are so much more quenched than similar molecules, for instance anthracene, observed as solutes. In the latter case, because of the great dilution, the energy cannot reach the quenching molecule by migration but only via material diffusion. If the latter is less effective than the migration mechanism, then it is understandable why molecules observed as solutes are less quenched than those observed as solvents. However, in solutions where xylene and methylnaphthalene were diluted with tri-n-butylphosphate to 30%, it was observed that energy transfer to the solute and to the quencher are only slightly changed. The quenching constant is reduced by about 15% when one goes from the pure solvent to a solvent which contains only 30% of the transferring solvent. The energy transfer parameter in equation 1 is under similar conditions changed by about 1. 5.

In order to discuss the above results the constants α and δ for the case of migration of energy in a solvent with H molecules per unit volume may be evaluated as follows. H comprises only those molecules which are capable of transferring energy. Consider N_0 excited molecules instantaneously created. $N_0(1 - e^{-VC})$ of these are immediately quenched or their excitation energy transferred to an acceptor when these processes occur instantaneously. V and C are the volume of the sphere of influence and concentration respectively of either acceptors or quenchers. For the remaining $N_0 e^{-VC}$ molecules, migration sets in. If τ is the average time required for an excited solvent molecule to jump to its nearest solvent neighbor, then this molecule migrates $\frac{\tau_B}{\tau}$ times to other solvent molecules. For each of these there is the chance of being transferred to an acceptor or quencher given by $(1 - e^{-VC})$. Therefore, the number of processes of transfer or quenching by migration is given by $N_0 e^{-VC}(1-e^{-VC})\,\tau_B/\tau$. Taking into account the fact that transfer to an acceptor or quenching in reality requires a certain time, τ_0, even if the excited solvent molecules are inside the respective sphere of influence, the probability becomes $(1 - e^{-VC})\frac{\tau}{\tau + \tau_0}$ and consequently the number U of quenching or transfer processes during the time τ_B is given by

$$U = N_0 \left[(1 - e^{-VC}) \frac{\tau}{\tau + \tau_0} + (1 - e^{-VC}) \frac{(e^{-VC}\ \tau_B}{\tau + \tau_0} \right]$$

XIV. 5.

The first term is the fraction of the excited molecules which transfer or are quenched before migration sets in. The second term is the fraction due to migration; it is the contribution during the time τ_B. (The contribution per unit time is obtained by dividing the second term of expression 5 by τ_B.) The probability of transfer $\frac{1}{\tau_A}$ or of quenching $\frac{1}{\tau_D}$ is then given by the following equation:

$$\frac{1}{\tau_A} \text{ or } \frac{1}{\tau_D} = \frac{(1 - e^{-VC})}{\tau + \tau_0} \qquad \text{XIV. 6.}$$

Since $\frac{1}{\tau_A}$ and $\frac{1}{\tau_D}$ are respectively given by αC_A and δC_D, the parameter α or δ can be determined from equation 6 after evaluating τ. The latter can be estimated from the relationship $\tau = \frac{h}{E}$, where h is Planck's constant and E is the interaction energy between an excited solvent molecule and its unexcited neighbor; $E = \mu_n{}^2/R^{n+1}$ where μ_n are the respective transition multipole moments with n = 2, 4, 8. . . for dipole, quadrupole, octapole, . . . transitions. $\frac{1}{R^3}$ is H, the total number of solvent molecules per unit volume, so that $\tau = h/\mu_n^2 H^{(n+1)/3}$. Introducing this relation in equation 6 allows the determination of α or δ as a function of H.

It is of special interest, in comparing with some experiments, to apply this formula to transfer from the dilute liquid transferring solvents discussed above and to a similar case for plastics to be described below. In such instances VC is considerably less than 1, and it directly follows from equation 6 that

$$\alpha \text{ or } \delta = \frac{V}{\tau + \tau_0} = \frac{V}{\dfrac{h}{\mu_n^2 H^{(n+1)/3}} + \tau_0} \qquad \text{XIV. 7.}$$

with V equal to the total volume of the spheres of influence for the α or δ process respectively. If τ_0, the actual transfer or quenching time, is small compared to the migration time τ, α or δ become proportional to H, $H^{5/3} \cdots$, for the dipole, quadrupole, ... transitions. For $\tau_0 >> \tau$, α and δ become practically independent of H. This latter case would mean that, even if the excited solvent molecule is close to an acceptor or quencher, transfer or quenching is unlikely. Strong transfer and quenching is, however, observed. Consequently, it is concluded that τ_0 is relatively short. (Thus, these processes occur with great probability if the excited solvent molecule is close to an acceptor or quencher.) Our results with diluted solvents can then be interpreted as an indication that material diffusion and not migration is responsible for the transport of excitation energy to the quencher and to the fluorescent solute molecule. This is at variance with the conclusion of the discussion given above and seems to show that material diffusion has been underestimated. In this regard, the calculation for rigid media given below shows that τ is even smaller than τ_B for a dilute transferring solvent. This result means that no migration occurs during the lifetime of the excited solvent molecule.

Equation 3 was also checked with respect to its dependence upon the solute concentration for all the systems investigated so far. For large C_A values considerable deviation from the expression $1/(Q + C_A)$ were observed. The reasons for these deviations have not yet been established.

14.3. EXPERIMENTS WITH RIGID MEDIA

Experiments similar to those with liquid systems have been carried out with the following rigid systems.

1. Polystyrene containing various amounts of different fluorescent solutes and various amounts of $\alpha, \alpha, \alpha, \alpha', \alpha', \alpha'$, -hexachloro-p-xylene as quencher. In liquid solution this compound acts in all respects very similarly to CCl_4.

2. Polystyrene systems as in 1 but with various amounts of naphthalene incorporated which serve as intermediate "solvent" to take over the energy from polystyrene.

3. Polyvinylnaphthalene containing various amounts of different solutes with various amounts of quencher added.

Experimental results for cases 1 and 2 have been published (Brown, Furst and Kallmann '59b) and are summarized in Table 1. Table 1 compares the fluorescence in the various plastics with those of liquid solutions and also the respective quenching.

The following conclusions can be drawn. In polystyrene, energy transport to the solute occurs with larger concentrations C_A required than in liquids. The law $PC_A/(Q + C_A)$ does not hold accurately. It is assumed that energy migration and energy transport by jumps to the solute are responsible for the strong fluorescence observed under high-energy excitation. The occurrence of migration is indicated by the strong quenching observed even in polystyrene containing considerable amounts of solute to which the energy is transferred; this quenching is too strong to be accounted for by instantaneous quenching according to equation 4. Furthermore the quenching is diminished by the addition of naphthalene; this result shows directly that the quenching is not instantaneous. In such a case a non-instantaneous quenching without migration of energy cannot account for the observed quenching of about 70%, since 70% of the polystyrene molecules are certainly not close enough to a quenching molecule. Moreover, even if this were the case, the excitation energy would be removed to some extent by energy transfer to the solute before quenching occurs. Thus, the

TABLE XIV. 1.

ENERGY TRANSFER AND QUENCHING*

Solvent	Molar Half Value Concentration $C_A 1/2$	Quencher Concentration (M) C_D
Polystyrene	0.02	0.12 quenched to 50%
Polystyrene + 0.2M Naphthalene	0.004	0.2 quenched to 70%
Polyvinylnaphthalene	0.005	0.15 quenched to 80%
Xylene	0.002	0.015 quenched to 50%

*Solute: PPO Quencher: $\alpha, \alpha, \alpha, \alpha', \alpha', \alpha'$-hexachloro-p-xylene

observed amount of quenching is a good indication that energy migrates in polystyrene.

When naphthalene is added in moderate amounts the situation is quite different. There is almost no quenching of naphthalene in this rigid system, although naphthalene is strongly quenched in liquid media. From this result it was concluded that no energy migration takes place in the naphthalene-polystyrene system (0.2 M naphthalene in polystyrene). This conclusion is in agreement with theoretical considerations. In 0.2 M naphthalene systems the interaction energy between naphthalene molecules which could bring about energy migration is of the order μ^2/R^3 where μ is the transition dipole moment. A reasonable estimate is that μ is not more than 10^{-20} in cgs units, which leads to an interaction energy of 10^{-20} erg for 0.2 M. This conclusion gives a τ of 6×10^{-7} sec which is probably longer than the lifetime τ_B of the naphthalene molecule. The result means that, in such a case, energy transport to an acceptor or a quencher is not increased by the term τ_B/τ in expression

5 and thus energy migration is negligible during time τ_B.

On the other hand one observes large energy transport to the acceptor in such systems with an acceptor concentration of only 0.01 M. Since energy migration seems to be ruled out at this naphthalene concentration, the conclusion is drawn that energy transfer to the solute occurs via jumps over distances of 10^{-7} cm according to the Förster mechanism ('51) in these polystyrene-naphthalene systems.

The next step in the investigation was to use a rigid system which contains very large concentrations of the naphthalene group. For this purpose, samples of polyvinylnaphthalene containing various amounts of fluorescent solutes and quenchers have been made.

Polyvinylnaphthalene is one of the best systems for energy transfer to the acceptor. Figure 1 describes the fluorescent light output as function of the acceptor concentration with a half-value concentration (the concentration at which half the energy is transferred to the solute) almost as low as those found in liquids. The curve can be represented fairly well by the formula $PC_A/(Q + C_A)$. The results indicate very good energy transport to the acceptor, and again the question arises as to whether it occurs via migration or via jumps or via both. A decision between these possibilities can again be made by quenching experiments, which are described in Figure 2. The important result is that the quenching is very small and almost independent of the acceptor concentration. This can be only interpreted as showing that no migration of energy takes place in this system. Thus, one comes to the conclusion that the high fluorescent light output at low acceptor concentration is due to energy-transport to the acceptor by jumps since it is the only remaining possibility.

It is interesting to see that polymers containing naphthalene or benzene groups behave so differently from liquid systems where both behave rather similarly for energy transfer; the main difference in liquids is that the naphthalene group is slightly less quenched than the benzene group. At present, no reason can be given for energy migration occurring in polystyrene but not in polyvinylnaphthalene.

There is, however, one possibility of avoiding this difficulty which should be mentioned. Thus it may be argued

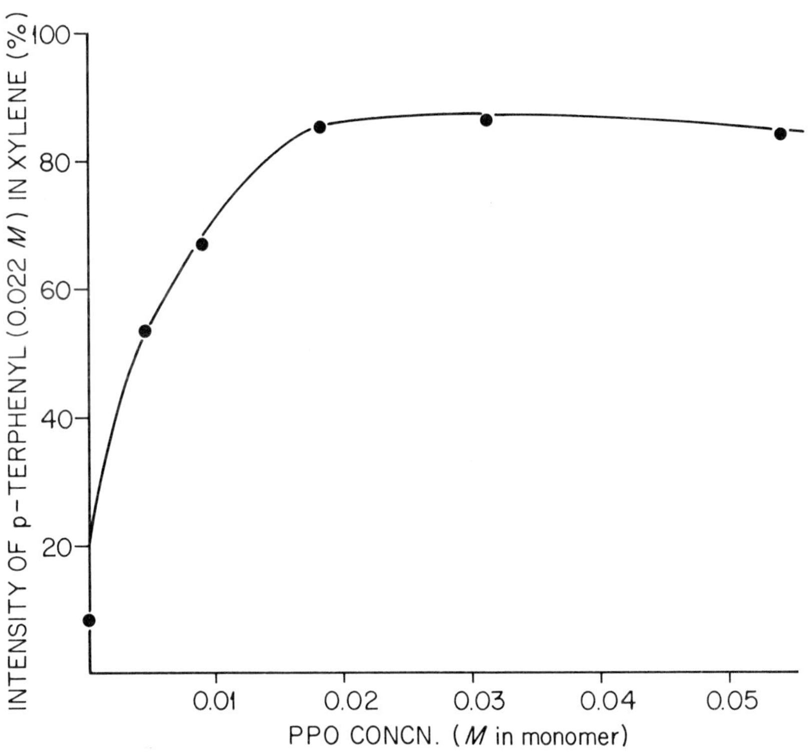

Fig. XIV. 1. Gamma-ray induced fluorescence of PVN/PPO.

that migration does not occur in polystyrene despite the above given considerations. Hexachloroxylene contains a benzene group as does polystyrene; thus an energy exchange is possible from polystyrene to the hexachloroxylene. No such exchange can occur from the naphthalene group to the hexachloroxylene since an excited naphthalene group has a smaller excitation energy than the benzene group. From this point of view, the polystyrene results can be explained by absence of migration, the quenching occurring instead by direct jumps to this quencher. When these jumps can occur over three or more atomic distances, then quenching sets in at relatively low values of C_D. In such a case quenching and energy transfer both would be due to jumps. The main

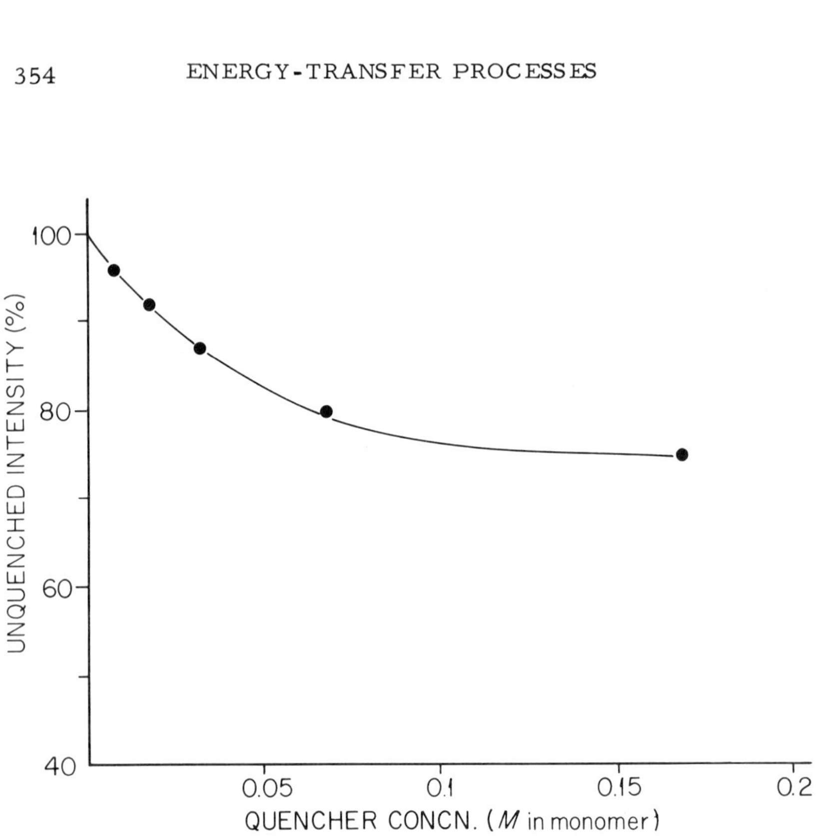

Fig. XIV. 2. Quenching of PVN/PPO (0.045 M) by *a*, *a*, *a*, *a*', *a*', *a*',-hexachloro-p-xylene.

objection to this possibility is the fact that polystyrene and hexachlorobenzene are both energetically so close that energy transfer should not take place to the quencher over large distances. The quenching by 70% of a polystyrene sample containing large amounts of solute to which the energy is also transferred would mean that energy transfer to hexachloroxylene is about half as effective as to the solute. This also seems unlikely because of the closeness of the energy levels of polystyrene and hexachloroxylene. Consequently, it is believed that migration in polystyrene does take place.

We hope this paper shows that many questions connected with transfer of excitation energy can be answered by investigating quenching effects in the systems of the type discussed, even if the quenching process itself is not connected with a resonance mechanism.

DISCUSSION

BURTON: (communicated): Professor Kallmann's interesting paper, itself a comment on Professor Förster's paper, was communicated after the Puerto Rico Conference at my personal request. One result is that there has been no opportunity for the extensive discussion which I am certain his presentation would have evoked. I would like to make the point, however, that evidence in support of migration of excitons (and against direct jumps over large distances) for certain liquid systems of considerable interest has been presented recently (Nosworthy, Magee and Burton '60, in press). According to this recent view there may, in certain cases, be two processes of interaction with quencher but only one with the scintillator solute. Furthermore, other recent work on similar systems (cf. Burton, Ghosh and Yguerabide '59) shows that the high rate constants found for excitation transfer rule out diffusion-controlled reactions between molecules of ordinary dimensions. Also, as Professor Kallmann states, the quenching operation clearly does not involve excitation "jumps". I am pleased to see that in regard to the difference between quenching constants of solvents and solutes Professor Kallmann's views accord with those I presented earlier in the discussion.

15

By the Editors

Summary

The Editors feel that the most profitable summing-up of a Conference such as this is to be found in a broad view of the entire field rather than a concise list of what we consider to be our specific accomplishments. At any time there are problems which have just been solved and new problems which have just come up. It is clear that rates of progress are always of interest in connection with assessments of status. There is also another factor which can be called "point of view", an important, undefinable thing which changes from time to time. In any case, it is clear that the historical background of the Conference is significant with respect to any summary we can make. Thus, for comparison, we consider the several conferences of the same sort with which we are familiar, the Oberlin Symposium, the several Highland Park Conferences, etc.

Even before the Oberlin Symposium there was a general feeling among physicists, chemists, and biologists that they had a research field of mutual interest. This feeling has increased and matured in the intervening ten years with the establishment of the Radiation Research Society as evidence of its sincerity.

One familiar with attempts of the various groups to discuss mutual problems has been aware of the naivite evidenced by each specialist as he leaves his field to apply his knowledge to other problems. We have the feeling that this effect is still dominating but that some progress is being made toward the development of a real appreciation of problems in other areas by a significant number of people.

During the last ten years there have been new developments in the theory of electronic energy levels of condensed systems. These developments were largely not inspired by the possibility of application to biological problems, but have their roots in the exciton theory which is thirty years old and have been made because of their importance in physics. This work is certain to have great influence in radiation chemistry and radiation biology in the next ten years. It changes many of the details of our interpretation of the interaction of high energy radiation with matter and the structure of the tracks of ionizing radiations in condensed molecular media. It furnishes mechanisms for transfer of electronic excitation to points remote from the locus of initial absorption. We believe that the most important message of Chapters 2, 5, 7 and 13 is the application of this theory.

During the last ten years there has been an impressive increase in understanding of the mechanism of radiolysis of water. The radical diffusion model has been studied experimentally and theoretically for its application in pure chemistry. An excellent summary of this work is presented in Chapter 8. There has been a very great impact of this work on the study of direct and indirect effects in radiation biology.

During the last ten years radiation chemists and physicists have increased their studies of mechanism tremendously. The recent work on mechanism in organic systems, such as presented in Chapters 9 and 11, will, in time, have application in radiationbiology. As an example of a growing interest in problems which are more definitely of immediate interest in biology, we can cite Chapter 12.

During the last ten years very impressive progress can be seen in certain areas of radiation biology, for example, in the field of photoreactivation. At the first Highland Park Conference speculations on possible mechanisms of the effect were made but at that time it was impossible to choose between mechanisms. The information presented in Chapter 4 shows how far we have progressed in this one area.

If we look back over the past ten years, it is easy to discern progress in our understanding of the action of ionizing, ultraviolet, visible and near infrared radiation. However, it is more important to look to the future. In each chapter of this volume more problems are raised than answered. It can only be hoped that the ability of physicists, chemists and biologists to cooperate will increase as time proceeds. We believe that there is some evidence that it is happening, but the situation is not clear and the pace is slow.

We hope that this volume will be useful for those scientists who are working in the area of radiation effects. In particular we hope that it will create a wider interest in some of the unsolved problems brought to light.

References

NOTE AS TO ARRANGEMENT:

The publications of a single author appear chronologically.

The publications of a multiplicity of authors appear chronologically with reference to the first author. In such case, alphabetical arrangment is not followed.

For the convenience of the reader, in the case of papers involving A. Pullman and B. Pullman in joint authorship, only chronological arrangement is followed regardless of which author appears first in the actual publication.

ALEXANDER, P., A. CHARLESBY and M. ROSS 1954. The degradation of solid polymethyl-methacrylate by ionizing radiation. Proc. Roy. Soc. A223: 392-404.

——, R. M. BLACK and A. CHARLESBY 1955. Radiation-induced changes in the structure of polyisobutylene. Proc. Roy. Soc. A232: 31-48.

—— and A. CHARLESBY 1957. Effect of x-rays and γ-rays on synthetic polymers in aqueous solution. J. Polymer Sci. 23: 355-357.

—— and D. J. TOMS 1958. Protection provided by added substances against the direct action of ionizing radiations. Radiation Research 9: 509-524.

—— and K. A. Stacey 1959. Fourth International Congress of Biochemistry. Vol. IX. Physical Chemistry of High Polymers of Biological Interest. Pergamon Press, London, p. 98.

ALLAN, J. T., E. M. HAYON and J. WEISS 1959. The chemical action of ionizing radiations in solution. XXIII. The action of Co γ-rays on aqueous solutions of ethanol in the liquid and frozen state. J. Chem. Soc. 3913-3919.

ALLEN, A. O. 1948. Radiation chemistry of aqueous solutions. J. Phys. Colloid Chem. 52: 479-490.

—— 1954. The yields of free H and OH in the irradiation of water. Radiation Research 1: 85-96.

——, C. J. HOCHANADEL, J. A. GHORMLEY and T. W. DAVIS 1952. Decomposition of water and aqueous solutions under mixed fast neutron and gamma radiation. J. Phys. Chem. 56: 575-586.

—— and H. A. SCHWARZ 1958. Decomposition of water under high-energy radiation. Proceedings of the Second United Nations International Conference on the Peaceful Uses of Atomic Energy 29: 30-37. Geneva.

ALLMAND, A. J. 1926. Einstein's law of photochemical equivalence. Trans. Faraday Soc. 21: 438-452.

ALPER, T. 1956. The modification of damage caused by primary ionization of biological targets. Radiation Research 5: 573-586.

ANDERSON, N. G. 1954. The sensitivity of DNA in crude fresh preparations to gamma and x-rays. Fed. Proc. 13: 3.

ARMSTRONG, D. H., E. COLLINSON and F. S. DAINTON 1959. Radical and molecular yields in solutions of acrylamide in light and heavy water irradiated with Co γ-rays. Trans. Faraday Soc. 55: 1375.

ARONSON, D. L., M. J. FRASER and C. L. SMITH 1956. Enzyme alteration by ionizing radiation. Radiation Research 5: 225-237.

AUGENSTINE, L. G. and B. R. RAY 1957. Trypsin monolayers at the air-water interface. II. Effect of x-ray and ultraviolet radiation upon enzymatic activity. J. Phys. Chem. 61: 1380-1384.

——— and ——— 1957. Trypsin monolayers at the air-water interface. III. Structural postulates on inactivation. J. Phys. Chem. 61: 1385-1388.

BÄCKSTRÖM, H. L. J. 1934. Der kettenmechanismus bei der autoxydation von aldehyden. Z. phys. Chem. B25: 99-121.

——— and K. SANDROS 1958. The quenching of the long-lived fluorescence of biacetyl in solution. Acta. Chem. Scand. 12: 823-832.

———, A. STENERYR and P. PERLMANN 1958. The photochemical reaction between nitric oxide and isopropanol, sensitized by benzophenone. Acta. Chem. Scand. 12: 8-23.

BACQ, Z. M. and P. ALEXANDER 1955. Fundamentals of radiobiology. Academic Press, New York.

BARBER, D. A., G. J. NEARY and R. S. RUSSELL 1957. Radiosensitivity of salt uptake in plants. Nature 180: 556-557.

BARR, N. F. and A. O. ALLEN 1959. Hydrogen atoms in the radiolysis of water. J. Phys. Chem. 63: 928-931.

——— and R. H. SCHULER 1959. The dependence of radical and molecular yields on linear energy transfer in the radiation decomposition of 0.8 N sulfuric acid solutions. J. Phys. Chem. 63: 808-812.

BARRON, E. S. G. 1954. The effect of x-rays on systems of biological importance. In, Radiation Biology. A. Hollaender Ed., Vol. I., pp. 283-314. McGraw-Hill, New York.

BARTON, D. H. R. 1959. Some photochemical rearrangements. Helv. Chim. Acta 52: 2604-2616.

BAWDEN, F. C. and A. KLECZKOWSKI 1953. The behavior of some plant viruses after exposure to ultraviolet radiation. J. Gen. Microbiol. 8: 145-156.

—— and —— 1959. Photoreactivation of nucleic acid from tobacco mosaic virus. Nature 183: 503-504.

BAXENDALE, J. H. and G. HUGHES 1958. The x-irradiation of aqueous methanol solutions, Part I. Reactions in H_2O, Part II. Reactions in D_2O. Z. phys. Chem. 14: 306-338.

BERKOWITCH, J., A. CHARLESBY and V. DESREUX 1957. Radiation effects on aqueous solutions of polyvinyl alcohol. J. Polymer Sci. 25: 490-492.

BESSMAN, M. J., I. R. LEHMAN, E. S. SIMMS and A. KORNBERG 1958. Enzymatic synthesis of deoxyribonucleic acid. II. General properties of the reaction. J. Biol. Chem. 233: 171-177.

BETHE, H. A. 1933. Quantenmechanik der Ein- und Zwei-Electronenprobleme. Handbuch der Physik Vol. 24, 1. pp. 273-560. See pp. 503-504.

BEUKERS, R., J. IJLSTRA and W. BERENDS 1958. The effect of ultraviolet light on some components of the nucleic acids. II. In rapidly frozen solutions. Rec. trav. chim. 77: 729-732.

——, —— and —— 1959. The effects of ultraviolet light on some components of the nucleic acids. III. Apurinic acid. Rec. trav. chim. 78: 247-251.

BILLEN, D. 1957. Modification of the release of cellular constituents by irradiated Escherichia coli. Arch. Biochem. Biophys. 67: 333-340.

BIONDI, M. A. and R. E. FOX 1958. Dissociative attachment of electrons in iodine. III. Discussion. Phys. Rev. 109: 2012-2014.

BIRTH, G.S. 1960. Dual-monochromator spectrophotometer for agricultural applications. Amer. Soc. Agr. Eng. (in press).

BOHM, D. 1951. Quantum Theory, Prentice-Hall, Inc., New York. Chapter 19: Degenerate Perturbations.

BORSTEL: see VON BORSTEL, R. C.

BORTHWICK, H. A., M. W. PARKER and S. B. HENDRICKS 1948. Wave length dependence and the nature of photoperiodism. Lotsya 1: 71-78.

———, S. B. HENDRICKS, M. W. PARKER, E. H. TOOLE and V. K. TOOLE 1952. A reversible photoreaction controlling seed germination. Proc. Natl. Acad. Sci. 38: 662-666.

———, ———, E. H. TOOLE and V. K. TOOLE 1954. Action of light on lettuce seed germination. Botan. Gaz. 115: 205-225.

———, ——— and M. W. PARKER 1956. Photoperiodism. In, Radiation Biology, A. Hollaender, Ed. Vol. III, pp. 479-517. McGraw-Hill, New York.

BOUBY, L. and A. CHAPIRO 1955. Radiolyse des solutions diluees de diphenylpicrylhydrazyle (DPPH) dans les solvents organiques. J. chim. phys. 52: 645-652.

BOVEY, F. A. 1958. Effects of ionizing radiation on natural and synthetic high polymers. Interscience, New York.

BRANDT, C. L. and A. C. GIESE 1956. Photoreversal of nuclear and cytoplasmic effects of short wave length ultraviolet radiation on Paramecium caudatum. J. Gen. Physiol. 39: 735-751.

BREALEY, G. J. and M. KASHA 1955. The role of hydrogen bonding in the $n \rightarrow \Pi^*$ blue-shift phenomenon. J. Am. Chem. Soc. 77: 4462-4468.

BREE, A. and L. E. LYONS 1956. The intensity of ultraviolet light absorption by monocrystals. Part II. Absorption and reflection by anthracene of plane-polarized light. J. Chem. Soc., 2662.

BRIDGE, N.K. and G. PORTER 1958. Primary photoprocesses in quinones and dyes. 1. Spectroscopic detection of intermediates. Proc. Roy. Soc. A244: 259-275.

BROWN, F.H., M. FURST and H. KALLMANN 1958. Transfert d'energie dans les systemes organiques liquides et rigides. J. chim. phys. 55: 688.

——, —— and —— 1959a. Light and high-energy induced energy transfer in liquid and rigid organic scintillators. Disc. Faraday Soc. 27: 43-56.

——, —— and —— 1959b. The processes involved in the improvement of plastic scintillators. In, Nuclear Electronics, pp. 15-26. International Atomic Energy Agency, Vienna.

BROWN, L.O. and N. MILLER 1955. On the light emitted during the irradiation of aqueous media with α-particles. Trans. Faraday Soc. 51: 1623-1632.

BUCHANAN, J.M., J.G. FLAKS, S.C. HARTMAN, B. LEVENBERG, L.N. LUKENS and L. WARREN 1957. The enzymatic synthesis of inosinic acid de novo. In, The Chemistry and Biology of Purines. A Ciba Foundation Symposium. London, 1957. pp. 233-252.

BUCHEL'NIKOVA, N.S. 1958. Effective capture cross-section of slow electrons by some halogen-containing molecules, oxygen and water. Zhur. Eksptl. i. Teoret. Fiz. 35: 1119-1130.

BUCHSBAUM, R. and R.E. ZIRKLE 1949. Shrinking and swelling after alpha irradiation of various parts of large erythrocytes. Proc. Soc. Exp. Biol. Med. 72: 27-31.

BÜCHER, TH. and J. KASPERS 1947. Photochemische Spaltung des Kohlenoxydmyoglobins durch Ultraviolette Strahlung. Biochim. et Biophys. Acta 1: 21.

BÜNAU, G. von, K. C. KURIEN, W. H. TAYLOR and L. M. THEARD 1959. Summary of Proceedings of an Informal Discussion of the Radiation Chemistry of Water. University of Notre Dame Press, Notre Dame, Indiana. p. 1-45.

BUNBURY, D. L., R. R. WILLIAMS, JR. and W. H. HAMILL 1956. Diffusion and hot radical kinetics in the photolysis of ethyl iodide. J. Am. Chem. Soc. 78: 6228-6233.

BUNN, D., F. S. DAINTON, G. A. SALMON and T. J. HARDWICK 1959. The reactivity of hydroxyl radicals in aqueous solutions. Trans. Faraday Soc. 55: 1-8.

BURCH, P. R. J. 1959. A theoretical interpretation of the effect of radiation quality on yield in the ferrous and ceric sulfate dosimeters. Radiation Research 11: 481-497.

BURTON, M., J. CHANG, S. LIPSKY and M. P. REDDY 1958. Radiation protection in cyclohexane. Radiation Research 8: 203-213.

——, W. H. HAMILL and J. L. MAGEE 1958. A consideration of elementary processes in radiation chemistry. Proceedings of the Second United Nations International Conference on the Peaceful Uses of Atomic Energy. Geneva. 29: 391-398.

—— and H. DREESKAMP 1959. On luminescence decay times and their relation to mechanisms of energy transfer in radiation chemistry of liquids. Disc. Faraday Society 27: 64-73.

—— and —— 1960. Energy transfer and decay times in radiation-induced luminescence of benzene solutions. Z. Elektrochem. 64: 165.

—— and K. C. KURIEN 1959. Effects of solute concentration in radiolysis of water. J. Phys. Chem. 63: 899-903.

BUTLER, J. A. V. 1956. Effects of x-rays and radiomimetic agents on nucleic acids and nucleoproteins. In, Ionizing Radiations and Cell Metabolism. A Ciba Foundation Symposium. J. A. Churchill, Ltd. London. p. 59-69.

—— 1959. Changes induced in nucleic acids by ionizing radiations and chemicals. Radiation Research Supplement 1: 403-416.

—— and B. E. CONWAY 1953. The action of photochemically generated radicals from hydrogen peroxide on deoxyribonucleic acid and simple model substances. Proc. Roy. Soc. B141: 562-580.

BUTLER, W. L., K. H. NORRIS, H. W. SIEGELMAN and S. B. HENDRICKS 1959. Detection, assay, and preliminary purification of the pigment controlling photoresponsive development of plants. Proc. Natl. Acad Sci. 45: 1703-1708.

CAFFREY, J. M., JR. and A. O. ALLEN 1958. Radiolysis of pentane adsorbed on mineral solids. J. Phys. Chem. 62: 33-37.

CHANTRENNE, H. 1958. Nucleic acid metabolism and induced enzyme formation. Rec. trav. chim. 77: 586-601.

CHAPIRO, A. 1950. Sur la polymerisation des composes vinyliques, amorcees par les rayons gamma. J. chim. phys. 47: 747-775.

CHARLESBY, A. 1952. Crosslinking of polythene by pile radiation. Proc. Roy. Soc. A215: 187-214.

—— 1953. Effect of high-energy radiation on long-chain polymers. Nature 171: 167.

—— 1954. Molecular-weight changes in the degradation of long-chain polymers. Proc. Roy. Soc. A224: 120-128.

—— 1960. Atomic Radiation and Polymers. Pergamon Press, New York.

—— and P. ALEXANDER 1955. Reticulation des Polymeres en Solution Aqueuse par les Rayons Gamma. J. chim. phys. 52: 699-708.

——, E. von ARNIM and L. GALLAGHAN 1958. Effects of crystallinity on radiation effects in polyethylene. Intern. J. Appl. Radiation and Isotopes 3: 226-231.

——, W. H. T. DAVISON and D. G. LLOYD 1959. Intensity and concentration dependence of some radiation-induced reactions in anthracene solutions. J. Phys. Chem. 63: 970-975.

—— and D. G. LLOYD 1959. Competitive reactions in the irradiation of anthracene + cyclohexane solutions. Proc. Roy. Soc. A249: 51-64.

—— and S. H. PINNER 1959. Analysis of the solubility behavior of irradiated polyethylene and other polymers. Proc. Roy. Soc. A249: 367-386.

CHEESMAN, D. and J. T. DAVIES 1954. Physiochemical and biological aspects of proteins at interfaces. Adv. Prot. Chem. 9: 439-501.

CHRISTENSEN, E. and A. C. GIESE 1954. Changes in absorption spectra of nucleic acids and their derivatives following exposure to ozone and ultraviolet radiations. Arch. Biochem. et Biophys. 51: 208-216.

CLAYTON, R. K. 1959. Permeability barriers and the assay of catalase in intact cells. Biochim. et Biophys. Acta 36: 35-39.

COLLINSON, E. and A. J. SWALLOW 1956. The radiation chemistry of organic substances. Chem. Rev. 56: 471-568.

COOPER, W. and W. H. STAFFORD 1958. Chemical changes induced in organic systems by ionizing radiations. Proceedings of the Second United Nations International Conference on the Peaceful Uses of Atomic Energy. Geneva. 29: 115-127.

CRAIG, D. P. and P. C. HOBBINS 1955. The polarized spectrum of anthracene. Part II. Weak transition and second-order crystal field perturbations. The polarized spectrum of anthracene. Part III. The system at 3800 Å. J. Chem. Soc. 2302-2308, 2309-2319.

DAINTON, F. S. 1954. Principles and mechanisms in electron transfer reactions. Chem. Soc. Special Pub. No. 1: 18-39.

—— 1958. Radiation chemistry in retrospect and prospect. Brit. J. Radiology 31: 645-659.

—— 1959. Chemical effects of radiation. Radiation Research Supplement 1: 1-22.

DALE, W. M. 1955. Modern trends in radiation-biochemistry. In, Aspects chimiques et biologiques des radiations, Vol. I. p. 205-243. M. Haissinsky Ed., Masson et Cie, Paris.

DALY, M. M., V. G. ALLFREY and A. E. MIRSKY 1950. Purine and pyrimidine contents of some desoxypentose nucleic acids. J. Gen. Phys. 33: 497-510.

DANIELLI, J. 1958. Surface chemistry and cell membranes. In, Surface Phenomena in Chemistry and Biology. J. Danielli, K. Pankhurst, and A. Riddiford, Editors, Pergamon Press, London. pp. 246-265.

DANNO, A. J. 1958. Gel formation of aqueous solution of polyvinyl alcohol irradiated by gamma rays from Cobalt-60. Phys. Soc. Japan 13: 722-727.

DAVISON, W. H. T., S. H. PINNER and R. WORRALL 1957. Polymerization of isobutene with high-energy radiation. Chem. and Ind. (London) 1274.

——, —— and —— 1959. The radiation-induced polymerization of isobutene. Proc. Roy. Soc. A252: 187-196.

DAVYDOV, A. S. 1948. Theory of absorption spectra of molecular crystals. Zhur. Exsptl. i. Teoret. Fiz. 18: 210.

DEBYE, P. and J. EDWARDS 1952. Long-lifetime phosphorescence and the diffusion process. J. Chem. Phys. 20: 236-239.

DEWHURST, H. A., A. H. SAMUEL and J. L. MAGEE 1954. A theoretical survey of the radiation chemistry of water and aqueous solutions. Radiation Research 1: 62-84.

DEXTER, D. L. 1953. A theory of sensitized luminescence in solids. J. Chem. Phys. 21: 836.

DIBELER, V. H. and R. M. REESE 1955. Selected positive and negative ions in the mass spectra of the monohalomethanes. J. Rcs. Natl. Bur. Stand. 54: 127-134

DOERING, W., von and L. H. KNOX 1954. The cycloheptatrienylium (tropilium) ion. J. Am. Chem. Soc. 76: 3203-3206.

DOLE, M. D. C. MILNER and T. F. WILLIAMS 1958. Irradiation of polyethylene. II. Kinetics of unsaturation effects. J. Am. Chem. Soc. 80: 1580-1588.

DONALDSON, D. M. and N. MILLER 1956. The action of α-particles on solutions containing ferrous and cupric ions. Trans. Faraday Soc. 52: 652-659.

DOTY, P. 1959. Configurations of biologically important macromolecules in solution. Revs. Mod. Phys. 31: 107-117.

——, H. DOEDTKER, J. R. FRESCO, B. D. HALL and R. HASELKORN 1959. Configurational studies of polynucleotides and ribonucleic acid. Ann. New York Acad. Sciences 81: 693-708.

DULBECCO, R. 1950. Experiments on photoreactivation of bacteriophage inactivated with ultraviolet radiation. J. Bacteriol. 59: 329-347.

DYNE, P. J. and J. M. KENNEDY 1958. The kinetics of radical reactions in the tracks of fast electrons: a detailed study of the Samuel-Magee model for the radiation chemistry of water. Can. J. Chem. 36: 1518-1536.

—— and —— 1960. A detailed study of the Samuel-Magee model for the radiation chemistry of water. Part II. Kinetics of radiation reactions in the tracks of densely ionizing particles. Can. J. Chem. 38: 61-76.

EKERT, B. and R. MONIER 1959. Structure of thymine hydroperoxide produced by x-irradiation. Nature 184: BA 58-59.

EPHRUSSI, B. and H. HOTTINGUER 1951. On an unstable cell state in yeast. Cold Spring Harbor Symposium. Quant. Biol. 15: 75-85.

EPSTEIN, L. M. 1957. Electron crosslinking of polyethylenes of two different densities. J. Polymer Sci. 26: 399-401.

ERRERA, M. 1952. Etude photochimique de l'acide desoxyribonucleique. I. Mesures energetiques. Biochim. et Biophys. Acta 8: 30-37.

—— 1955. Mechanisms of biological action of ultraviolet and visible radiations. Progress in Biophys. 3: 88-130.

EULER, H. V. and R. BLIX 1919. Verstärkung der Katalässewirkung in Hefezellen. Z. Physiol. Chem. 105: 83-114.

FANO, U. 1956. Atomic theory of electromagnetic interactions in dense materials. Phys. Rev. 103: 1202-1218.

—— 1960. Normal modes of a lattice of oscillators with many resonances and dipolar coupling. Phys. Rev. 118: 451-455.

FARADAY SOCIETY DISCUSSION 1926. Photochemical reactions in liquids and gases. Trans. Faraday Soc. 21: 437-656.

FARMER, J. B., I. H. S. HENDERSON, F. P. LOSSING and D. G. H. MARSDEN 1956. Free radicals by mass spectrometry. IX. Ionization potentials of CF_3 and CCl_3 radicals and bond dissociation energies in some derivatives. J. Chem. Phys. 24: 348-352.

FELSENFELD, G. and A. RICH 1957. Studies on the formation of two- and three-stranded polyribonucleotides. Biochim. et Biophys. Acta 26: 457-468.

FIRESTONE, R. F. 1957. The radiation chemistry of water vapor. The indirect effect on deuterium and the exchange of D atoms with water molecules. J. Am. Chem. Soc. 79: 5593-5598.

FISHER, W. D., N. G. ANDERSON and K. M. WILBUR 1959. Studies on nuclei. II. Effects of x-rays on deoxyribonucleoprotein from rat thymus. Exptl. Cell Research 18: 481-493.

FLANDERS, D. A. and H. FRICKE 1958. Application of a high-speed electronic computer in diffusion kinetics. J. Chem. Phys. 28: 1126-1129.

FLINT, R. H. and E. D. MACALISTER 1937. Wave length of radiation in the visible spectrum promoting the germination of light sensitive lettuce seed. Smithsonian Inst. Misc. Coll. 96: 1-8.

FÖRSTER, TH. 1946. Energiewanderung und Fluoreszenz. Naturwiss. 33: 166.

—— 1948. Zwischenmolekulare Energiewanderung und Fluoreszenz. Ann. d. Physik. 2:55.

—— 1951. Fluoreszenz organisher Verbindungen. Vandenhoeck and Ruprecht, Göttingen, 312 pp.

—— 1959. Transfer mechanisms of electronic excitation. Disc. Faraday Soc. 27: 7.

—— and E. KÖNIG 1957. Absorptionsspektren und Fluoreszenzeigenschaften konzentrierter Lösungen organischer Farbstoffe. Z. Elektrochem. 61: 344.

FONER, S. N. and R. L. HUDSON 1953. Detection of the HO_2 radical by mass spectrometry. J. Chem. Phys. 21: 1608-1609.

FORSSBERG, A. and R. NOVAK 1958. Early growth rate and biochemical changes caused by small x-ray doses in Phycomyces Blakesleeanus. Radiation Research 9: 115-116.

FORSYTH, P. F., E. N. WEBER and R. H. SCHULER 1954. Radiation-induced reaction between iodine and heptane. J. Chem. Phys. 22: 66-70.

——, —— and —— 1955. Radical production in the radiolysis of the hydrocarbons. Radiation Research 3: 68-75.

FOX, D. and O. SCHNEPP 1955. Theory of the lower excited electronic states of the benzene crystal. J. Chem. Phys. 23: 767.

FRANCK, J. and E. TELLER 1938. Migration and photochemical action of excitation energy in crystals. J. Chem. Phys. 6: 861-872.

FRENKEL, J. 1931. The transformation of light into heat in solids. Phys. Rev. 37: 17-44, 1276-1294.

FRICKE, H., C. A. LEONE and W. LANDMANN 1957. Role of structural degradation in the loss of serological activity of ovalbumin irradiated with gamma rays. Nature 180: 1423-1425.

GARNER, W. W. and H. A. ALLARD 1920. Effect of the relative length of day and night and other factors of the environment on growth and reproduction in plants. J. Agr. Research 18: 553-606.

GATES, F. L. 1928. On nuclear derivatives and the lethal action of ultraviolet light. Science 68: 479-480.

—— 1930. A study of the bactericidal action of ultraviolet light. III. The absorption of ultraviolet light by bacteria. J. Gen. Physiol. 14: 31-42.

GHORMLEY, J. A. 1956. Lifetime of intermediates in water subjected to electron irradiation. Radiation Research 5: 247-251.

——— and A. C. STEWART 1956. Effects of γ-irradiation on ice. J. Am. Chem. Soc. 78: 2934-2939.

GOLDENBERG, H. and A. D. McLAREN 1951. Effect of ultraviolet light on the specific activity of chymotrypsin and trypsin. J. Am. Chem. Soc. 73: 1131-1133.

GOODGAL, S. H., C. S. RUPERT and R. M. HERRIOTT 1957. Photoreactivation of hemophilus influenzae transforming factor for streptomycin resistance by an extract of Escherichia coli B. In, The Chemical Basis of Heredity, W. D. McElry and H. B. Glass, Editors. The Johns Hopkins Press, Baltimore. pp. 341-343.

GORDY, W. 1959. Electron spin resonance of free radicals in irradiated biochemicals. Radiation Research Supplement 1: 491-510.

GRANICK, S. and L. MICHAELIS 1940. The radical of tri-p-tolylamine. J. Am. Chem. Soc. 62: 2241-2242.

GREENBERG, G. R. and L. JAENICKE 1957. The activation of the one-carbon unit for the biosynthesis of purine nucleotides. In, The Chemistry and Biology of Purines. A Ciba Foundation Symposium. London. pp. 204-232.

HAMILL, W. H. 1960. Direct and indirect initial effects on chemical systems. This symposium, Chapter IX.

HARDWICK, T. J. 1959. Measurement of the recombination of H and OH in γ-irradiated aqueous systems. J. Chem. Phys. 31: 226-231.

HART, E. J. 1955. Radiation chemistry of aqueous ferrous sulfate-cupric sulfate solutions. Effect of γ-rays. Radiation Research 2: 33-46.

—— 1957. Chemical effects of ionizing radiations on aqueous solutions. J. Chem. Ed. 34: 586-593.

—— 1958. Recent studies on reactions in irradiated water. Proceedings of the Second United Nations International Conference on the Peaceful Uses of Atomic Energy. Geneva. 29: 5-12.

—— 1959. Water subexcitation electrons in aqueous formic acid radiolysis. J. Am. Chem. Soc. 85: 6885.

——, S. GORDON and D. A. HUTCHISON 1952. Free radical initiated 0^{16} 0^{18}-H_2O^{16} exchange reaction in aqueous solutions. J. Am. Chem. Soc. 74: 5548.

——, W. R. McDONELL and S. GORDON 1956. The decomposition of light and heavy water boric acid solutions by nuclear reactor radiations. Proceedings of the International Conference on the Peaceful Uses of Atomic Energy. Geneva. 7: 593-598.

HAUPT, W. 1958. Hellrot-Dunkelrot-Antogonismus bei der Auslösung der Chloroplastenbewegung. Naturwiss. 45: 273-274.

HELLER, W. R. and A. MARCUS 1951. A note on the propagation of excitation in an idealized crystal. Phys. Rev. 84: 809.

HELLMANN, H. 1937. Einführung in die Quantenchemie. Deuticke, Leipzig and Vienna. pp. 76-78.

HEMS, G. 1958. Effect of ionizing radiation on aqueous solutions of guanylic acids and guanosine. Nature 181: 1721-1722.

—— and M. L. EIDINOFF 1958. Effect of x-radiation on aqueous solutions of adenosine diphosphate. Radiation Research 9: 305-311.

HENDRICKS, S. B 1960. The photoreactions controlling photoperiodism and related responses. In, Symposium on Comparative Biochemistry of Photoreactive Pigments. M. B. Allen, Ed. Academic Press.

—— and H. A. BORTHWICK 1955. Photoresponsive growth. In, Aspects of Synthesis and Order in Growth. Dorothea Rudnick, Ed. Princeton University Press. pp. 149-169.

——, —— and R. J. DOWNS 1956. Pigment conversion in the formative responses of plants to radiation. Proc. Natl. Acad. Sci. 42: 19-26.

—— and —— 1959. Photocontrol of plant development by the simultaneous excitations of two interconvertible pigments. Proc. Natl. Acad. Sci. 45: 344-349.

—— and —— 1959. Photocontrol of plant development by the simultaneous excitations of two interconvertible pigments. II. Theory and control of anthocyanin synthesis. Botan. Gaz. 120: 187-193.

HENGLEIN, A. and C. SCHNEIDER 1958. Abbau und Vernetzung des Polystyrols in Lösung durch Co-60-Gamma-Strahlen. Z. phys. Chem. 18: 56-73.

—— and —— 1958. Degradation and crosslinking of polystyrene in solution by γ-radiation. Radiation Research 9: 128.

—— and —— 1959. Die γ-Strahlenchemie des Polyisobutylens in Lösung. Z. phys. Chem. 19: 367-376.

HICKHAM, W. M. and D. BERG 1958. Negative ion formation and electric breakdown in some halogenated gases. J. Chem. Phys. 29: 517-523.

HIRSHBERG, Y. and E. FISHER 1954. Photochromism and reversible multiple internal transitions in some spiropyrans at low temperatures. Parts I and II. J. Chem. Soc. 297-303, 3129-3137.

HOCHANADEL, C. J. 1952. Effects of cobalt γ-radiation on water and aqueous solutions. J. Phys. Chem. 56: 587-594.

—— 1955. The photolysis of dilute hydrogen peroxide in the presence of dissolved hydrogen and hydrogen plus oxygen. Radiation Research 3: 234.

—— 1956. The radiation-induced reaction of hydrogen and oxygen in water at 25° C to 250° C. Proceedings of the International Conference on the Peaceful Uses of Atomic Energy. Geneva. 7: 521-525.

HODES, M. E. and E. CHARGAFF 1956. The apurinic acids from deoxyribonucleic acid fractions. Biochim. et Biophys. Acta 22: 348-360.

HORGAN, V. J. and J. S. PHILPOT 1953. Attempted estimation of organic peroxides in x-irradiated mice. Trans. Faraday Soc. 49: 324-325.

HOWARD-FLANDERS, P. 1958. Physical and chemical mechanisms in the injury of cells by ionizing radiations. Advanc. biol. med. Phys. VI: 553-603.

HUBBARD, R. 1956. Retinene isomerase. J. Gen. Physiol. 39: 935-962.

HUTCHINSON, F. 1952. Inactivation of monomolecular layers of bovine serum albumin with deuterons. Arch. Biochem. et Biophys. 41: 317-324.

—— 1954. Energy requirements for the inactivation of bovine serum albumin by radiation. Radiation Research 1: 43-52.

—— 1954. A review of the properties of films at oil-water interfaces. In, Monomolecular Layers, H. Sobotka, Ed. AAAS, Washington. pp. 161-174.

—— 1957. The distance that a radical formed by ionizing radiation can diffuse in a yeast cell. Radiation Research 7: 473-483.

—— and E. R. MOSBURG 1954. Deuteron inactivation of adsorbed monolayers of bovine serum albumin. Biochim. et Biophys. Acta 51: 436-443.

——, A. PRESTON and B. VOGEL 1957. Radiation sensitivity of enzymes in wet and dry yeast cells. Radiation Research 7: 465-472.

—— and J. ARENA 1960. Destruction of the activity of DNA in irradiated cells. Radiation Research (in press).

—— and C. NORCROSS 1960. Inactivation by ionizing radiation of Coenzyme A in various cells. Radiation Research 12: 13-19.

JAGGER, J. 1958. Photoreactivation. Bacteriol. Revs. 22: 99-142.

—— and R. LATARJET 1956. Spectra d'action de la photorestauration chez E. coli B/r. Ann. Inst. Pasteur 91: 858-873.

JAYSON, G. G., G. SCHOLES and J. WEISS 1957. Chemical action of ionizing radiations in solution. Part XX. Action of x-rays (220 kv) on ethanol in aqueous solution. J. Chem. Soc. 1358-1368.

JONES, A. S. and D. S. LATHAM 1954. Studies in the determination of nucleotide sequence in deoxypentosenucleic acids. Biochim. et Biophys. Acta 14: 438-440.

—— and —— 1956. The nucleotide sequence in deoxypentosenucleic acids. Part I. The action of mercaptoacetic acid on calf-thymus deoxyribonucleic acid. J. Chem. Soc. 2573-2578.

KAPLAN, J. G. 1955. The alteration of intracellular enzymes. I. Yeast catalase and the Euler effect. Experimental Cell Research 8: 305-328.

—— and M. J. FRASER 1952. The effect of ultraviolet radiation on monomolecular films of egg albumin. Biochim. et Biophys. Acta 9: 585-596.

KASHA, M. 1950. Characterization of electronic transitions in complex molecules. Disc. Faraday Soc. 9: 14.

—— 1960a. Paths of molecular excitation. Radiation Research Supplement II (in press).

—— 1960b. In, Symposium on light and life (McCollum Pratt Institute), Johns Hopkins University Press.

KELNER, A. 1949. Effect of visible light on the recovery of streptomyces griseus conidia from ultraviolet irradiation injury. Proc. Natl. Acad. Sci. 35: 73-79.

KIMBALL, R. F., N. GAITHER and S. M. WILSON 1959. Recovery in stationary-phase Paramecium from radiation effects leading to mutation. Proc. Natl. Acad. Sci. 45: 833-839.

KIRBY-SMITH, and M. L. RANDOLPH 1960. Production and lifetime of radiation-induced free radicals in some molecules of biological importance. Suppl. Intern. J. Radiation Biol. (in press).

KORGAONKAR, K. S. and A. M. DESAI 1959. Monolayer technique to study effects of ionizing radiation on histone. Radiation Research 11: 625-635.

KREBS, H. A. and H. L. KORNBERG 1957. A survey of energy transformations in living matter. Ergb. d. Physiol. 49: 212-298.

LAIDLER, K. J. 1954. Elementary processes in the radiation chemistry of water. J. Chem. Phys. 22: 1740-1745.

LASSETTRE, E. N. 1959. Collision cross-section studies on molecular gases and the dissociation of oxygen and water. Radiation Research Supplement I: 530-547.

LATARJET, R., H. EPHRUSSI-TAYLOR and N. REYBEYROTTE 1959. On the target size of a transforming factor based on x-ray inactivation. Radiation Research Supplement I: 417-430.

LAUTSCH, W., W. BROSER and V. GODICKE 1957. Experiments in energy transfer. Z. Naturforsch. 12b: 303.

LAVOREL, J. 1957. Influence of concentration on the absorption spectrum and the action spectrum of fluorescence of dye solution. J. Phys. Chem. 61: 1600.

LAWLEY, P. D. 1957. The relative reactivities of deoxyribonucleotides and of the bases of DNA towards alkylating agents. Biochim. et Biophys. Acta 26: 450-451.

—— 1957. The hydrolysis of methylated deoxyguanylic acid at pH 7 to yield 7-methylguanine. Proc. Chem. Soc. 290-291.

LAWTON, E. J., A. M. BRIECHE and J. S. BALWIT 1953. Irradiation of polymers by high-energy electrons. Nature 172: 76-77. London.

LEA, D. E. 1947. The actions of radiations on dilute aqueous solutions: the spatial distribution of H and OH. Brit. J. Radiol. Suppl. 1: 59.

—— 1955. Actions of radiations on living cells. 2nd Ed. Cambridge University Press, Cambridge, England.

LEFORT, M. 1955. Chimie des radiations des solutions aqueouses-aspect actuel des resultats experimentaux, in Actions Chimique et Biologiques des Radiations. Vol. I, edited by M. Haissinsky, Masson et Cie., Paris. 93-204.

LENNOX, E. S., S. E. LURIA and S. BENZER 1954. On the mechanism of photoreactivation of ultraviolet inactivated bacteriophage. Biochim. et Biophys. Acta 15: 471-474.

LERMAN, L. and L. TOLMACH 1959. Genetic transformation. II. The significance of damage to the DNA molecule. Biochim. et Biophys. Acta 33: 371-387.

LEWIS, G. N. and D. LIPKIN 1942. Reversible photochemical processes in rigid media: The dissociation of organic molecules into radicals and ions. J. Am. Chem. Soc. 64: 2801-2808.

—— and M. KASHA 1944. Phosphorescence and the triplet state. J Am. Chem. Soc. 66: 2100.

LINSCHITZ, H., M.G. BERRY and D. SCHWEITZER 1954. The identification of solvated electrons and radicals in rigid solutions of photooxidized organic molecules; recombination luminescence in organic phosphors. J. Am. Chem. Soc. 76: 5833-5839.

——, J. RENNERT and T.M. KORN 1954. Symmetrical semiquinone formation by reversible photooxidation and photoreduction. J. Am. Chem. Soc. 76: 5839-5842.

LIPSKY, S. and M. BURTON 1959. Comparison of high-energy and ultraviolet radiation-induced luminescence in liquid systems. J. Chem. Phys. 31: 1221-1226.

LIVINGSTON, Ralph 1959. Physical observation of radiation-induced intermediates. J. Chem. Ed. 36: 349-353.

—— 1957. Intermolecular transfer of electronic excitation. J. Phys. Chem. 61: 860-864.

LIVINGSTON, Robert 1955. Photochemistry. In, Radiation Biology, Vol. II. A. Hollaender, Ed. McGraw-Hill, New York. pp. 1-40.

——, H. ZELDES and E.H. TAYLOR 1954. Atomic hydrogen hyperfine structure in irradiated acids. Phys. Rev. 94: 725.

——, —— and —— 1955. Paramagnetic resonance studies of atomic hydrogen produced by ionizing radiations. Disc. Faraday Soc. 19: 166-173.

LOSSING, F., P. KEBARLE and J.B. DESOUSA 1959. Ionization potentials of alkyl and halogenated alkyl free radicals. In, Advances in Mass Spectrometry, J.D. Waldron, Ed. Pergamon Press, New York. pp. 431-441.

LUCKE, B., A.R. RICCA and A.K. PARPART 1951. Differential effects of roentgen rays on cell permeability and cell cleavage. J. Natl. Cancer Inst. 11: 1007-1130.

LYMAN, H., H. T. EPSTEIN and J. SCHIFF 1959. Ultraviolet inactivation and photoreactivation of chloroplast development in Euglena without cell death. J. Protozool. 6: 264-265.

LYONS, L. E. and J. W. WHITE 1958. Low temperature quenching of energy transfer. J. Chem. Phys. 29: 233.

McCLURE, D. S. 1959. Electronic spectra of molecules and ions in crystals. Part I. Molecular crystals. In, Solid State Physics, Vol. 8. Academic Press, New York and London.

McNULTY, W. P., JR. and F. HUTCHINSON 1954. Deuteron inactivation of dry bovine serum albumin. Arch. Biochim. et Biophys. 50: 92-103.

McRAE, E. G. and M. KASHA 1958. Enhancement of phosphorescence ability upon aggregation of dye molecules. J. Chem. Phys. 28: 721-722.

McWEENY, R. and T. E. PEACOCK 1957. The electronic structure and spectra of some nitrogen heterobenzenes. Proc. Phys. Soc. (London) A70: 41-50.

MAGEE, J. L. 1953. Mechanisms of energy degradation and chemical change: Effects of secondary electrons. In, Basic Mechanisms in Radiobiology. NRC Pub. 305, Washington, D. C. 51-ff.

——— 1953. Radiation chemistry. Ann. Rev. Nuclear Sci. 3: 171-192.

——— 1960. Elementary processes in action of ionizing radiation. This symposium, Chapter VII.

——— and M. BURTON 1951. Elementary processes in radiation chemistry. II. Negative ion formation by electron capture in neutral molecules. J. Am. Chem. Soc. 73: 523-532.

MAHLMAN, H. A. 1956. Radiation-induced nitrite formation from concentrated nitrate solutions. Oak Ridge National Laboratory Report ORNL-2208.

—— 1958. Activity concept in radiation chemistry. J. Chem. Phys. 31: 993-995.

—— and J. W. BOYLE 1958. Primary cobalt-60 radiolysis yields in heavy water. J. Am. Chem. Soc. 80: 773-774.

MANION, J. P. and M. BURTON 1952. Radiolysis of hydrocarbon mixtures. J. Phys. Chem. 56: 560-569.

MANN, M. M., H. HUSTRULID and J. T. TATE 1940. The ionization and dissociation of water vapor and ammonia by electron impact. Phys. Rev. 58: 340-347.

MARCOVICH, H. 1958. Activite radioprotectrice de la glycerine ches les bacteries. In, Les Peroxides Organiques en Radiobiologie. R. Latarjet, Ed. Masson et Cie., Paris. pp. 116-124.

MARMUR, J. and P. DOTY 1959. Heterogeneity in deoxyribonucleic acids. Nature 183: 1427-1429.

MARTON, L., L. B. LEDER and H. MENDLOWITZ 1955. Characteristic energy losses of electrons in solids. In, Advances in Electronics and Electron Physics. Vol. 7. Academic Press, New York. pp. 183-239.

——, et al. 1956. Conference on Quantum Interactions of the Free Electron. Rev. Mod. Phys. 28: 171-276.

MATAGA, N. 1957. Note on the polymerization of dyes in solution. Bull. Chem. Soc. Japan 30: 375.

MATHESON, M. S. 1958. Photochemical contributions to radiation chemistry. Proceedings of the Second United Nations International Conference on the Peaceful Uses of Atomic Energy, Geneva. 29: 385-390.

—— and B. SMALLER 1955. Paramagnetic species in gamma-irradiated ice. J. Chem. Phys. 23: 521-528.

MAZIA, D. and G. BLUMENTHAL 1948. Inactivation of enzyme-substrate films by small doses of x-rays. Proc. Natl. Acad. Sci. 34: 328-336.

—— and —— 1950. Inactivation of enzyme-substrate film by small doses of x-rays. J. Cellular Comp. Physiol. 35: Supplement 1: 171-186.

MEAD, J. F. 1952. The irradiation-induced autoxidation of linoleic acid. Science 115: 470-472.

MERYMAN, H. T. 1957. Physical limitations of the rapid freezing method. Proc. Roy. Soc. B147: 452-459.

MESHITSUKA, G. and M. BURTON 1959. Hydrogen iodide production in radiolysis of organic liquids containing dissolved iodine: cyclohexane, n-hexane, benzene. Radiation Research 10: 499-506.

MICHAELIS, L., M. P. SCHUBERT and S. GRANICK 1939. The free radicals of the type of Wurster's salts. J. Am. Chem. Soc. 61: 1981-1992.

MILLER, N. 1957. Effects of ionizing radiation on aqueous media. Reviews of Pure and Appl. Chem. (Australia) 7: 123-154.

MITCHELL, J. S. and E. K. RIDEAL 1938. Photochemical reactions in monolayers. Proc. Roy. Soc. (London) 167: 342-367.

MOH, C. C. and R. B. WITHROW 1958. Non-ionizing radiant energy as an agent in altering the incidence of x-ray induced chromatid aberrations. Radiation Research 10: 13-19; 11: 18-25.

MOHR, H. 1956. Die Beeinflussung der Keimung von Farnsporen durch Licht und andere Faktoren. Planta (Berlin) 46: 534-551.

MULLIKEN, R.S. 1935. Electronic structures of polyatomic molecules. VII. Ammonia and water type molecules and their derivatives. J. Chem. Phys. 3: 506-514.

NAKAYAMA, S. 1958. Studies on the dark process in the photoperiodic response of Pharbitis seedlings. Scientific Report, Tohoku Univ. (Japan) 24: 137-183.

NIIRA, K. 1952. On the abnormal rotation of OH radicals. J. Phys. Soc., Japan 7: 193-199.

NOVICK, A. and L. SZILARD 1949. Experiments on light reactivation of ultraviolet inactivated bacteria. Proc. Natl. Acad. Sci. 35: 591-600.

OKADA, S. and G. FLETCHER 1960. Radiation action on DNase in dry and hydrated states. Radiation Research (in press).

OLDENBERG, O. and H. A. FROST 1936. Kinetics of OH radicals as determined by their absorption spectrum. I. The electric discharge through water vapor. J. Chem. Phys. 4: 642-648.

—— and F. F. RIECHE 1939. Kinetics of OH radicals as determined by their absorption spectrum. V. A spectroscopic determination of a rate constant. J. Chem. Phys. 7: 485-492.

OSTER, G. 1956. Crosslinking of polyethylene with selective wave lengths of ultraviolet light. J. Polymer Sci. 22: 185.

—— and O. SHIBATA 1957. Graft copolymer of polyacrylamide and natural rubber produced by means of ultraviolet light. J. Polymer Sci. 26: 233-234.

——, G. K. OSTER and H. MOROSON 1959. Ultraviolet induced crosslinking and grafting of solid high polymers. J. Polymer Sci. 34: 671-684.

PARISER, R. 1956. Theory of the electronic spectra and structure of the polydecenes and of alternant hydrocarbons. J.

Chem. Phys. 24: 250-268.

PARKER, M. W., S. B. HENDRICKS, H. A. BORTHWICK and N. J. SCULLY 1946. Action spectra for photoperiodic control of floral initiation in short-day plants. Botan. Gaz. 108: 1-26.

PARR, R. G. and R. PARISER 1955. On the electronic structure and spectra of ethylene-like molecules. J. Chem. Phys. 23: 711-725.

PATHAK, M. 1958. Biochemistry of furocoumarins. M. S. Thesis, University of Oregon Medical School, Portland, Oregon.

PAULI, W. 1933. Die allgemeinen Prinzipien der Wellenmechanik. Handbuch der Physik. 2nd ed., Vol. XXIV/1: p. 160. Julius Springer, Berlin.

PEIERLS, R. 1932. Zur Theorie der Absorptionsspektren fester Körper. Ann. d. Physik. 13: 905-952.

PIERCE, S. and A. C. GIESE 1957. Photoreversal of ultraviolet injury to frog and crab nerves. J. Cell. Comp. Physiol. 49: 303-317.

PIETTE, L. H., R. C. REMPEL and H. E. WEAVER 1959. EPR studies of electron irradiated ice and solid hydrogen. J. Chem. Phys. 30: 1623-1624.

PINES, DAVID 1955. Electron interaction in metals. In, Solid Physics. Vol. 1. Academic Press, New York. pp. 367-450.

PITTMAN, D., B. RANGANATHAN and F. WILSON 1959. Photoreactivation studies in yeasts. II. Photoreactivation of the ultraviolet damage producing respiratory deficiency in haploid and tetraploid yeasts. Exptl. Cell. Res. 17: 368-377.

——, J. WEBB, A. ROSHANMANESH and L. COKER 1960. Evidence for the genetic control of photoreactivation. Genetics (in press).

PLATZMAN, R. L. 1953. Energy transfer from secondary electrons to matter. In, Basic Mechanisms in Radiobiology. NRC Publ. 305, Washington, D. C. 22-ff.

—— 1955. Subexcitation electrons. Radiation Research 2: 1-7.

—— 1958. The physical and chemical basis of mechanisms in radiation biology. In, Radiation Biology and Medicine. Ed. by Claus, Addison-Wesley. pp. 15-72.

PLOTNIKOW, J. 1920. Allgemeine Photochemie. Vereinigung Wiss., Berlin. p. 607.

POLLARD, E. C. 1959. Radiation inactivation of nucleic acids and phage particles. Rev. Mod. Phys. 31: 273-281.

——, W. F. POWELL and S. H. REAUNE 1952. The physical inactivation of invertase. Proc. Natl. Acad. Sci. 38: 173-180.

—— and N. BARRETT 1959. The radiation sensitivity and molecular weight of dry β-galactosidase. Radiation Research 11: 781-792.

PORTER, G. and E. STRACHAN 1958. The electronic spectra of benzyl. Spectrochim. Acta 12: 299.

POTTIE, R. F., R. BARKER and W. H. HAMILL 1959. Ion-molecule reaction of methyl and ethyl iodides. Radiation Research 10: 644-670.

POWERS, E. L., R. B. WEBB and C. F. EHRET 1959. An oxygen effect in dry bacterial spores and its temperature dependence. Exp. Cell. Res. 17: 550-554.

——, —— and —— 1960. Storage, transfer and utilization of energy from x-rays in dry bacterial spores. (in press).

PREISS, J. W. 1958. The localization of invertase in the yeast cell with low-voltage electrons. Arch. Biochim. et Biophys. 75: 186-195.

—— 1959. The effects of range-restricted electrons on intracellular yeast catalase. Arch. Biochim. et Biophys. 79: 261-268.

PREVOT-BERNAS, A., A. CHAPIRO, C. COUSIN, Y. LANDLER and M. MAGAT 1952. The radiolysis of some organic liquids. Disc. Faraday Soc. 12: 98-109.

PROSKURNIN, M. A. and Y. M. KOLOTYRKIN 1958. Studies in the radiation chemistry of aqueous solutions. Second United Nations International Conference on the Peaceful Uses of Atomic Energy. Geneva. 29: 52-61.

PULLMAN, B. 1959. Electronic structure, chemical reactivity and basicity of purines and pyrazolopyrimidines. J. Chem. Soc. 1621-1623.

—— and A. PULLMAN 1952. Les theories electroniques de la chimie organique. Masson et Cie, Paris.

—— and —— 1954. Sur les transformations metaboliques des hydrocarbures cancerogenes. Bull. soc. chim. (France) 21: 1097-1104.

—— and —— 1955. Electronic structure and carcinogenic activity of aromatic molecules. Adv. Cancer. Res. 3: 117-169.

—— and —— 1955. Cancerisation par les substances chimiques et structure moleculaire. Masson et Cie., Paris.

——, —— and G. BERTHIER 1958a. Recherches sur la structure electronique des composes actifs en chimiotherapie due cancer. Compt. rend. 243: 380-383.

—— and —— 1958b. Structure electronique des intermediaires fondamentaux dans la synthese de novo et la degradation metabolique des acides nucleiques. Compt. rend. 246: 1613-1616.

—— and —— 1958c. Free valence in conjugated organic molecules. Progress in Organic Chemistry. Vol. 4: 31-71.

—— and —— 1958d. Recherches sur la structure electronique des constituants cellulaires fondamentaux et des composes actifs en chimiotherapie anticancereuse. II. Les bases purique fondamentales. Bull. soc. chim. (France): 766-772.

—— and —— 1959. Elements of a general theory of enzymatic hydrolysis. Proc. Natl. Acad. Sci. 45: 1572-1582.

—— and —— 1959a. Recherches sur la structure electronique des constituants cellulaires, fondamentaux et des composes actifs en chimotherapie anticancereuse. VI. Les pyrimidines naturelles fondamentales. Bull. soc. chim. (France) 594-597.

RAUT, C. 1954. Heritable non-genic changes induced in yeast by ultraviolet light. J. Cell. Comp. Physiol. 44: 463-475.

—— and W. L. SIMPSON 1955. The effect of x-rays and of ultraviolet light of different wave lengths on the production of cytochrome deficient yeasts. Arch. Biochim. et Biophys. 57: 218-228.

RIESZ, P. and E. J. HART 1959. Absolute rate constants for H atom reactions in aqueous solutions. J. Phys. Chem. 63: 858-861.

RIGG, T., G. STEIN and J. WEISS 1952. The action of x-rays on ferrous and ferric salts in aqueous solution. Proc. Roy. Soc. (London) A211: 375-398.

ROERSCH, A., R. BEUKERS, J. IJLSTRA and W. BERENDS 1958. The effect of UV light on some components of the nucleic acids. I. Uracil, thymine. Rec. trav. chim. 77: 423-429.

ROSENBERG, A. M. 1958. Ph. D. Thesis, University of Pennsylvania. Philadelphia, Pennsylvania.

ROSENKRANZ, H. S. and A. BENDICH 1959. Studies on the effect of heat on deoxyribonucleic acid. J. Am. Chem. Soc. 81: 6255-6259.

ROSS, M. and A. CHARLESBY 1953. The effect of pile radiation on polymethyl methacrylate ("Perspex") - I. Atomics 4: (8) 189-194

ROTHSCHILD, W.G. and A.O. ALLEN 1958. Studies in the radiolysis of ferrous sulfate solutions. III. Air free solutions at high pH. Radiation Research 8: 101-110.

ROTHSTEIN, A. 1959. Biochemical and physiological changes in irradiated yeast. Radiation Research Supplement 1: 357-371.

RUPERT, C.S. 1960. Photoreactivation of transforming DNA by an enzyme from baker's yeast. J. Gen. Physiol. 43: 573-595.

——, S.H. GOODGAL, and R.M. HERRIOTT 1958. Photoreactivation in vitro of ultraviolet inactivated Hemophilus influenzae transforming factor. J. Gen. Physiol. 41: 451-471.

SAMUEL, A.H. and J.L. MAGEE 1953. Theory of radiation chemistry. II. Track effects in radiolysis of water. J. Chem. Phys. 21: 1080-1087.

SCHOLES, G., G. STEIN and J. WEISS 1949. Action of x-rays on nucleic acids. Nature 164: 709-710.

—— and J. WEISS 1950. Chemical action of ionizing radiation on nucleic acids in aqueous systems. Nature 166: 640-642.

—— and —— 1959. Oxygen effects and formation of peroxides in aqueous solutions. Radiation Research Supplement 1: 177-189.

SCHULER, R.H. 1957. The effect of solutes on the radiolysis of cyclohexane. J. Phys. Chem. 61: 1472-1476.

—— and A.O. ALLEN 1956. Radiation chemistry studies with cyclotron beams of variable energy: Yields in aerated ferrous sulfate solution. J. Am. Chem. Soc. 79: 1565-1572.

SCHUMACHER, K. 1958. Über die Veränderung von HD- und ND-polyäthylen durch Röntgenbestrahlung. Kolloid-Z. 157: 16-27.

SCHWARZ, H. A. 1955. Effects of solutes on molecular yields in the radiolysis of aqueous solutions. J. Am. Chem. Soc. 77: 4960-4964.

——, J. M. CAFFREY and G. SCHOLES 1959. Radiolysis of neutral water by cyclotron produced deuterons and helium ions. J. Am. Chem. Soc. 81: 1801-1809.

SENVAR, C. B. and E. J. HART 1958. Decomposition of aqueous solutions by alpha particles. Proceedings of the Second United Nations International Conference on the Peaceful Uses of Atomic Energy. Geneva. 29: 19-23.

SETLOW, R. B. 1955. A relation between cystine content and ultraviolet sensitivity of proteins. Biochim. et Biophys. Acta 16: 444-445.

—— 1957. Action spectroscopy. In, Advances in Biological and Medical Physics. Vol. V. J. H. Lawrence and C. A. Tobias, Editors. Academic Press, New York, pp. 37-74.

——, G. WATTS and C. DOUGLAS 1959. Inactivation of proteins by vacuum ultraviolet radiation. In, Proceedings of the First National Biophysics Conference. H. Quastler and H. J. Morowitz, Editors. Yale University Press, New Haven. p. 174-183.

SHAPIRA, R. 1959. Adsorption of ribonuclease on glass. Biochemical and Biophysical Research Comm. 1: 235-237.

SHAPIRO, H. S. and E. CHARGAFF 1957. Studies on the nucleotide arrangement in deoxyribonucleic acids. I. The relationship between the production of pyrimidine nucleoside 3', 5'-diphosphates and specific features of nucletide sequence. Biochim. et Biophys. Acta 26: 596-608.

SHULTZ, A. R. and F. A. BOVEY 1956. Electron irradiation of polyacrylates. J. Polymer Sci. 22: 485-494.

SHEPPARD, C. W. and M. STEWART 1952. The direct effect of radiation on erythrocytes. J. Cellular Comp. Physiol. 39: Supplement 1: 189-215.

SHEPPARD, S. E. 1927. Antagonism of radiations in photochemical and photographic reactions. Chem. Revs. 4: 319-353.

SHUGAR, D. and K. L. WIERZCHOWSKI 1958. Photochemistry of nucleic acids, nucleic acid derivatives, and related compounds. Postepy Biochem. 4: 243-296.

SIEGEL, A., W. GINOZA and S. G. WILDMAN 1957. The early events of infection with tobacco mosaic virus nucleic acid. Virology 3: 554-559.

SIEGELMAN, H. W. and S. B. HENDRICKS 1957. Photocontrol of anthocyanin formation in turnip and red cabbage seedlings. Plant Physiol. 32: 393-398.

—— and —— 1958. Photocontrol of anthocyanin synthesis in apple skin. Plant Physiol. 33: 185-190.

SIMPSON, W. T. and D. L. PETERSON 1957. Coupling strength for resonance force transfer of electronic energy in Van der Waals solids. J. Chem. Phys. 26: 588-593.

SJOSTRAND, F. S. 1959. Fine structure of cytoplasm: the organization of membranous layers. Rev. Mod. Phys. 31: 301-318.

SKREB, Y. and M. ERRERA 1957. Action des rayons U. V. sur des fragments nuclees et enuclees d'amibes. Exptl. Cell. Res. 12: 649-656.

SMITH, C. L. 1954. The inactivation of monomolecular films of protein and its relation to the lifetime of active radicals formed in water by x-radiation. Arch. Biochim. et Biophys. 50: 322-336.

SMYTH, H. D. and D. W. MUELLER 1933. The ionization of water vapor by electron impact. Phys. Rev. 43: 116-120.

SPARRMAN, B., L. EHRENBERG and A. EHRENBERG 1959. Scavenging of free radicals and radiation protection by nitric oxide in plant seeds. Acta chem. Scand. 13: 199-200.

STAPLETON, G. E. and C. W. EDINGTON 1956. Temperature dependence of bacterial inactivation by x-rays. Radiation Research 5: 39-45.

STUY, J. H. 1960. Studies on the irradiation inactivation of microorganisms. VI. X-ray induced breakdown of deoxyribonucleic acid in Haemophilus influenzae and in other bacteria. J. Bacteriol. (in press).

SUTTON, H. C. and J. ROTBLAT 1957. Dose-rate effects in radiation-induced chemical reactions. Nature 180: 1332-1333.

SWALLOW, A. J. 1960. The radiation chemistry of organic compounds. Chapter 3. Pergamon Press, London. (in press).

SWANSON, C. P. 1944. X-ray and ultraviolet studies on pollen tube chromosomes. I. The effect of ultraviolet (2537 Å) on x-ray induced chromosomal aberrations. Genetics 29: 61-68.

——— 1949. Further studies on the effect of infrared radiation on x-ray induced chromatid aberrations in Tradescantia. Proc. Natl. Acad. Sci. 35: 237-244.

SWORSKI, T. J. 1954. Yields of hydrogen peroxide in the decomposition of water by cobalt γ-radiations. I. Effect of bromide ion. J. Am. Chem. Soc. 76: 4687-4692.

——— 1957. Photoreduction of ceric ion in sulfuric acid solutions. Effect of thallous ion. J. Am. Chem. Soc. 79: 3655-3657.

SYMPOSIUM ON THE STRUCTURE AND REACTIVITY OF ELECTRONICALLY EXCITED SPECIES 1958. Canadian J. Chem. 36: 1-146.

TAKEMURA, S. 1959. Hydrazinolysis of nucleic acids. I. The formation of deoxyriboapyrimidinic acid from herring sperm deoxyribonucleic acid. Bull. Chem. Soc. Japan 32: 920-926.

—— and M. MIYAZAKI 1959. Hydrazinolysis of nucleic acids. II. The formation of riboapyrimidinic acid from yeast ribonucleic acid. Bull. Chem. Soc. Japan 32: 926-931.

TAMM, C., M. E. HODES and E. CHARGAFF 1952. The formation of apurinic acid from the desoxyribonucleic acid of calf thymus. J. Biol. Chem. 195: 49-63.

TERENIN, A. and V. ERMOLAEV 1952. Sensitized phosphorescence of organic molecules at low temperatures. Intermolecular energy transfer with excitation of the triplet level. Doklady Akad. Nauk. 85: 547.

—— and —— 1956. Sensitized phosphorescence in organic solutions at low temperature. Energy transfer between triplet states. Trans. Faraday Soc. 52: 1042.

TING, T. P. and R. E. ZIRKLE 1940. The nature and cause of the hemolysis produced by x-rays. J. Cell. Comp. Physiol. 16: 189-195.

—— and —— 1950. The kinetics of the diffusion of salts into and out of x-irradiated erythrocytes. J. Cell. Comp. Physiol. 16: 197-206.

TOOLE, E. H., V. K. TOOLE, H. A. BORTHWICK and S. B. HENDRICKS 1955. Photocontrol of lepidium seed germination. Plant Physiol. 30: 15-21.

——, —— and —— 1959. Photocontrol of plant development by the simultaneous excitation of two interconvertible pigments. III. Control of seed germination and axis elongation. Botan. Gaz. 121: 1-8.

TRUMBORE, C. N. and E. J. HART 1959. γ-Ray oxidation of ferrous sulfate in 0.4 M sulfuric acid solutions. The effect of 0 to 0.4 M oxygen. J. Phys. Chem. 63: 867-872.

UBER, F. and A. D. MCLAREN 1941. A photochemical yield for the inactivation of crystalline trypsin. J. Biol. Chem. 141: 231-237.

UEHLING, E. A., Editor. 1960. Penetration of charged particles in matter. Publication 752, National Academy of Science — National Research Council. Washington, D. C.

URI, N. 1952. Reactions of atoms and free radicals in aqueous solution with particular reference to their photochemical formation. Research Council Israel, Spec. Publ. No. 1: 193-204. Chem. Rev. 50: 375-454.

U.S.S.R. SYMPOSIUM 1957. Proceedings of the First All Union Conference on Radiation Chemistry, Moscow. Part II. Radiation Chemistry of Aqueous Solutions. English Trans. Consultants Bureau, New York (1959). 80 pp.

VAVILOV, S.I. 1943. The theory of the influence of concentration on the fluorescence of solutions. J. Phys. (U.S.S.R.) 7: 141-152.

VON BORSTEL, R. C. and S. WOLFF 1955. Photoreactivation experiments on the nucleus and cytoplasm of the Habrobracon egg. Proc. Natl. Acad. Sci. U.S. 41: 1004-1009.

VON BÜNAU, (see Bünau, G. von).

WADDINGTON, F. B. 1958. The electron crosslinking of polypropylene and high density polyethylenes. J. Polymer Sci. 31: 221-225.

WALD, G. 1953. The biochemistry of vision. Ann. Rev. Biochem. 22: 497-526.

WALL, L. A. 1955. Factors influencing the behavior of polymers exposed to high-energy radiation. J. Polymer Sci. 17: 141-142.

—— and M. MAGAT 1953. Effects of atomic radiation on polymers. Mod. Plast. 30: 176-178.

WANG, S. Y. 1958. Photochemistry of nucleic acids and related compounds. I. The first step in the ultraviolet irradiation of 1, 3-dimethyluracil. J. Am. Chem. Soc. 80: 6196-6198.

—— 1959. Ultraviolet irradiation of 1, 3-dimethylthymine. Nature 184: 59-61.

—— 1959. Photoautomerization of cytosine derivatives on ultraviolet irradiation. Nature 184: 184-185.

——, M. APICELLA and B. R. STONE 1956. Ultraviolet irradiation of 1, 3-dimethyluracil. J. Am. Chem. Soc. 78: 4180.

WARBURG, O. 1948. Schwermetalle als Wirkungsgruppen von Fermenten. W. Saenger, Berlin.

WATANABE, K. and M. ZELIKOFF 1953. Absorption coefficients of water vapor in the vacuum ultraviolet. J. Opt. Soc. Amer. 43: 753-755.

WAWILOW, S. I. 1954. Die Mikrostruktur des Lichtes (German translation) Akademie Verlag, Berlin.

WEISS, J. 1958. Some effects of oxygen and the radiation-induced formation of hydroperoxides from nucleic acids and related compounds. In, Les peroxides organiques en radiobiologie, M. Haissinsky, Ed. Masson et Cie., Paris. p. 42-45.

WEST, W. and B. PAUL 1932. The quantum yield in the photodecomposition of the alkyl halides in non-polar solvents. Trans. Faraday Soc. 28: 688-697.

WHELAND, G. W. 1955. Resonance in organic chemistry. John Wiley and Sons, Inc., New York. pp. 281-289.

WIERZCHOWSKI, K. L. and D. SHUGAR 1957. Photochemistry of cytosine and nucleosides and nucleotides. Biochim. et Biophys. Acta 25: 355-364.

WILLIAMS, R. R., JR. and W. H. HAMILL 1954. Chemical effects of electron capture by solutes in hydrocarbons during gamma irradiation. Radiation Research 1: 158-164.

WILSON, D. E. 1959. The effect of ionizing radiation on the enzyme desoxyribonuclease in the dry state. Int. J. Rad. Biol. 1: 360-365.

WOLF, H. C. 1959. The electronic spectra of aromatic molecular crystals. In, Solid State Physics, Vol. 9. Academic Press, New York and London.

WOOD, T. H. 1959. Inhibition of cell division. Radiation Research Supplement 1: 332-346.

WORRALL, R. and A. CHARLESBY 1958. Effect of additives on the radiation-induced polymerization of isobutene. Int. J. Appl. Rad. Isotopes 4: 84-88.

ZELLE, M. R. and A. HOLLAENDER 1955. Effects of radiations on bacteria. In, Radiation Biology, Vol. II. A. Hollaender, Ed. McGraw-Hill, New York. pp. 365-430.

ZIMMER, K. G. 1959. Evidence for free radical production in living cells exposed to ionizing radiation. Radiation Research Supplement 1: 519-529.

ZIRKLE, R. E. and C. A. TOBIAS 1953. Effects of ploidy and linear energy transfer on radiobiological survival curves. Arch. Biochim. et Biophys. 47: 282-306.

Discussant Index

Author Index

Subject Index

In the use of this INDEX, recall that this book is concerned with radiation effects in general. Thus, all references to specific substances or systems are directly or indirectly concerned with radiation effects.

E